Astrophysics and Space Science Proceedings

Volume 46

More information about this series at http://www.springer.com/series/7395

Josep M. Trigo-Rodríguez • Maria Gritsevich
Herbert Palme
Editors

Assessment and Mitigation of Asteroid Impact Hazards

Proceedings of the 2015 Barcelona Asteroid Day

Springer

Editors
Josep M. Trigo-Rodríguez
Campus UAB, 2a planta
Institute of Space Sciences
Bellaterra Barcelona
Spain

Herbert Palme
Sektion Meteoritenforschung
Senckenberg, Naturmuseum und Forschungsinstitut
Frankfurt, Germany

Maria Gritsevich
Department of Physics
University of Helsinki
Helsinki, Finland

Ural Federal University, Institute of Physics and Technology
Ekaterinburg, Russia

National Land Survey of Finland, Finnish Geospatial Research Institute
Masala, Finland

ISSN 1570-6591 ISSN 1570-6605 (electronic)
Astrophysics and Space Science Proceedings
ISBN 978-3-319-46178-6 ISBN 978-3-319-46179-3 (eBook)
DOI 10.1007/978-3-319-46179-3

Library of Congress Control Number: 2016959006

Printed on acid-free paper

This Springer imprint is published by Springer Nature
The registered company is Springer International Publishing AG
The registered company address is: Gewerbestrasse 11, 6330 Cham, Switzerland

Preface: Impact Hazard, a Russian Roulette

Science is never ending, and the more we learn, the more we see how much we ignore. Descartes said that he would give everything he knew, if he would get half of what he ignored. What a good business for him, who died when Newton was still eight!

After three centuries of riding the knowledge wave created by the latter, and now the new one by Einstein, even we would be quite well off if we made that swap.

Let's take the planets outside our Solar system as an example: until 2009, using various means, we discovered about 50 per year. That year, the specialized space telescope Kepler was launched, and speed was multiplied threefold. Now, reviewing the database of its measurements till 2013 (when it malfunctioned) and using some degraded modes of operation, around 1000 are added per year. However, Kepler has only studied about 4 % of the sky. Making simple calculations about the number of stars in that area and the number of planets found, we can be sure that most stars do have them: there must be millions, and not too far, still unknown! ESA's PLATO mission will study a million stars over the whole sky. Let's see.

Now we even observe planetary systems in formation, and just the leftover bodies of planetary formation are the asteroid and comets currently forming part of our planetary system. These so-called minor bodies are very relevant because they allow us to date the origin of our solar system and the processes that sculpted planets. Because of their importance, a number of missions of exploration have been dedicated to study these bodies recently. There are literally millions, and we keep discovering lots of them every day. Until 1990, we knew about 15,000 objects. In 2000 they were already 125,000, and we are already at 700,000. A NASA program managed to catalog virtually all the very large ones over 1 km, of which at the end of April, 2016, we know about 875. On February 15, 2013, an object just 20 m in size caused destruction in Chelyabinsk, in Siberia, and the world woke up concerning this risk. Would one of those fall on us? Last year a comprehensive study demonstrated that there are 10 million such objects close enough to Earth, and very few are actually known. As if the growing number of near-Earth asteroids wasn't enough, there is also a population of extinct comets, with highly eccentric and much higher impact power, that are extremely dark and difficult to detect. An example was the object, 600 m in diameter, called Great Pumpkin (officially 2015 TB145), which approached to only 480,000 km of Earth for Halloween 2015. This extinct comet was discovered with only 3 weeks of margin, exemplifying the need of multiplying the resources dedicated to detecting these dangerous objects that result from the aging and final disruption of periodic comets.

Data revealed in 2013 from the military infrared observatories confirms exactly the numbers about the flux of these objects, and we have even recent evidence that the flux of Chelyabinsk-size projectiles (few tens of meters) could be higher than we thought. We need to take action. It is imperative to give a large impulse to specialized space telescope missions like the Sentinel of the B612 Foundation, in order to find these million objects and have early warning. And in parallel we need to work on deflection techniques like the Don Quijote project. Time to take action in front of a threat of this caliber might be certainly more valuable than gold.

However, space agencies have not yet decided to assign the necessary resources to those missions, although we may be getting close. The new version of Don Quijote, called AIDA, a joint NASA/ESA project, is still in study phase on both sides. NEOCam, a NASA proposal, is also awaiting funding. In this framework, the international outreach initiative called Asteroid Day is a magnificent way to inform the public about our serious concern. In such framework the international workshop "Assessing the nature of asteroid impact hazard to transform it into technological opportunity" was an initiative of the Institute of Space Sciences (CSIC-IEEC), supported by the Academia Europaea, Parc de Recerca UAB (PRUAB), Asteroid Day, Elecnor-Deimos y SENER.

We would like to warmly thank the scientific organizing committee (SOC) who helped us shape the content of this workshop:

Maria Gritsevich (co-chair, Finnish Geodetic Institute—Ural University—Russian Academy of Sciences—MIIGAiK, Finland and Russia)
Valery Lupovka (Moscow State University of Geodesy and Cartography, MIIGAiK, Russia)
Herbert Palme (Academia Europaea/Forschung Institute Senckenberg, Germany)
Toni Santana-Ros (Astronomical Observatory, Adam Mickiewicz University, Poland)
Josep M. Trigo-Rodríguez (chair, Institute of Space Sciences, CSIC-IEEC, Spain)

The success of this workshop and the outreach conference not had been possible without the local support team (LOC) help concerning the organizational issues:

Anna Bertolin (IEEC)
Carles E. Moyano-Cambero (CSIC-IEEC)
Marina Martínez-Jiménez (IEEC)
Pilar Montes (IEEC)
Manuel Moreno-Ibañez (CSIC-IEEC)
Eva Notario (IEEC)

Finally we would like to remark that the Association of Space Explorers, the world's astronauts, is promoting within our own Agencies, with the public, and with the United Nations, the idea that it is time to take action and move from concept to development. This year 2016 both NASA and ESA are due to take decisions about their future programs, and we promote and hope that both Agencies realize that the returns of such missions may be enormous and that it is time to stop playing *Russian Roulette* in the hostile environment through which our common spaceship Earth moves day in and day out.

Cologne, Germany
Barcelona, Spain
April 2016

Pedro Duque
Josep M. Trigo-Rodríguez

Contents

Barcelona Asteroid Day 2015: Revisiting the Threat by Asteroid and Comet Impact

Josep M. Trigo-Rodríguez, Herbert Palme, and Maria Gritsevich

Abstract The general context in which was organized the Barcelona Asteroid Day 2015 workshop is introduced, together with a brief description of the contributions to this proceedings book. To promote the study of asteroids, comets and meteorites was one of the goals of the Barcelona event, an initiative born as part of the international Asteroid Day movement that tries to popularize the Tunguska event (on June 30th, 1908), the most energetic contemporary impact detected so far. This book compiles the contributions presented to this research workshop organized by the Institute of Space Sciences (IEEC-CSIC) at the UAB campus in Bellaterra, together with an outreach event led by astronaut Pedro Duque that conveyed the main issues to the general public. This workshop was coordinated with other initiatives around the world and intended to offer to the general public, and astronomy enthusiasts an environment for public awareness and open discussion. The event was intended to present the most recent science, to promote international cooperation and to raise the scientific interest of studying asteroids to the general public and the media.

J.M. Trigo-Rodríguez (✉)
Institute of Space Sciences (IEEC-CSIC), Meteorites, Minor Bodies and Planetary Sciences Group Campus UAB, Carrer de Can Magrans, s/n 08193 Cerdanyola del Vallés (Barcelona), Catalonia, Spain
e-mail: trigo@ice.csic.es

H. Palme
Forschungsinstitut und Naturmuseum Senckenberg, Senckenberganlage 25, Frankfurt am Main D-60325, Germany

M. Gritsevich
Department of Geodesy and Geodynamics, Finnish Geospatial Research Institute (FGI), National Land Survey of Finland, Geodeentinrinne 2, Masala FI-02431, Finland

Department of Physical Methods and Devices for Quality Control, Institute of Physics and Technology, Ural Federal University, Mira street 19, Ekaterinburg 620002, Russia

Department of Computational Physics, Russian Academy of Sciences, Dorodnicyn Computing Centre, Valilova 40, Moscow 119333, Russia

J.M. Trigo-Rodríguez et al. (eds.), *Assessment and Mitigation of Asteroid Impact Hazards*, Astrophysics and Space Science Proceedings 46,
DOI 10.1007/978-3-319-46179-3_1

1 Introduction: The Impact Hazard in Perspective

The recent discovery by telescopic surveys of hundreds of potentially hazardous asteroids and the popularization of their encounters with Earth increases the public concern about the hazard associated with the impact of asteroids and comets. Several world-wide initiatives began with the goals of collecting accurate information and improving the current knowledge of the near-Earth environment. Included in these initiatives are the European Space Situational Awareness programme (Koschny et al. 2010), dedicated NEO-surveys, such as NEOWISE (Mainzer et al. 2014), PanSTARRS (Denneau et al. 2013), the Catalina Sky Survey (Larson et al. 2007), the International Asteroid Warning Network (IAWN) which was established as part of the United Nations effort to mitigate the NEO impact threat, as well as others. Indeed, as a result of these large investments into NEO tracking, the past decade has brought impressive improvements in large-scale asteroid discovery. So far, the initiatives have resulted with orbital data obtained on nearly 700,000 known asteroids (most of which lie in the main belt, and of which approximately 13,000 are NEOs), with over 1/5 of these having other additional physical characterization measurements collected as well (DeMeo et al. 2015). This new knowledge greatly improves our early warning awareness and preparedness.

In contrast to these advances, a comprehensive model capable of tracking a hazardous object from its current location in orbit to its intersection with Earth, accounting for all possible scenarios and resulting in reliable forecasts of regional environmental consequences is still missing. The Asteroid Day initiative (http://asteroidday.org/) wants to inform the general public about the initiatives taken by scientists and space agencies in order to increase our capabilities to forecast and palliate the effects of such cosmic disasters.

These issues outlined before are some of the main topics that motivate us to organize this scientific workshop. It counted with the participation of 40 people coming from all around Europe, plus more than 100 additional people participating in the Parc de Recerca UAB (PRUAB) outreach talk given by the famous Spanish astronaut Pedro Duque. In that event also participated Prof. Jordi Isern, as director of the Institute of Space Sciences, Julia Palma of Parc de Recerca of the UAB (PRUAB), and Prof. Buenaventura Guamis representing the Universitat Autònoma de Barcelona (UAB).

2 Chapter Contributions in This Book

Almost 1 year was necessary to compile the book contributions, after ongoing accurate blinded peer-reviews. The proceedings start with a general overview of our current knowledge on impact hazard in a chapter written by Trigo-Rodríguez et al. (2017). The main asteroid belt is generally assumed to be the main source of meter-sized projectiles. But there are other important sources for delivering dangerous projectiles to Earth. One of them is the Jupiter Family Comet (JFC) source. Another direct pathway is associated with the disruption of potentially hazardous asteroids (PHAs) in near-Earth space.

A chapter by Palme and Zipfel (2017) deals with fundamental questions about the bulk chemistry of solar system objects. Tiny sub-micrometer interplanetary dust particles, sub-millimeter to millimeter sized micro-meteorites, meter-sized chondritic meteorites, bulk planetesimals and large planets including the Earth have largely the same bulk chemical composition resembling the heavy element composition of the Sun. But there are differences in stable isotopes of these objects. Some key questions are addressed: to what extent are solar system bodies chemically and isotopically uniform? What is the relationship between meteorites, asteroids, comets and dust? Are meteorites building blocks of the Earth and other terrestrial planets?

It is well known that asteroids are small bodies with star-like appearance. It is then challenging to understand them as we need to obtain basic physical parameters like e.g. the diameter, shape, and bulk density, among others. The Santana-Ros et al. (2017) chapter describes the relevance of ground based observations, and the state-of-the-art techniques used to solve the lightcurve inversion problem, and obtain asteroid shape models. Although the most abundant type of observational data comes from light curves, models of asteroids can be obtained using various other data obtained by a variety of techniques. Santana Ros et al. describe the way to combine the different techniques in some detail: radar imaging, stellar occultation timings, and disk-integrated photometry to unveil the physical properties of asteroids.

Martínez-Jiménez et al. (2017) explore the bulk composition of undifferentiated meteorites, focusing on the potential economic significance of asteroids. Is asteroid mining feasible and economically worthwhile? There is not doubt that chondritic asteroids contain large amounts of valuable resources such as platinum group metals, iron-nickel-metal and to a lesser extent, rare earth elements (REE), and water. They report results of analyses of small fractions of chondrites which we expect to be returned from future missions. The technique used is inductively coupled plasma mass spectrometry (ICP-MS), which allows precise analyses of tiny fractions of extraterrestrial material.

A chapter by Rimola et al. (2017) deals with a fundamental process that occurred in hydrated asteroids. At the earliest time after the accretion of asteroids, some water was released and soaked their interiors. These authors explore the effects of the aqueous alteration: elemental mobilization, formation of secondary minerals and changes in the primordial chemistry and mineralogy. Among the key processes is the transformation of mafic silicates initially the major constituents of chondrules and matrix fragments, into phyllosilicates. With the aim to have atomic-scale in-sights of these water alteration processes, in this chapter we present our first atomistic simulations related to the water alteration of silicates obtained by means of quantum chemical calculations.

Usually we talk about impact hazard associated with sub-km to km-sized asteroids or comets, but what about meter-sized meteoroids released from them? Entry events caused by these objects, take place regularly and bombard the Earth at the rate of more than 100 t a day. The object causing the fireball can penetrate the atmosphere down to the height of around 20 km in its luminous phase, and in specific cases, even lower. Recent evidence from crater excavation in the Carancas fall, suggests the existence of monolithic and high-strength meter-sized rocks that might be dangerous to humans, and may even excavate craters. The Moreno-Ibáñez et al. (2017) chapter deals with these questions and describes the recent progress made in this regards. Certainly, to properly evaluate impact hazards it is also needed to develop physical models capable to forecast the ability of the terrestrial atmosphere to protect us from interplanetary projectiles. It is well known that the atmosphere shields us from the direct impact of these rocks but certainly the development of new methods to quantify the effects and discard any hazard to humans is necessary.

From the scientific point of view, fireballs (or bolides) determine the linkages between meteorites and their parent bodies. When a meteorite is successfully recovered on the ground, the detailed atmospheric flight path (as recorded by cameras and other instruments) is an excellent tool to test and improve the existing entry models assuring the robustness of their future applications. It is crucial to rapidly locate fresh meteorite falls for several reasons; it yields the range of the initial parameters defining the hazardous object's entry into the atmosphere, its subsequent flight through the atmosphere, as well as the physical properties of the object itself. The chapter presented by Gritsevich et al. (2017) develops, improves, and merges together models to better interpret the instrumentally recorded atmospheric entry of meteoroids, which occur much more frequently than larger extraterrestrial impacts. The authors solve inverse problem to determine the pre-atmospheric parameters of large meteoroids and illustrate their approaches using the observational data available for the recently recovered Kosice meteorite.

Blanch et al. (2017) describe recent progress made in the study of daylight bolides that are being tracked in the framework of the Spanish Meteor and fireball Network (SPMN) initiative. Daylight fireball studies are rare in the literature because of the intrinsic difficulty to calibrate the often biased trajectory descriptions given by eyewitnesses. Modern video detectors can be adapted for monitoring the sky during broad daylight. Some encouraging results of a new meteor video-detection station at the Ebre Observatory (URL-CSIC) and Montseny and Folgueroles (IEEC-CSIC) stations are presented.

The Tapia and Trigo-Rodríguez (2017) chapter deals with the effects associated with the entrance of large meteoroids or small asteroids into the Earth's atmosphere. Not only meter-sized projectiles with certain entry geometries and slow geocentric velocities can be a source of hazard to humans, but some small asteroids may deposit enough energy in the atmosphere to produce destructive phenomena as, for example, exemplified by the Tunguska and Chelyabinsk events. Hazards produced in these events may adopt challenging forms: sonic booms, airbursts, hot plasma or even crater excavation and earthquakes.

A new age in space research started with sample return missions from minor bodies of the solar system, with, for example, the Stardust (NASA), Hayabusa (JAXA) and Rosetta (ESA) missions. Future space missions are aimed to achieve the sample return from Near Earth Asteroids (Osiris-REx and Hayabusa 2). Additional space missions are envisioned to test impact deflection techniques, to be used in future hypothetical encounters of our planet with Potentially Hazardous Asteroids (PHA). A chapter by Moyano-Cambero et al. (2017) deals with laboratory studies of chondrites to better understand the mechanical and reflective properties of shocked asteroids that might be representative of PHAs. Such bodies are targets of future missions as, for instance, the Asteroid Impact and Deflection Assessment (AIDA) a joint ESA-NASA mission to test the impact deflection technique on binary asteroid 65803 Didymos. That asteroid system consists of a main asteroid of about 800 m in diameter and its satellite with a diameter of 150 m. The mission will be carried out by two spacecraft: the Asteroid Impact Mission (AIM) lead by ESA and the Double Asteroid Redirection Test (DART) of NASA. It is planned that the DART spacecraft will impact the smaller component of the Didymos system in order to test and perhaps quantify our ability to deflect asteroids with a projectile. AIM will study the impact from close orbit with an innovative payload, including instruments on board of cubesats.

Finally, the last chapter by Trigo-Rodríguez et al. (2017) describes the different studies of minor bodies at Montsec Astronomical Observatory. Using the 0.8 m Joan Oró Telescope a Near Earth Asteroids and Potentially Hazardous Asteroids campaign is being performed in the context of Euro-NEAR program. Another campaign to obtain photometric data of selected main belt asteroids which aims to enhance the Solar System science coming out from *Gaia* space mission. Additionally, a continuous monitoring of large fireballs over Catalonia is being performed.

3 Putting the Barcelona AD Workshop in Context

The scientific meeting that has led to these proceedings was organized by the Institute of Space Sciences (CSIC) and the Institut d'Estudis Espacials de Catalunya (IEEC) with the collaboration of the Scientific Secretariat of the Institute for Catalan Studies (IEC). Spanish Ministry of Science funded the research project AYA2015-67175-P in which context of this book was published. The workshop took place on June 30, 2015 in the Institute of Space Sciences (CSIC) building (Fig. 1). A memorable outreach talk given by astronaut Pedro Duque took place in the Parc de Recerca de la Universitat Autònoma de Barcelona (PRUAB) before lunch time (see Fig. 2). All participants agreed that the meeting location was a perfect place to discuss many interdisciplinary issues concerning the nature of impact hazard. Consequently, this meeting resulted in a rewarding experience for all participants (Fig. 3).

Fig. 1 A general view of workshop participants in front of the Institute of Space Sciences (IEEC-CSIC) building in Cerdanyola del Vallès, Barcelona (IEEC-CSIC)

Fig. 2 A picture of the outreach talk given at PRUAB by Pedro Duque (PRUAB)

Fig. 3 Meeting organizers and book editors, relaxed and satisfied at the end of the Barcelona Asteroid Day 2015 workshop. From *right* to *left*: Marina Martínez-Jiménez, Josep M. Trigo-Rodríguez, Herbert Palme, Manuel Moreno-Ibáñez, and Maria Gritsevich

4 The Proceedings Book

This interdisciplinary book is a result of a common effort of the authors, who share their science, and achievements in a really enthusiastic environment provided by the Barcelona Asteroid Day event. The book is published with Institut d'Estudis Espacials de Catalunya (IEEC) funds associated with the Meteorites, Minor Bodies and Planetary Sciences group.

Finally, the editors of this volume are also especially grateful to all distinguished authors for preparing texts of high quality despite having many duties ahead in their agendas. We also thank the kind and valuable additional effort made by the reviewers, who often read and improved the book contributions in shorter time than usual for peer-review. We are grateful to the reviewers who timely contributed to assure blinded peer-review, and also provided significant feedback to guarantee scientific quality, appropriated citation of previous literature, and prose clarity. The chapter reviewers were: Thomas Burbine, Robert Hawkes, Peter Jenniskens, Jordi Llorca, Andy Rivkin, Eleanor Sansom, Elizabeth Silber, Paolo Tanga, Iwan P. Williams, Mike Zolensky, and some additional anonymous reviewers.

The authors and the referees deserve all the credit in the production of this comprehensive book of proceedings that compiles most of the work presented during the Asteroid Day 2015 workshop of Barcelona. We sincerely hope that all this common effort is useful for having many readers keeping the track on this fascinating research area.

Acknowledgments Workshop attendance of some participants was financially supported by Institut d'Estudis Espacials de Catalunya (IEEC), Parc de Recerca de la Universitat Autònoma de Barcelona (PRUAB) and Academia Europaea. Finally, the Institute of Space Sciences (CSIC) and the IEEC provided the needed logistics. MG acknowledges travel support provided by the Magnus Ehrnrooth Foundation. Iwan P. Williams is also acknowledged by his involvement in the edition tasks. This book has been prepared in the framework of the research lines studied by Spanish Excellence project AYA 2015-67175-P (P.I. Trigo-Rodríguez).

References

Blanch, E., Trigo-Rodríguez, J.M., Madiedo, J.M., Lyytinen, E., Moreno-Ibáñez, M., Gritsevich, M., Altadill, D.: Detection of nocturnal and daylight bolides from Ebre Observatory in the framework of the SPMN fireball network. In: Trigo-Rodríguez, J.M., Gritsevich, M., Palme H. (eds.) Assessment and Mitigation of Asteroid Impact Hazards, pp. 185–197. Springer, New York (2017)

DeMeo, F.E., Alexander, C.M. O'D., Walsh, K.J, Chapman, C. R., Binzel, R.P.: The compositional structure of the Asteroid Belt. In: Asteroids IV in the Space Science Series, pp. 13–41 (2015)

Denneau, L., Jedicke, R., Grav, T., et al.: The Pan-STARRS moving object processing system. PASP **125**, 357–395 (2013)

Gritsevich, M., Dmitriev, V., Vinnikov, V., Kuznetsova, D., Lupovka, V., Peltoniemi, J., Mönkölä, S., Brower, J., Pupyrev, Y.: Constraining the pre-atmospheric parameters of large meteoroids: Košice, a case study. In: Trigo-Rodríguez, J.M., Gritsevich, M., Palme H. (eds.) Assessment and Mitigation of Asteroid Impact Hazards, pp. 153–183. Springer, New York (2017)

Koschny, D., Drolshagen, G., Bobrinsky, N., Gritsevich, M.: The European Space Situational Awareness programme. In: European Planetary Science Congress 2010, abstract EPSC2010-686 (2010)

Larson, S.: Near Earth Objects, our Celestial Neighbors: Opportunity and Risk. In: Valsecchi, G.B., Vokrouhlicky, D., Milani, A. (eds.) Proceedings if IAU Symposium 236, pp. 323–328. Cambridge University Press, Cambridge (2007)

Mainzer, A., Bauer, J., Cutri, R.M., Grav, T., Masiero, J., Beck, R., Clarkson, P., Conrow, T., Dailey, J., Eisenhardt, P., Fabinsky, B., Fajardo-Acosta, S., Fowler, J., Gelino, C., Grillmair, C., Heinrichsen, I., Kendall, M., Kirkpatrick, J.D., Liu, F., Masci, F., McCallon, H., Nugent, C.R., Papin, M., Rice, E., Royer, D., Ryan, T., Sevilla, P., Sonnett, S., Stevenson, R., Thompson, D.B., Wheelock, S., Wiemer, D., Wittman, M., Wright, E., Yan, L.: Initial performance of the NEOWISE reactivation mission. Astroph. J. **792**(30), 14 (2014)

Martínez-Jiménez, M., Moyano-Cambero, C.E., Trigo-Rodríguez, J.M., Alonso-Azcárate, J., Llorca, J.: Asteroid mining: mineral resources in undifferentiated bodies from the chemical composition of carbonaceous chondrites. In: Trigo-Rodríguez, J.M., Gritsevich, M., Palme H. (eds.) Assessment and Mitigation of Asteroid Impact Hazards, pp. 73–101. Springer, New York (2017)

Moreno-Ibáñez, M., Gritsevich, M., Trigo-Rodríguez, J.M.: Measuring the terminal heights of bolides to understand the atmospheric flight of large asteroidal fragments. In: Trigo-Rodríguez, J.M., Gritsevich, M., Palme H. (eds.) Assessment and Mitigation of Asteroid Impact Hazards, pp. 129–151. Springer, New York (2017)

Moyano-Cambero, C.E., Trigo-Rodríguez, Pellicer, E., Martínez-Jiménez, M., Llorca, J., Metres, N., and Sort, J. Chelyabinsk meteorite as a proxy for studying the properties of Potentially Hazardous Asteroids and impact deflection strategies. In: Trigo-Rodríguez, J.M., Gritsevich, M., Palme H. (eds.) Assessment and Mitigation of Asteroid Impact Hazards, pp. 219–241. Springer, New York (2017)

Palme, H. and Zipfel, J. The chemistry of Solar System Materials: Sun, Planets, Asteroids, Meteorites and dust. In: Trigo-Rodríguez, J.M., Gritsevich, M., Palme H. (eds.) Assessment and Mitigation of Asteroid Impact Hazards, pp. 33–53. Springer, New York (2017)

Rimola, A., and Trigo-Rodríguez, J.M. Atomistic Simulations of Aqueous Alteration Processes of mafic silicates in carbonaceous chondrites. In: Trigo-Rodríguez, J.M., Gritsevich, M., Palme H. (eds.) Assessment and Mitigation of Asteroid Impact Hazards, pp. 103–127. Springer, New York (2017)

Santana-Ros, T., Dudzinski, G. and Bartczak, P. Shape models and physical properties of asteroids. In: Trigo-Rodríguez, J.M., Gritsevich, M., Palme H. (eds.) Assessment and Mitigation of Asteroid Impact Hazards, pp. 55–71. Springer, New York (2017)

Tapia, M., Trigo-Rodríguez, J.M.: Natural hazard associated to shock waves of meter-sized meteoroids. In: Trigo-Rodríguez, J.M., Gritsevich, M., Palme H. (eds.) Assessment and Mitigation of Asteroid Impact Hazards, pp. 199–217. Springer, New York (2017)

Trigo-Rodríguez, J.M., Williams, I.P.: Dynamic sources of contemporary hazard from meteoroids and small asteroids. In: Trigo-Rodríguez, J.M., Gritsevich, M., Palme H. (eds.) Assessment and Mitigation of Asteroid Impact Hazards, pp. 11–32. Springer, New York (2017)

Trigo-Rodríguez, J.M., Santana-Ros, T., Moreno-Ibáñez, M., Rodríguez, D., Sanz, J., Lloro, I., Sánchez, A.: Asteroids, comets and meteorite-dropping bolides studied from the Montsec Astronomi-cal Observatory. In: Trigo-Rodríguez, J.M., Gritsevich, M., Palme H. (eds.) Assessment and Mitigation of Asteroid Impact Hazards, pp. 243–256. Springer, New York (2017)

Dynamic Sources of Contemporary Hazard from Meteoroids and Small Asteroids

Josep M. Trigo-Rodríguez and Iwan P. Williams

Abstract Ground-based observations of meteors and fireballs increase our data and statistics on meter-sized events entering the Earth's atmosphere. Impacts by larger bodies are less frequent and telescopic surveys to find potentially hazardous objects are still crucial to infer the flux of these over long timescales. Telescopic surveys provide significant data on Near Earth Asteroids of few tens or hundreds of meters in diameter that can be only detected when these bodies are close to the Earth. Statistically, bodies with a diameter from a few meters up-to about a 100 m can be considered as the most direct source of contemporary hazard. Of course, larger bodies will do more damage, but impact less frequently. The behaviour of stony bodies interacting with the atmosphere is reasonably well known, but little is known about the either the flux or the behaviour of materials from dormant comets that are often associated with meteoroid streams and small Near Earth Objects. We will introduce some examples that meter-sized meteoroids following high-inclination, and eccentric orbits are not necessarily fragile, and can trace the existence of hazardous objects: dormant comets or Damocloids being an example. From all the available data, a better understanding of the rate at which asteroids impact the Earth can be derived. If meteoroids of cometary origin are included the flux of objects into the Earth's atmosphere will be increased (Space Sci Rev 84(3/4): 327–471, 1998). However, the typical strengths of such meteoroids are too low to survive ablation in the upper atmosphere, so that they are unlikely to impact the ground. However, events such as that over Tunguska in 1908, where an air burst caused considerable damage over a large area, indicates that we should not underestimate fragile bodies as potentially hazardous sources. New missions aimed at returning samples of Near Earth Asteroids to Earth for analysis (Osiris-REx and Hayabusa 2) are very important because they will deliver to our laboratories materials probably

J.M. Trigo-Rodríguez (✉)
Institute of Space Sciences (IEEC-CSIC), Meteorites, Minor Bodies and Planetary Science Group, Campus UAB, Carrer Can Magrans, s/n, Cerdanyola del Vallés, Barcelona 08193, Spain
e-mail: trigo@ice.csic.es

I.P. Williams
School of Physics and Astronomy, Queen Mary, University of London, Mile End Rd, London E1 4NS, UK

J.M. Trigo-Rodríguez et al. (eds.), *Assessment and Mitigation of Asteroid Impact Hazards*, Astrophysics and Space Science Proceedings 46,
DOI 10.1007/978-3-319-46179-3_2

non-sampled in meteorite collections. A better understanding of the composition of Near Earth Objects will allow the most efficient deflection techniques to be developed so that they present no hazard to human beings.

1 Introduction: NEO Fragments as a Source of Hazard to the Earth

For several decades now, there has been a concern over the threat posed to the Earth from impacts by Near Earth Objects, originating in both asteroids and comets and networks set up to detect them (see for example Gehrels et al. 1972; Chapman et al. 1982, 1989). In 1996 a Resolution of the Council of Europe was issued to promote the development of world-wide programs for the search and physical characterization of Near Earth Objects (NEOs). These bodies have orbits with perihelion distance $q < 1.3$ astronomical units (AU) and aphelion distance $Q > 0.983$ AU. As of December 24, 2015, there are about 13,600 objects known. Amongst these, many belong to well-known groups, 7292 are Apollos, 979 are Atens and 16 are Atiras (Table 1). Members of these groups are typically rocky bodies with diameters from ten meters up to few kilometers, and are usually called Near Earth Asteroids (NEAs). Near-Earth space is dominated by a population of thousands asteroids, compared with about 100 (104 at the current time) dark objects that have dynamical and physical properties suggesting that they are dormant or extinct comets (Jenniskens et al. 2007). The composition of these dark objects is still poorly known. If we need to deflect them, we must make every effort to determine their nature through studying their physical and reflective properties (Trigo-Rodríguez et al. 2014).

Despite the large number of recent discoveries, there is an urgent need to complete the current NEO population surveys so that all bodies larger than a few tens of meters have been discovered. This is not the case at present as is illustrated by two nearly simultaneous, but apparently unrelated events. On February 15, 2013 the scientific community correctly predicted the close approach, within 27,700 km, of the roughly 30-m near-Earth asteroid 367943 Duende (previous temporary designation 2012DA14), but failed to predict the far more dramatic direct collision of an 18-m in diameter asteroid just southwest of the Russian city of Chelyabinsk about 16 h earlier. More than a ton of meteorites reached the ground and were recovered. This meteorite, being named Chelyabinsk, exemplifies the shocked nature of materials reaching the near-Earth space (Fig. 1), and the damage caused by air burst (Wasson 2003; Boslough and Crawford 2008). As the program Protec-2-2014 from the Horizon 2020 is entitled "PROTECTION OF EUROPEAN ASSETS IN AND FROM SPACE-2014-LEIT SPACE" and has as specific goals to create knowledge about these bodies in view of their future exploration and mining potential, this event was of interest to this project.

The near-Earth population is diverse, but it is dominated by chondritic asteroids mostly formed out of silicates and metals. Such undifferentiated bodies are a source

Table 1 The definitions of the orbits of the various Near Earth Object groups and their respective acronyms (Adapted from NEO JPL)

Group	Description	Orbital definition
NECs	**N**ear-**E**arth **C**omets	$q < 1.3$ AU, $P < 200$ years
NEAs	**N**ear-**E**arth **A**steroids	$q < 1.3$ AU
Atiras	On similar orbits to asteroid 163693 Atira	$a < 1.0$ AU, $Q < 0.983$ AU
	Orbits are contained entirely with the orbit of the Earth	
Atens	On similar orbits to asteroid 2062 Aten	$a < 1.0$ AU, $Q > 0.983$ AU
	Earth-crossing NEAs with semi-major axes smaller than Earth's	
Apollos	On similar orbits to asteroid 1862 Apollo	$a > 1.0$ AU, $q < 1.017$ AU
	Earth-crossing NEAs with semi-major axes larger than Earth's	
Amors	On similar orbits to asteroid 1221 Amor	:$a > 1.0$ AU, $1.017 < q < 1.3$ AU
	Orbits exterior to Earth's but interior to Mars'	
PHAs	**P**otentially **H**azardous **A**steroids: NEAs whose Minimum Orbit Intersection Distance (MOID) with the Earth is 0.05 AU or less and whose absolute magnitude (H) is 22.0 or brighter	$MOID <= 0.05$ AU, $H <= 22.0$

Fig. 1 A specimen of the Chelyabinsk meteorite that was recovered by an international expedition. This fragment can now be found in the IEEC-CSIC meteorite collection. The classical light lithology of ordinary chondrites is diversified by the presence of a dark lithology with significant amounts of shocked minerals, impact-generated veins and opaques (J.M.Trigo/CSIC-IEEC)

of precious elements that are rare in the Earth's crust and other planetary bodies (see Martínez-Jiménez et al., 2017). An influential book in this area was *Mining the Sky: Untold Riches from the Asteroids, Comets, and Planets*, written by John S. Lewis (1997). Before its publication, asteroid and comet mining was confined to the realm of science fiction. Today the required technology is available to make this a reality. In fact, such direct application of asteroidal materials as mineral resources is the key point for transforming risk assessment from an issue of public safety into an opportunity for technical and, industrial development. Seen in this way, justifying asteroid research and exploration in the near-Earth space is relatively simple. We should not forget that meteorite collections contain a quite biased sample set of diverse materials but the diversity of the sample can be increased by promoting the robotic exploration of these Near Earth Objects (see e.g. Trigo-Rodríguez and Blum 2009a, b).

Despite the difficulties of getting samples of most of these NEO's, sometimes *Mother Nature* gives us a helping hand. Large fireballs sometimes result in the fall of a meteorite that can be recovered, while often meteoritic material can be recovered without the fireball being witnessed. In either case, meteorites provide a direct source of materials from asteroids reaching the Earth vicinity, and are not seen as dangerous because not clear casualties are historically reported (Yau et al. 1994). This chapter will describe the challenging diversity of the nature of bodies moving in the near-Earth environment stresses the needed to bring together several different research lines in order to gain insight about their real nature and to inferring ways to minimize the hazard resulting from a direct Earth impact. We will also identify here the main sources of impact hazard to Earth, and the existence of quite unexpected populations of dangerous bodies following high-inclination orbits.

2 Meteorite Falls: Clues on the Origin of Hazardous Meteoroids

The most primitive materials arriving to Earth come from undifferentiated comets and asteroids (Nesvorny et al. 2003, 2006). If a stony meteoroid is to survive the passage through the Earth's atmosphere and produce a detectable meteorite on the ground, the initial body should exceed about half a meter. Such a meteoroid would generate a very bright *fireball* or even *superbolides* with brightness greater than magnitude-16. Such bright fireballs are very rare and can occur anywhere at any time. The reliability or otherwise of the observations is a matter of luck and will vary from event to event. Fireball networks around the world have observed thousands of orbits of fireballs during the last decades and determined their orbits, but in most cases the meteoroid did not survive passage through the atmosphere so that the recovery of a meteorite following a fireball is a very rare event. The first was the Pribram event on 7 April 1959 (For a review of this topic, see e.g. Trigo-Rodríguez et al. 2015). At the time of writing, 21 meteorites exist where the heliocentric orbit has been calculated from observations of the fireball generated by the passage of the meteoroid through the Earth's atmosphere (Table 2). About half of these meteorite recoveries were achieved because the luminous phase was imaged by a fireball network carrying out routine monitoring. The remainder came through casual and fortuitous observations. The reliability or otherwise of the observations is a matter of luck and will vary from event to event. Some are relatively good while others are less trustworthy, depending on the type, quality and number of records. Some events were imaged by accident by untrained observers. However, modern digital cameras have allowed casual images to be obtained even in day-time, producing valuable records of the luminous fireball phase that might be calibrated (see e.g. Trigo-Rodríguez et al. 2006). Even though the fireballs are extremely bright, the progenitor meteoroids are still less than a few meters across, and so are usually too small to be recognized by telescopic monitoring programs searching for potentially hazardous objects. However, there are some exceptions such as 2008 TC3, a small asteroid about 5-m across that disrupted over the Nubian desert and produced the Almahatta Sita meteorite (Jenniskens et al. 2009).

Collisions within the main asteroid belt (hereafter MB) have often been invoked as the cause for delivering small to moderate-sized asteroids to Earth (see e.g. Chapman et al. 1989; Davis et al. 2002). It is obvious that a collisional cascade of events produce smaller bodies that are subjected to non-gravitational forces. Among them, the Yarkovsky effect is a subtle force due to the asymmetric absorption and re-radiation of solar energy of the asteroid in space. It produces a progressive loss of energy and an inward migration of their orbits. Consequently, it was first realized that small asteroids and meter-sized meteoroids are crossing resonances mainly by the Yarkovsky effect, not by collisions (Farinella and Vokrouhlický 1999). Also the so-called YORP effect is a key force that changes the spins of small bodies, sometimes spinning them so fast that they break apart. These non-gravitational effects are also producing significant effects on the dynamic evolution of meter-sized and smaller meteoroids (Jenniskens 1998; Ceplecha et al. 1998). Morbidelli and Nesvorný (1999) demonstrated that numerous resonances are present in the MBA delivery of meteoroids and small asteroids to the near-Earth space. Another clear source of meteoroids reaching the Earth's atmosphere is of comets. The close relationship between comets and meteoroid streams was established in the latter part of the nineteenth century (see Williams 2011), well before the nature of comets was known. Comet ices sublimate as they approach the Sun and the gas outflow drags dust particles onto heliocentric orbits that are slightly different from that of the parent comet, forming coherent meteoroid streams (see Williams 2004a, b). Though the orbits are affected by planetary perturbations, they typically remain as coherent streams for tens of thousands of years. These are seen as meteors when the Earth passes through such a stream. Given sufficient time, streams loose their coherence due to a number of factors, close planetary encounters, Pointing—Robinson drag and collisions. These then form a part of the general interplanetary dust background and are seen as sporadic meteors when they meet the Earth's atmosphere (see Williams 1995) these sporadic particles move on orbits that are unrelated to the parent body orbit and so classifying them as cometary or asteroidal

Table 2 List of recovered meteorites with reliable orbital information by chronologic order

						Orbital elements			
Meteorite name	Year of fall	Type	Vg (km/s)	q (AU)	1/a (AU^{-1})	e	i (°)	ω (°)	Ω (°)
Příbram	1959	H5	17.43	0.78951	0.416	0.6711	10.482	241.75	17.79147
Lost City	1970	H5	14.2	0.967	0.602	0.417	12.0	161.0	283.0
Innisfree	1977	L5	14.2	0.986	0.534	0.4732	12.27	177.97	316.80
Peekskill	1992	H6	14.7	0.886	0.671	0.41	4.9	308	17.030
Tagish Lake	2000	C2-ung	15.8	0.884	0.505	0.55	2.0	224.4	297.9
Morávka	2000	H5	19.6	0.9823	0.541	0.47	32.2	203.5	46.258
Neuschwanstein	2000	EL6	20.95	0.7929	0.417	0.670	11.41	241.20	16.82664
Park Forest	2003	L5	16.1	0.811	0.395	0.680	3.2	237.5	6.1156
Villalbeto de la Peña	2004	L6	16.9	0.860	0.435	0.63	0.0	132.3	283.6712
Bunburra Rockhole	2007	Eucrite	13.4	0.6428	1.175	0.245	9.07	209.87	297.59528
Almahata Sitta	2008	Ureilite-an	12.42	0.8999	0.7644	0.31206	2.5422	234.448	194.10114
Buzzard Coulee	2008	H4	18.0	0.961	0.8130	0.22	25.5	212.0	238.9
Maribo	2009	CM2	28.5	0.481	0.45	0.8	0.26	99.0	117.64
Grimsby	2009	H5	20.9	0.9817	0.490	0.518	28.07	159.865	182.9561
Jesenice	2009	L6	13.78	0.9965	0.571	0.431	9.6	190.5	19.196
Mason Gully	2010	H5	14.53	0.98240	0.405	0.6023	0.832	18.95	203.2112
Košice	2010	H5	10.3	0.957	0.369	0.647	2.0	204.2	340.072
Sutter's Mill	2012	C	28.6	0.456	0.386	0.824	2.38	77.8	32.774
Novato	2012	L6	13.67	0.9880	0.478	0.526	5.51	347.35	24.9900
Chelyabinsk	2013	LL5	19.03	0.738	0.581	0.571	4.98	107.67	326.459
Annama	2014	H5	24.2	0.634	0.503	0.69	14.7	264.8	28.611

The uncertainty in each orbital element is not given here for simplicity, but it is implicit in the last figure given
For a full list of references see Trigo-Rodríguez et al. (2015)

in origin is harder (see Jopek and Williams 2013; Williams and Jopek 2014). The flux of low-strength aggregates associated with comets are preferentially disrupted in the upper atmosphere (Trigo-Rodríguez and Llorca 2006, 2007; Trigo-Rodríguez and Blum 2009a, b). Some arrive with the right geometry and velocity to survive as Interplanetary Dust Particles, but most are ablated and or disaggregated in the upper atmosphere as a consequence of their aerodynamic deceleration.

The Minimum Orbit Intersection Distance (MOID) can be used for evaluating the possibility of Earth encounters with an NEO. However, it does not indicate whether or not the two objects actually have an encounter at this minimum distance (Binzel et al. 2010). These objects usually have significant errors in their published orbital parameters and additionally non-gravitational effects usually make that the uncertainty in its location increases much more rapidly than the knowledge of the orbit itself. Thus, we are forced to continuous monitor these objects using ground-based telescopes, especially for large PHAs. Despite of these efforts, and mainly due to current surveys' observational limitations, we expect a very short notice for smaller asteroids or large meteoroids as was exemplified by Chelyabinsk or Almahata Sitta events (Brown et al. 2013; Jenniskens et al. 2009).

In summary, the different chondrite classes (Fig. 2) are basically conglomerates of fine dust (a mixture of silicates, oxides, metal, sulfides and organic constituents, see Fig. 3), chondrules, and refractory or mafic inclusions (Weisberg et al. 2006). Some of them, e.g. ordinary chondrite groups, were significantly metamorphosed in moderately large asteroid progenitors. The primordial water content in the different groups is difficult to assess, but aqueous alteration minerals have been identified even in ordinary chondrites. In general, chondrites containing Fe-metal are indicative of being anhydrous in origin. We also know that many carbonaceous chondrite groups are unprocessed because they exhibit unequilibrated minerals, and also contain interstellar grains with isotopic anomalies that survived processing in the solar nebula and accretionary processes when incorporated to these rocks (Huss et al. 2006; Trigo-Rodríguez and Blum 2009b). Aqueous alteration plays against pristinity as there is clear evidence that some isotopic ratios (like e.g. D/H) are altered during parent body aqueous alteration processes (Alexander et al. 2012; Marty 2012).

Achondritic meteorites are not so common, but provide significant information about the composition and evolution of planetary bodies. From the inferred orbits of achondrites we could expect a significant breakthrough in our understanding of the main dynamic pathways for their delivery to the near-Earth space. In any case, Table 2 clearly shows that we are still far from such a goal as we only know the orbit of an eucrite called Bunburra Rockhole (Bland et al. 2009), and an anomalous ureilite called Almahata Sitta (Jenniskens et al. 2009). We have also recently identified that a small Apollo asteroid designed as 2012XJ112 is the source of bright bolides whose emission spectra seem to be clearly achondritic (Madiedo et al. 2014a).

Achondritic asteroids are also common in the MB. A good example is asteroid 2867 Steins that was studied by Rosetta (ESA) mission during a fly-bye approach. On the basis of these observations, Barucci et al. (2005) classified Steins as an

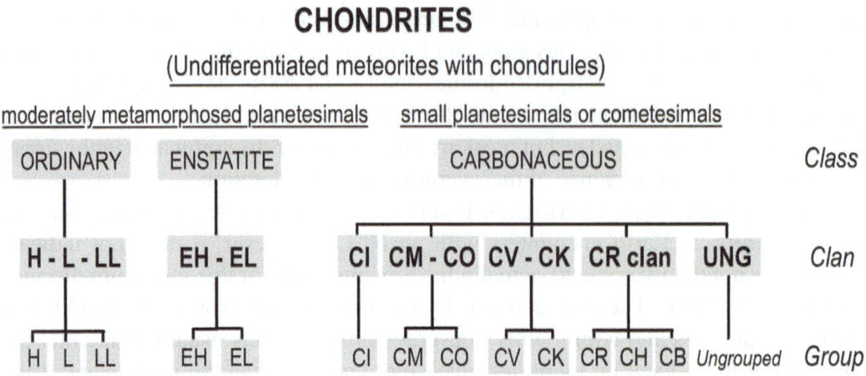

Fig. 2 Main classes, clans and groups of chondritic meteorites. A tentative origin is suggested, but not with idea of being exhaustive (Adapted from Trigo-Rodríguez 2015)

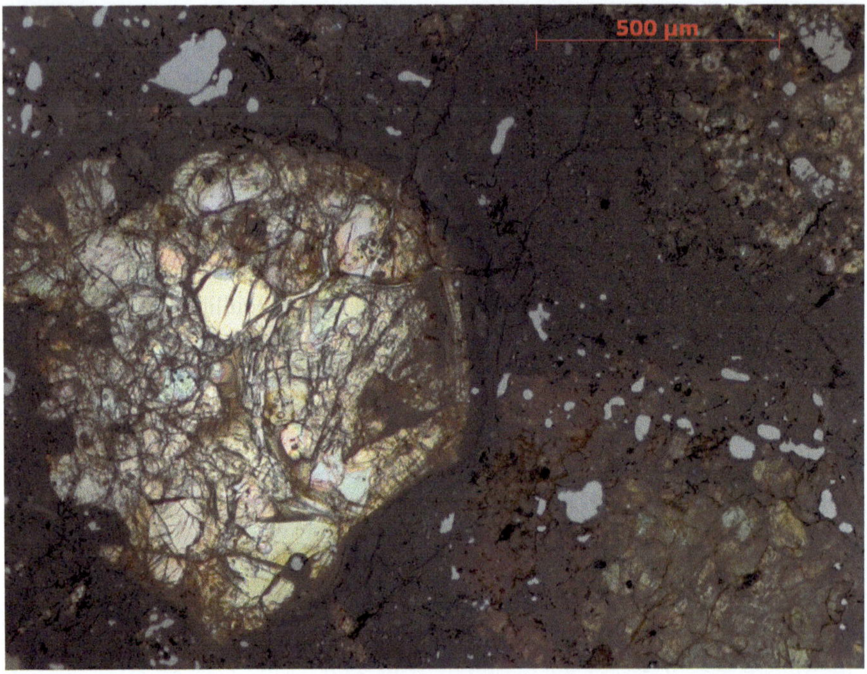

Fig. 3 A thin section of the GRA 95229 CR chondrite from NASA Antarctic collection seen in reflected light with a Zeiss petrographic microscope. Several mm-sized porphyritic olivine-rich chondrules are dominant in the image. Fe-Ni metal grains in grey colour are distributed in the chondrules or isolated in the fine-grained dark matrix (J.M.Trigo/IEEC-CSIC)

E-type asteroid based on visual and near-infrared spectra. E-type asteroids may be similar to enstatite achondrite meteorites (aubrites) and display only grey to moderately red colors. 2867 Steins reflectance spectrum exhibits an unusually strong 0.50 μm feature and a significantly redder spectral slope than previously studied E-types, or their terrestrial meteorite analogs (Weissman et al. 2008).

3 Space Missions Relevant to the Study of Asteroids

Several space missions are currently either ongoing or envisaged for the next decades that are particularly relevant to the study of asteroids. There is a real need to understand the nature of many of the asteroids in the MB, particularly those that feed the near-Earth region through planetary resonances. Because of the intrinsic difficulty in comparing asteroid and meteorite spectra, we need to increase telescopic observations and discovery surveys in order to increase our knowledge on these bodies. We also require sample-return dedicated missions.

The main goal of ESA's Rosetta mission was to rendezvous with periodic comet *67P/Churyumov-Gerasimenko* (67P/C-G) and spend approximately 2 years studying that comet. In addition the Philae landed on the surface providing some date on the internal structure and near-surface environment En route to the comet, Rosetta flew by two MB asteroids, *2867 Steins*, on September 5, 2008, and *21 Lutetia* on July 10, 2010, and carried out observations of them.

A particular mission that should lead to a significant progress in our global knowledge about asteroids is *Gaia*. It was launched on 19 December 2013, and, after an extended commissioning phase started scientific operations in September 2015. *Gaia* performs very accurate astrometry, but also obtains spectral data of the sources through a continuous scanning of the sky. Moreover, the fully automated selection of the sources also ensure that nearly all Solar System objects brighter than +20 visual magnitude will be observed by this unique space observatory (Tanga and Mignard 2012). Gaia has two telescopes with a rectangular aperture of 1.45×0.5 m^2 observing simultaneously two fields separated by an angle of 106.5°. Both telescopes have a common focal plane composed of 106 CCD detectors: 77 of them will perform the astrometry and the photometry of the sources, while 14 CCDs will measure the light dispersed by two prisms, in order to obtain the color of all sources observed by the astrometric focal plane. Finally, another 12 CCDs will collect the data coming from the high resolution spectrometer (RVS) to measure the radial velocity of the sources, while the remaining CCDs are used for calibration purposes. During its 5 year life, it is expected to observe about 400,000 asteroids on average 60 times, discovering many new asteroids in the process. From the reflectance spectra planned, it should also produce a breakthrough in our understanding of the different minor bodies populations.

A good example of the importance of sample-return initiatives to understand the nature and evolution of asteroids come from the Japanese Space Agency (JAXA) Hayabusa mission (Fujiwara et al. 2006). Asteroid Itokawa (Fig. 4) is our best

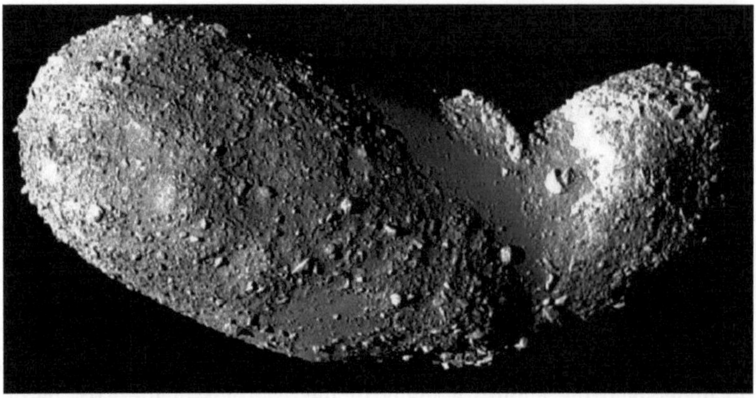

Fig. 4 Image of asteroid 25143 Itokawa our best known example of rubble pile (Hayabusa/JAXA)

studied rubble pile example, necessarily created from a complex collisional and re-aggregation history that exemplifies how evolved are the bodies dominating the near-Earth environment (see e.g. Chapman et al. 1989). The smallest asteroids typically delivered by mean motion planetary resonances have suffered catastrophic disruptions, and collisional cascade processes as envisioned by Michel et al. (2001, 2002). In fact, catastrophic disruptions are behind the formation of asteroidal cluster or families (Nesvorný et al. 2002). Then, it is not so surprising after all that shock effects in different degree are quite commonly reported in meteorites delivered to Earth.

There are two currently ongoing sample-return missions with significant cosmochemical interest addressing pristine asteroids associated with carbonaceous chondrites. The *Hayabusa 2* mission is a Japanese Space Agency (JAXA) initiative focused on returning samples from Apollo asteroid *162173 Ryugu* (also known by the provisional designation *1999* JU_3). This body belongs to a rare Cg spectral type that is defining a C-type asteroid presumably formed by carbonaceous chondrites, but exhibiting hydrated features in its reflectance spectrum like those observed in G-type asteroids represented by *1 Ceres*. The return of samples to Earth is expected to occur in December 2020 (Nakamura et al. 2015).

The *OSIRIS-Rex* NASA mission is focused in returning samples of asteroid (*101955*) *Bennu*. This body likely originated as a discrete asteroid in the inner MB approximately 0.7–2 Gyr ago as a fragment from the catastrophic disruption of a pristine carbonaceous asteroid (Lauretta et al. 2015).

Both sample-return missions will make history in returning primitive materials for study in terrestrial laboratories. Another proposed relevant mission is the *NASA Asteroid Redirect Mission* (*ARM*). This will be the first-ever robotic mission to visit a large near-Earth asteroid, collect a multi-ton boulder from its surface, and use it in an enhanced gravity tractor asteroid deflection demonstration. The spacecraft will then redirect the multi-ton boulder into a stable orbit around the moon, where astronauts will explore it and return with samples in the mid-2020s (Mazanek et al. 2013).

Unfortunately, a European sample-return mission called Marco Polo-R, initially pre-selected by the European Space Agency (ESA), failed final selection but promoted a significant scientific European cooperation in this field (Barucci et al. 2012). Undoubtedly such initiatives promote scientific cooperation in the context of the planned Asteroid Impact Deflection Asteroid mission (AIDA).

4 Hazardous Meteoroids Originating from Near Earth Asteroids

The possibility that meteorite-dropping bolide complexes associated with near-Earth asteroids could exist was first proposed by Halliday (1987). Trigo-Rodríguez et al. (2007) also found dynamic associations between large meteoroids and Near Earth Objects (NEOs). Many asteroids are rubble piles and so probably do not require a collision in order to be disrupted (Trigo-Rodríguez et al. 2009a, b; Madiedo et al. 2014a, b, c). The fragmentation process is likely to produce many meter-sized rocks as well as larger boulders and rubble pile asteroids that could form a complex of asteroidal fragments once disrupted all initially moving on nearly identical orbits. Detecting such families or associations may not be easy because the temporal stability of such orbital complexes is quite short (few tens of thousands of years) as consequence of planetary perturbations (Pauls and Gladman 2005), except perhaps for those cases exhibiting orbits with high inclination, where life-times can be considerably higher (Jones and Williams 2008). Disruptive and collisional processes also cause a divergence in the orbits (Bottke et al. 2002). Significant brecciation, and shock-induced darkening has been found e.g. in the Almahata Sitta and the Chelyabinsk meteorites (Kohout et al. 2014; Bischoff et al. 2010) indicating that collisions played a role in their evolution.

Pravec et al. (2012) studied a sample of 583 MB asteroids and NEAs and found that the typical albedo for carbonaceous asteroids (Tholen/Bus/DeMeo spectral classification: C/G/B/F/P/D) is 6 %, while the associated with ordinary chondrites (spectral classes S/A/L) is about a 20 %. It is not so surprising then that the NEA population discovered so far is dominated by the latter ones.

Other mechanisms for delivering meter-size meteoroids to Earth include tidal fracturing caused by close encounters with planets and fast rotation (Trigo-Rodríguez et al. 2007, 2008; Chapman 2010). Meteoroids ejected by a fast rotator could have quite low de-coherence timescales if we consider the YORP effect. Alternatively, a catastrophic disruption produce fragments in which typically the escape velocity is considerably smaller than the orbital velocity, so a large amount of the mass is ejected away at escape velocity (Bottke et al. 2005) which is considerably smaller than the orbital velocity. Consequently, a significant number

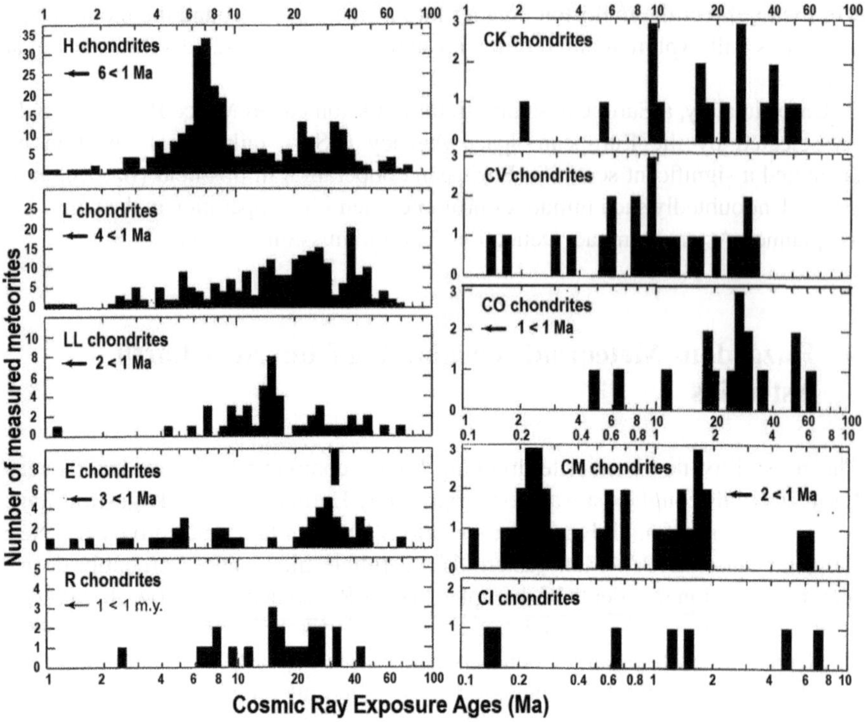

Fig. 5 Typical CREAs for chondritic meteorites (Adapted from Eugster et al. 2006)

of meter-sized pebbles or even boulders are released forming a stream of asteroidal fragments moving on nearly identical orbits (Williams 2004a, b; Jenniskens 2006; Trigo-Rodríguez et al. 2007). This is a likely source of meteorites to Earth, and it is more probable that such delivery from NEOs is dominated by low compressive strength materials: well fractured ordinary chondrites or well fragile carbonaceous chondrites (Trigo-Rodríguez and Blum 2009a, b).

While we cannot recover a more significant number of meteorites that have orbital information, some clues can be obtained from the study of Cosmic Ray Exposure Ages (CREAs) in meteorites inferred from the solar wind noble gases implanted in them during the movement around the Sun of the meter-sized progenitor meteoroids. It is generally accepted that the inferred CREAs corroborate that most meteorites reach the Earth after timescales of tens of Ma (see e.g. the review by Marti and Graf 1992). Such periods of time are consistent with meteorites delivered from the MB through dynamic resonances. A more exhaustive and recent review by Eugster et al. (2006) identified some chondrites exhibiting smaller CREAs (see Fig. 5).

If this hypothesis concerning the existence of a small, but significant fraction of meteorites being delivered from NEAs is correct, it is likely that a significant fraction of small meteorites found in Antarctica (where the minimum meteorite mass recovered is about one order of magnitude smaller than in the rest of locations) exhibit CREAs of few million years or even lower (Trigo-Rodríguez 2015). This is the expected timescale in which NEOs are crossing the near-Earth space before returning to the MB, and if they contribute significantly we should see at least some meteorites having such CREAs or even lower. First, some meteorites belonging to the L and H groups of ordinary chondrites exhibit CREAs lower than 1 Ma. Some examples are some finds: the H4 chondrite Cullison (CREA $= 1.4 \pm 0.3$ Ma), the H5 chondrite Grove Mountains 98,004 (0.05 Ma), or the L6 chondrite Ladder Creek (CREA $= 0.9 \pm 0.1$ Ma). For further details on these CREAs and references for the data outlined here see Eugster et al. (2007) compilation.

Many grouped and ungrouped carbonaceous chondrites remain to be studied and may contain fascinating clues on disruptive processes in their parent bodies like for example Elephant Moraine (EET) 96026, which that has an extremely low CRE age of 0.21 ± 0.03 Ma (Ma et al. 2002). Such CREAs can be perfectly consistent with a quick delivery from disrupted bodies (C-rich asteroids and extinct comets) located in the near-Earth space. It is important to mention that the typical CREAs inferred for ordinary chondrites is higher than for carbonaceous ones (see e.g. Fig. 5). In order to confirm these ideas to complete CREA data of small Antarctic chondrites seems essential, particularly these of carbonaceous nature.

5 Cometary Sources of Impact Hazard

Centaurs comprise a transitional population deriving from cometary reservoirs beyond Neptune that are crossing the orbits of giant planets (Jewitt 2008). Due to gravitational perturbations and the decay associated with their envisioned fragility, some of them could reach Earth-crossing orbits like it was found for comet 95P/Chiron (Hahn and Bailey 1990; Asher and Steel 1993). Recent claims have proposed a more careful study of the dynamic evolution and physical decay of these large objects (Napier et al. 2015).

This scenario invoking the importance of cometary bodies in impact hazard is not new. It is currently accepted that trans-Neptunian objects (TNOs) and centaurs, in chaotic orbits, evolve towards the inner planetary region (Duncan et al. 1995; Levison and Duncan 1997) at a significant rate via the Jupiter Family Comets (JFCs). These comets can be considered fragments of TNOs (Jewitt 2008) and participating as a significant source of impact hazard. Dynamic studies have found that comets forming part of such population become Earth crossers (Ipatov et al. 2007). They follow highly-eccentric orbits that can bring them to Earth vicinity at much higher relative velocities than typical asteroidal orbits, and usually are km-sized bodies. Although Earth encounters with JFCs are estimated to be one order of magnitude less frequent that with asteroids, the energy released can be several orders of magnitude because of their larger masses and much higher velocity. In a similar way that the collisional processing of asteroids produces most of the small asteroids crossing the near-Earth space, the origin of JFCs was explained by the progressive decay and subsequent inward orbital movement of TNOs (Emel'yanenko et al. 2004; Jewitt 2008).

By using visual observations reported 150 years ago in the Comptes Rendus de l'Académie des Sciences de Paris the atmospheric trajectory of the Orgueil CI chondrite has been reconstructed. Despite the intrinsic uncertainty of visual observations and estimating the duration of the luminous phase of that bolide, it was found that this rare meteorite could have been delivered from a JFC (Gounelle et al. 2006). There is growing evidence suggesting that meteoroids following high-inclination orbits are not necessarily fragile and might exist (Jones and Williams 2008). A meteorite fall occurred on July 6, 2007 over Cali, Colombia and visual observations of the bolide were compiled. The recovered meteorite was characterized as a rare H/L chondrite interloper that was identified as coming from the MB (Trigo-Rodríguez et al. 2009a, b). A potential meteorite dropping bolide called SPMN110708 "Bejar" appeared on July 11, 2008 with a pre-atmospheric orbit that indicated a plausible association with the debris of disrupted comet C/1919 Q2 Metcalf. The meteoroid was about one meter in diameter, it indicated that a cometary disruption can be source of meteorites (Trigo-Rodríguez et al. 2009; Williams 2011). Events in successive years close to the same time suggested that they should be investigated further, and other orbital solutions for the Cali event were explored. It is well known that visual observations introduce significant observational uncertainties and whether a unique solution can be obtained probably depends on the data set considered. We realized that another solution coming from the most extremely long trajectory observations was consistent with a grazing bolide with a radiant consistent with that estimated for Bejar. Is it possible that a high-inclination stream of meter-sized meteoroids originating from the disruption of comet C/1919 Q2 Metcalf exists?

The answer to the above question is probably positive. H/L ordinary chondrite falls other than Cali, occurred on similar dates but separated by several decades each (see Table 3). This cannot be casual given the small amount of H/L chondrites falls (4 so far) compared with the number of falls associated with H (366), L (417) or LL (99) group chondrites, usually coming from the MB (Meteoritical Bulletin Database 2016). The fall of these chondrite groups of chondrites **are** scattered all over the year (Grady 2000), while the H/L chondrites are concentrated between May and July (Table 3). To have the 4 H/L chondrites falls occurred in about a 3 months period corresponds to a probability of about a 6 % ($1/(4 \times 4)$, for each trimester chance). We should also remember that there are 4 H/L chondrite falls, compared with 882 chondrites belonging to another groups so the statistic is limited.

This evidence is interesting because it suggests that an important flux of meter-sized rocks currently reaching the Earth could come from high-inclination orbits that we are not taking into account, particularly in view that only a small amount of meteorites that fall on Earth are recovered.

The existence of meteorite streams was first suggested by Halliday et al. (1990) based on a number of fireballs in the Meteorite Observation and Recovery Program (MORP) and Prairie network databases with exceedingly similar orbits. The detailed data for 259 fireballs observed with MORP can be found in Halliday et al. (1996). Wolf and Lipschutz (1995) and Lipschutz et al. (1997), have argued that there is evidence from meteorite falls for the existence of meteorite streams.

Table 3 Compilation of H/L meteorite falls

Meteorite	Fall date (UTC)	Group & petrological subtype	TKM (kg)
Bremervörde	May 13, 1855	H/L3.9	7.25
Tieschitz	Jul 15, 1878	H/L3.6	28
Cali	July 6, 2007 21h32min	H/L4	0.478
Famenin	June 27, 2015 4h30min	H/L3.8–3.9	0.630

References: [1] Meteoritical Bulletin Database: http://www.lpi.usra.edu/meteor/index.php

We previously mentioned the recently found evidence of meteorite-dropping bolides associated with NEOs. A dynamic association between the orbit of NEO 2002NY40 and several bolides recorded over Spain and Finland was found (Trigo-Rodríguez et al. 2007). Babadzhanov et al. (2008) also found meteor showers dynamically associated with 2003EH1. Two meteorite-dropping bolides recorded over Spain were also clearly linked with PHA 2007LQ19 (Madiedo et al. 2014b). In addition, we also found a plausible dynamic link between the H5 chondrite Annama and PHA 2014UR116 (Trigo-Rodríguez et al. 2015). Also recently a dynamic link has been also claimed between comet 2P/Encke, the Taurid complex NEOs and the Maribo and Sutter's Mill meteorites, though the spectra of the bodies obtained was inconclusive (Tubiana et al. 2015).

The relative absence of NECs compared with other groups of NEOs known so far seems to indicate that our surveys could miss a significant number of extinct or inactive comets. The recent discovery of 2015 TB145, with only 3 weeks of margin to take action in case of a direct encounter, is a fact that should not leave us indifferent. A major concern in our strategy should arise from the evidence of low-albedo comets with such a type of high-eccentricity orbits. The danger associated with this kind of objects is well exemplified with the encounter velocity of 35 km/s during the closest approach of 2015 TB145 is slightly farther away than the Earth-Moon distance. The low-reflectivity also made that the original 400 m in diameter estimate was lately reassessed into 600 m, once studied the radar images obtained from Arecibo and Goldstone. There is no doubt that the future study of this primitive object will take a scientific interest.

Another superbolide recorded from Spain on July 13, 2012 provided evidence that meter-sized bodies with high tensile strength can be associated with rare and hazardous Damocloids (Madiedo et al. 2013a). These bodies are named after asteroid 5335 Damocles and exhibit Halley-family or long-period highly eccentric orbits, but without showing a cometary coma or tail. It is evidence that some comets are formed by high-strength materials, also being a significant source of hazard lying in high inclination orbits that might escape to most asteroid surveys.

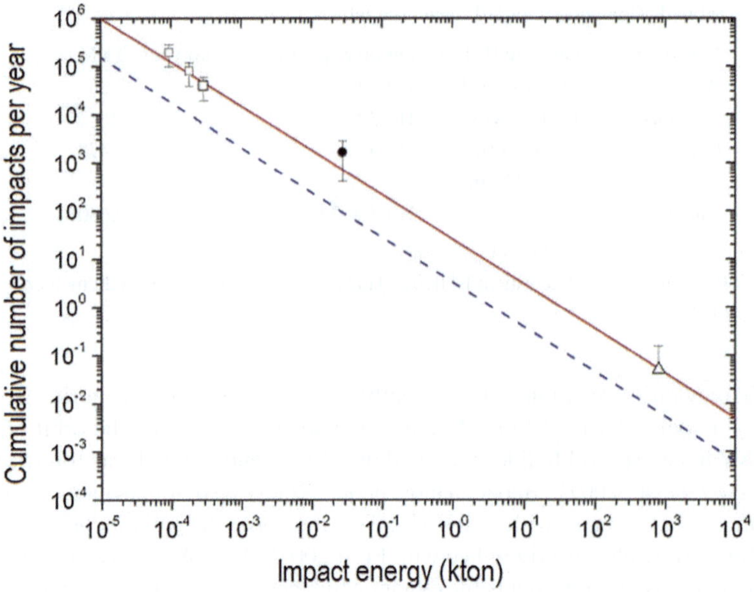

Fig. 6 Cumulative number of impacts per year as a function of the impact energy of the projectile. For details about the symbols please see the text (Adapted from Madiedo et al. 2014c)

Recent studies of lunar flares produced by sporadic meteoroid impacts have also confirmed that the flux of large meteoroids is higher than previously suspected (Ortiz et al. 2006). The cumulative frequency of impacts with the Earth as a function of the energy of the impact is shown in Fig. 6. The dashed line corresponds to the rate of impacts obtained by Brown et al. (2002), while the squares represent the results obtained by Ortiz et al. (2006) based on the analysis of flashes of impacts on the Moon. The continuous line corresponds to the frequency obtained from Ortiz et al. (2006). The result corresponds to the great impact on the Moon detected by Madiedo et al. (2014c). It is represented by the black circle, while the triangle corresponds to the impacts flow calculated by Brown et al. (2013) based on an analysis of the event of Chelyabinsk. There is no doubt that the evidence provided by infrasound and satellite flash detection of meter-sized projectiles is clearly relevant to flux estimates and it is outlined in (Tapia and Trigo-Rodríguez 2017)

6 Discussion and Conclusions

Fireball network evidence together with the information from recovered meteorites support the existence of a large diversity in the bodies contributing to Earth's contemporary impact hazard. Probably the relative absence of destructive events in the literature, but perhaps Tunguska and Chelyabinsk, is due to our relatively short written history. In the last century, some events have been clearly identified with asteroid or comet impacts, and significant progress is being made in our knowledge when meteorites are recovered from such rare events. This is a good reason to promote international cooperation in this regard, and try to decipher all routes of meteorites reaching our planet.

Consequently, the main conclusions of this review are:

1. Telescopic surveys are increasing the number of NEOs over the size range of tens to hundreds of meters, and the study of their physical and dynamic properties requires an interdisciplinary approach involving international cooperation. It opens the way of finding evidence of disruptions in the near-Earth environment, producing streams that can be source of meteorites over short-time scales.
2. The population of bodies crossing the Earth's orbit is extremely diverse. It is probably dominated by chondritic asteroids, but the presence of carbonaceous bodies is probably largely underestimated, particularly for small NEOs. This is because carbonaceous bodies have much lower albedos and following eccentric orbit escape to our ground-based biased monitoring techniques.
3. The collisional and dynamic pathways followed by NEOs until reaching our planet are biasing their populations towards high-strength materials that are the final products of successive collisions.
4. From the previous point, it is obvious that meteorite collections are not fully representative of the materials available in the Near Earth environment. There is a clear bias concerning low-strength materials that rarely survive to atmospheric entry or produce a crater, but likely produce air bursts and tektites.
5. Future sample return missions like OSIRIS Rex and Hayabusa 2 will develop new technology, and promote the study of asteroid samples in clean laboratories. Collecting new, previously non-sampled, materials from NEOs we will increase our capacity to develop palliative efforts, like e.g. much more efficient ways of deflecting asteroids.
6. Cometary samples are probably among the most fragile materials available in the solar system. Future cryogenic missions returning cometary samples could promote amazing discoveries on the first materials forming the protoplanetary disk.
7. Meter-sized meteoroids producing superbolides and following high-inclination, and eccentric orbits are not necessarily fragile, and can trace the existence of hazardous objects: dormant comets or Damocloids being an example.
8. H/L chondrites could have originated from disrupted comet C/1919 Q2 Metcalf. This was probably a rubble pile consisting of pieces of different petrologic nature from different parents, and a good example of the complex structure of a transitional body.

Acknowledgments This work was financially supported by AYA2015-67175-P.

References

Alexander, C.M., Bowden, R., Fogel, M.L., Howard, K.T., Herd, C.D.K., Nittler, L.R.: The provenance of asteroids, and their contributions to the volatile inventories of the terrestrial planets. Science **337**(6095), 721–723 (2012). doi:10.1126/science.1223474

Asher, D.J., Steel, D.I.: Orbital evolution of the large outer solar system object 5145 Pholus. Mon. Not. R. Astron. Soc. **263**, 179–190 (1993)

Babadzhanov, P.B., Williams, I.P., Kokhirova, G.I.: Meteor showers associated with 2003EH1. Mon. Not. R. Astron. Soc. **368**, 2271–2277 (2008)

Barucci, M.A., Fulchignoni, M., Fornasier, S., Dotto, E., Vernazza, P., Birlan, M., Binzel, R.P., Carvano, J., Merlin, F., Barbieri, C., Belskaya, I.: Asteroid target selection for the new Rosetta mission baseline: 21 Lutetia and 2867 steins. Astron. Astrophys. **430**, 313–317 (2005)

Barucci, M.A., Cheng, A.F., Michel, P., Benner, L.A.M., Binzel, R.P., Bland, P.A., Böhnhardt, H., Brucato, J.R., Campo, B.A., Cerroni, P., Dotto, E., Fitzsimmons, A., Franchi, I.A., Green, S.F., Lara, L.-M., Licandro, J., Marty, B., Muinonen, K., Nathues, A., Oberst, J., Rivkin, A.S., Robert, F., Saladino, R., Trigo-Rodríguez, J.M., Ulamec, S., Zolensky, M.: MarcoPolo-R near earth asteroid sample return mission. Exp. Astron. **33**, 645–684 (2012)

Binzel, R.P., Morbidelli, A., Merouane, S., DeMeo, F.E., Birlan, M., Vernazza, P., Thomas, C.A., Rivkin, A.S., Bus, S.J., Tokunaga, A.T.: Earth encounters as the origin of fresh surfaces on near-Earth asteroids. Nature **463**, 331–334 (2010)

Bischoff, A., Horstmann, M., Pack, A., Laubenstein, M., Haberer, S.: Asteroid 2008 TC3-Almahatta Sitta: a spectacular breccias containing many different ureilitic and chondritic lithologies. Meteorit. Planet. Sci. **45**, 1638–1656 (2010)

Bland, P.A., Spurný, P., Towner, M.C., Bevan, A.W.R., Singleton, A.T., Bottke Jr., W.F., Greenwood, R.C., Chesley, S.R., Shrbený, L., Borovička, J., Ceplecha, Z., McClafferty, T.P., Vaughan, D., Benedix, G.K., Deacon, G., Howard, K.T., Franchi, I.A., Hough, R.M.: An anomalous basaltic meteorite from the innermost main belt. Science **325**, 1525–1527 (2009)

Boslough, M., Crawford, D.A.: Low-altitude airbursts and the impact threat. Int. J. Impact Eng. **35**, 1441–1448 (2008)

Bottke Jr., W.F., Vokrouhlický, D., Rubincam, D.P., Broz, M.: The effect of Yarkovsky thermal forces on the dynamical evolution of asteroids and meteoroids. In: Bottke Jr., W.F., Cellino, A., Paolicchi, P., Binzel, R.P. (eds.) Asteroids III, pp. 395–408. University of Arizona Press, Tucson (2002)

Bottke, W.F., Durda, D.D., Nesvorný, D., Jedicke, R., Morbidelli, A., Vokrouhlický, D., Levison, H.F.: Linking the collisional history of the main asteroid belt to its dynamical excitation and depletion. Icarus **179**(1), 63–94 (2005)

Brown, P.G., Whitaker, R.W., ReVelle, D.O., Tagliaferri, E.: Multi-station infrasonic observations of two large bolides: signal interpretation and implications for monitoring of atmospheric explosions. Geophys. Res. Lett. **29**, 14-1–14-4 (2002). CiteID 1636

Brown, P.G., Assink, J.D., Astiz, L., Blaauw, R., Boslough, M.B., Borovička, J., Brachet, N., Brown, D., Campbell-Brown, M., Ceranna, L., Cooke, W., de Groot-Hedlin, C., Drob, D.P., Edwards, W., Evers, L.G., Garces, M., Gill, J., Hedlin, M., Kingery, A., Laske, G., Le Pichon, A., Mialle, P., Moser, D.E., Saffer, A., Silber, E., Smets, P., Spalding, R.E., Spurný, P., Tagliaferri, E., Uren, D., Weryk, R.J., Whitaker, R., Krzeminski, Z.: A 500-kiloton airburst over Chelyabinsk and an enhanced hazard from small impactors. Nature **503**(7475), 238–241 (2013)

Ceplecha, Z., Borovicka, J., Elford, J.G., Revelle, D.O., Hawkes, R.L., Porubcan, V., Simek, M.: Meteor phenomena and bodies. Space Sci. Rev. **84**(3/4), 327–471 (1998)

Chapman, C.R.: Stripped on passing by Earth. Nature **463**, 305–306 (2010)

Chapman, C.R., Davis, D.R., Greenberg, R.: Apollo asteroids: relationships to main belt asteroids and meteorites. Meteoritics **17**, 193–194 (1982)

Chapman, C.R., Paolicchi, P., Zappalà, V., Binzel, R.P., Bell, J.F.: Asteroid families: physical properties and evolution. In: Binzel, R.P., Gehrels, T., Matthews, M.S. (eds.) Asteroids II, pp. 386–415. University of Arizona, Tucson (1989)

Davis, D.R., Durda, D.D., Marzari, F., Campo Bagatin, A., Gil-Hutton, R.: Collisional evolutions of small body populations. In: Bottke Jr., W.F., Cellino, A., Paolicchi, P., Binzel, R.P. (eds.) Asteroids III, pp. 545–558. University of Arizona Press, Tucson (2002)

Duncan, M.J., Levison, H.F., Budd, S.M.: The dynamical structure of the Kuiper belt. Astron. J. **110**, 3073 (1995)

Emel'yanenko, V.V., Asher, D.J., Bailey, M.E.: High-eccentricity trans-Neptunian objects as a source of Jupiter-family comets. Mon. Not. R. Astron. Soc. **350**, 161–166 (2004)

Eugster, O., Herzog, G.F., Marti, K., Caffee, M.W.: Irradiation records, cosmic-ray exposure ages, and transfer times of meteorites. In: Lauretta, D., McSween, H.Y. (eds.) Meteorites and the early solar system II, pp. 829–851. The University of Arizona Press, Tucson (2006)

Eugster, O., Lorenzetti, S., Krähenbühl, U., Marti, K.: Comparison of cosmic-ray exposure ages and trapped noble gases in chondrule and matrix samples of ordinary, enstatite, and carbonaceous chondrites. Meteorit. Planet. Sci. **42**, 1357–1371 (2007)

Farinella, P., Vokrouhlický, D.: Semimajor axis mobility of asteroidal fragments. Science **283**, 1507–1510 (1999)

Fujiwara, A., et al.: The rubble-pile asteroid Itokawa as observed by Hayabusa. Science **312**, 1330–1334 (2006)

Gehrels, T., McMillan, R., Frecker, J., Roland, E., Stoll, C., Doose, L., Shoemaker, E., Nozette, S., Boesgaard, H.: Progress report on the Spacewatch camera. Bull. Am. Astron. Soc. **14**, 728 (1972)

Gounelle, M., Spurný, P., Bland, P.A.: The orbit and atmospheric trajectory of the Orgueil meteorite from historical records. Meteorit. Planet. Sci. **41**, 135–150 (2006)

Grady, M.: Catalogue of Meteorites. Cambridge University Press, Cambridge (2000). 689pp

Hahn, G., Bailey, M.E.: Rapid dynamical evolution of giant comet Chiron. Nature **348**, 132–136 (1990)

Halliday, I.: Detection of a meteorite 'stream'—observations of a second meteorite fall from the orbit of the Innisfree chondrite. Icarus **69**, 550–556 (1987)

Halliday, I., Blackwell, A.T., Griffin, A.A.: Evidence for the existence of groups of meteorite-producing asteroidal fragments. Meteoritics **25**, 93–99 (1990)

Halliday, I., Griffin, A.A., Blackwell, A.T.: Detailed data for 259 fireballs from the Canadian camera network and inferences concerning the influx of large meteoroids. Meteorit. Planet. Sci. **31**, 185–217 (1996)

Huss, G.R., Rubin, A.E., Grossman, J.N.: Thermal metamorphism in chondrites. In: Lauretta, D.S., McSween, H.Y. (eds.) Meteorites and the Early Solar System II, pp. 567–586. University of Arizona Press, Tucson (2006)

Ipatov S.I., Mather J.C.: Migration of comets to the terrestrial planets. In: Milani, A., Valsecchi, G.B.. Vokrouhlický, D (eds.) Near Earth Objects, our Celestial Neighbors: Opportunity and Risk, Proceedings of the IAU Symposium 236, pp. 55–64. University Press, Cambridge (2007)

Jenniskens, P.: On the dynamics of meteoroid streams. Earth Planets Space **50**, 555–567 (1998)

Jenniskens, P.: Meteor showers and their parent comets. Cambridge University Press, Cambridge (2006). 790 pp

Jenniskens, P.: (Mostly) dormant comets in the NEO population and the meteoroid streams that they crumble into. In: Milani, A., Valsecchi, G.B., Vokrouhlický, D. (eds.) Near Earth Objects, our Celestial Neighbors: Opportunity and Risk, Proceedings of the IAU Symposium 236, pp. 87–94. University Press, Cambridge (2007)

Jenniskens, P., Shaddad, M.H., Numan, D., Elsir, S., Kudoda, A.M., Zolensky, M.E., Le, L., Robinson, G.A., Friedrich, J.M., Rumble, D., Steele, A., Chesley, S.R., Fitzsimmons, A., Duddy, S., Hsieh, H.H., Ramsay, G., Brown, P.G., Edwards, W.N., Tagliaferri, E., Boslough, M.B., Spalding, R.E., Dantowitz, R., Kozubal, M., Pravec, P., Borovička, J., Charvat, Z., Vaubaillon, J., Kuiper, J., Albers, J., Bishop, J.L., Mancinelli, R.L., Sandford, S.A., Milam, S.N., Nuevo, M., Worden, S.P.: The impact and recovery of asteroid 2008 TC3. Nature **458**, 485–488 (2009)

Jewitt, D.: Kuiper belt and comets: an observational perspective. In: Altwegg, K., Benz, W., Thomas, N. (eds.) Trans-Neptunian Objects and Comets, pp. 1–78. Springer, Berlin (2008)

Jones, D., Williams, I.P.: High inclination meteorite streams can exist. Earth Moon Planets **102**, 35–46 (2008)

Jopek, T.J., Williams, I.P.: Stream and sporadic meteoroids associated with near-Earth objects. Mon. Not. R. Astron. Soc. **430**, 2377–2389 (2013)

Kohout, T., Gritsevich, M., Grokhovsky, V.I., Yakovlev, G.A., Haloda, J., Halodova, P., Michallik, R.M., Penttilä, A., Muinonen, K.: Mineralogy, reflectance spectra, and physical properties of the Chelyabinsk LL5 chondrite—insight into shockinduced changes in asteroid regoliths. Icarus **228**, 78–85 (2014)

Lauretta, D.S., Bartels, A.E., Barucci, M.A., Bierhaus, E.B., Binzel, R.P., Bottke, W.F., Campins, H., Chesley, S.R., Clark, B.C., Clark, B.E., Cloutis, E.A., Connolly, H.C., Crombie, M.K., Delbó, M., Dworkin, J.P., Emery, J.P., Glavin, D.P., Hamilton, V.E., Hergenrother, C.W., Johnson, C.L., Keller, L.P., Michel, P., Nolan, M.C., Sandford, S.A., Scheeres, D.J., Simon, A.A., Sutter, B.M., Vokrouhlický, D., Walsh, K.J.: The OSIRIS-REx target asteroid (101955) Bennu: constraints on its physical, geological, and dynamical nature from astronomical observations. Meteorit. Planet. Sci. **50**, 834–849 (2015)

Levison, H.F., Duncan, M.J.: From the Kuiper belt to Jupiter-family comets: the spatial distribution of ecliptic comets. Icarus **127**, 13–32 (1997)

Lewis, J.S.: Mining the Skies: Untold Riches from the Asteroids, Comets, and Planets. Perseus Publishing, Reading (1997)

Lipschutz, M.E., Wolf, S.F., Dodd, R.T.: Meteoroid streams as sources for meteorite falls: a status report. Planet. Space Sci. **45**, 517–523 (1997)

Ma, Y., Williams, I.P., Chen, W.: On the ejection velocity of meteoroids from comets. Mon. Not. R. Astron. Soc. **337**, 1081–1086 (2002)

Madiedo, J.M., Trigo-Rodríguez, J.M., Zamorano, J., Ortiz, J.L., de Miguel, A.S., Ocaña, F., Izquierdo, J., Castro-Tirado, A.J., Morales, N., Galadí, D., de Guindos, E., Lacruz, J., Organero, F., Ana-Hernández, L., Fonseca, F., Tapia, M., Gallego, F., Cabrera-Caño, J.: Analysis of a superbolide from a Damocloid observed over Spain on 2012 July 13. Mon. Not. R. Astron. Soc. **436**, 3656–3662 (2013a)

Madiedo, J.M., Trigo-Rodríguez, J.M., Zamorano, J., Ana-Hernández, L., Izquierdo, J., Ortiz, J.L., Castro-Tirado, A.J., Sánchez de Miguel, A., Ocaña, F., Pastor, S., de los Reyes, J.A., Galadí, D., de Guindos, E., Organero, F., Fonseca, F., Cabrera-Caño, J.: Trajectory, orbit, and spectroscopic analysis of a bright fireball observed over Spain on April 13, 2013. Astron. Astrophys. **569**, 8pp (2014c) id.A104

Madiedo, J.M., Trigo-Rodríguez, J.M., Williams, I.P., Konovalova, N., Ortiz, J.L., Castro-Tirado, A.J., Pastor, S., de los Reyes, J.A., Cabrera-Caño, J.: Near-Earth object 2012XJ112 as a source of bright bolides of achondritic nature. Mon. Not. R. Astron. Soc. **439**, 3704–3711 (2014a)

Madiedo, J.M., Trigo-Rodríguez, J.M., Ortiz, J.L., Castro-Tirado, A.J., Cabrera-Caño, J.: Bright fireballs associated with the potentially hazardous asteroid 2007LQ19. Mon. Not. R. Astron. Soc. **443**, 1643–1650 (2014b)

Madiedo, J.M., Ortiz, J.L., Morales, N., Cabrera-Caño, J.: A large Lunar impact blast on September 11th 2013. Mon. Not. R. Astron. Soc. **439**, 2364–2369 (2014c)

Marti, K., Graf, T.: Cosmic-ray exposure history of ordinary chondrites. Ann. Rev. Earth Planet. Sci. **20**, 221–243 (1992)

Martínez-Jiménez, M., Moyano-Cambero, C.E., Trigo-Rodríguez, J.M., Alonso-Azcárate, J., Llorca, J.: Asteroid mining: mineral resources in undifferentiated bodies from the chemical composition of carbonaceous chondrites. In: Trigo-Rodríguez, J.M., Gritsevich, M., Palme H. (eds.) Assessment and Mitigation of Asteroid Impact Hazards, pp. 73–101. Springer, New York (2017)

Marty, B.: The origins and concentrations of water, carbon, nitrogen and noble gases on Earth. Earth Planet. Sci. Lett. **313–314**, 56–66 (2012)

Mazanek, D.D., Brophy, J.R., Merrill, R.G.: Asteroid Retrieval Mission Concept—Trailblazing Our Future in Space and Helping to Protect Us from Earth Impactors. In Planetary Defense Conference 2013, IAA-PDC13-04-14 (2013)

Meteoritical Bulletin Database. http://www.lpi.usra.edu/meteor/. (2016)

Michel, P., Benz, W., Tanga, P., Richardson, D.C.: Collisions and gravitational reaccumulation: forming asteroid families and satellites. Science **294**, 1696–1700 (2001)

Michel, P., Benz, W., Tanga, P., Richardson, D.C.: Formation of asteroid families by catastrophic disruption: simulations with fragmentation and gravitational reaccumulation. Icarus **160**, 10–23 (2002)

Morbidelli, A., Nesvorný, D.: Numerous weak resonances drive asteroids toward terrestrial planets orbits. Icarus **139**, 295–308 (1999)

Nakamura T., Iwata, T., Kitasato, K., Abe, M., Osawa, T., Matsuoka, M., Nakauchi, Y., Arai, T., Komatsu, M., Hiroi, T., Imae, N., Yamaguchi, A., Kojima, H.: Reflectance Spectra Measurement of Various Carbonaceous Chondrites Using Hayabusa-2 Near Infrared Spectrometer. In: 78th Annual Meeting of the Meteoritical Society, LPI Contribution No. 1856, p. 5206 (2015)

Napier, B., Asher, D., Bailey, M., Steel, D.: Centaurs as a hazard to civilization. Astron. Geophys. **56**, 6.24–6.30 (2015)

Nesvorný, D., Bottke, W.F., Dones, L., Levison, H.F.: The recent breakup of an asteroid in the main-belt region. Nature **417**, 720–771 (2002)

Nesvorný, D., Bottke, W.F., Levison, H.F., Dones, L.: Recent origin of the solar system dust bands. Astrophys. J. **591**, 486–497 (2003)

Nesvorný, D., Enke, B.L., Bottke, W.F., Durda, D.D., Asphaug, E., Richardson, D.C.: Karin cluster formation by asteroid impact. Icarus **183**, 296–311 (2006)

Ortiz, J.L., Aceituno, F.J., Quesada, J.A., Aceituno, J., Fernández, M., Santos-Sanz, P., Trigo-Rodríguez, J.M., Llorca, J., Martín-Torres, F.J., Montañés-Rodríguez, P., Pallé, E.: Detection of sporadic impact flashes on the moon: implications for the luminous efficiency of hypervelocity impacts and derived terrestrial impact rates. Icarus **184**, 319–326 (2006)

Pauls, A., Gladman, B.: Decoherence time scales for "meteoroid streams". Meteorit. Planet. Sci. **40**, 1241–1256 (2005)

Pravec, P., Harris, A.W., Kusnirak, P., Galad, A., Hornoch, K.: Absolute magnitudes of asteroids and a revision of asteroid albedo estimates from WISE thermal observations. Icarus **221**, 365–387 (2012)

Tanga, P., Mignard, F.: The solar system as seen by Gaia: the asteroids and their accuracy budget. Planet. Space Sci. **73**, 5–9 (2012)

Tapia, M., Trigo-Rodríguez, J.M.: Natural hazard associated to shock waves of meter-sized meteoroids. In: Trigo-Rodríguez, J.M., Gritsevich, M., Palme H. (eds.) Assessment and Mitigation of Asteroid Impact Hazards, pp. 199–217. Springer, New York (2017)

Trigo-Rodríguez, J.M.: Aqueous alteration in chondritic asteroids and comets from the study of carbonaceous chondrites. In: Lee, M.R., Leroux, H. (eds.) Planetary Mineralogy, EMU Notes in Mineralogy, vol. 15, pp. 67–87. European Mineralogical Union and the Mineralogical Society of Great Britain and Ireland, London (2015). Chapter 3

Trigo-Rodríguez, J.M., Blum, J.: Tensile strength as an indicator of the degree of primitiveness of undifferentiated bodies. Planet. Space Sci. **57**, 243–249 (2009a)

Trigo-Rodríguez, J.M., Blum, J.: The effect of aqueous alteration and metamorphism in the survival of presolar silicate grains in chondrites. Publ. Astron. Soc. Australia **26**, 289–296 (2009b)

Trigo-Rodríguez, J.M., Llorca, J.: The strength of cometary meteoroids: clues to the structure and evolution of comets. Mon. Not. R. Astron. Soc. **372**, 655–660 (2006)

Trigo-Rodríguez, J.M., Llorca, J.: Erratum: The strength of cometary meteoroids: clues to the structure and evolution of comets. Mon. Not. R. Astron. Soc. **375**, 415 (2007)

Trigo-Rodríguez J. M., Bottke, W.F., Campo Bagatin, A., Tanga, P., Llorca, J., Jones, D. C., Williams, I., Madiedo, J.M., Lyytinen, E.: Is Asteroid 2002NY40 a Rubble Pile Gravitationally Disrupted? LPI Contribution No., 1391, 39th Lunar and Planetary Science Conference, p. 1692 (2008)

Trigo-Rodríguez, J.M., Rubin, A.E., Wasson, J.T.: Non-nebular origin of dark mantles around chondrules and inclusions in CM chondrites. Geochim. Cosmochim. Acta **70**, 1271–1290 (2006)

Trigo-Rodríguez, J.M., Lyytinen, E., Jones, D.C., Madiedo, J.M., Castro-Tirado, A.J., Williams, I., Llorca, J., Vítek, S., Jelínek, M., Troughton, B., Gálvez, F.: Asteroid 2002NY40 as a source of meteorite-dropping bolides. Mon. Not. R. Astron. Soc. **382**, 1933–1939 (2007)

Trigo-Rodríguez, J.M., Llorca, J., Rubin, A.E., Grossman, J.N., Sears, D.W.G., Naranjo, M., Bretzius, S., Tapia, M., Guarín Sepúlveda, M.H.: The Cali meteorite fall: a new H/L ordinary chondrite. Meteorit. Planet. Sci. **44**, 211–220 (2009a)

Trigo-Rodríguez, J.M., Madiedo, J.M., Williams, I.P., Castro-Tirado, A.J., Llorca, J., Vítek, S., Jelínek, M.: Observations of a very bright fireball and its likely link with comet C/1919 Q2 Metcalf. Mon. Not. R. Astron. Soc. **394**, 569–576 (2009b)

Trigo-Rodríguez, J.M., Lyytinen, E., Gritsevich, M., Moreno-Ibáñez, M., Bottke, W.F., Williams, I., Lupovka, V., Dmitriev, V., Kohout, T., Grokhovsky, V.: Orbit and dynamic origin of the recently recovered Annama's H5 chondrite. Mon. Not. R. Astron. Soc. **449**, 2119–2127 (2015)

Trigo-Rodríguez, J.M., Moyano-Cambero, C.E., Llorca, J., Fornasier, S., Barucci, M.A., Belskaya, I., Martins, Z., Rivkin, A.S., Dotto, E., Madiedo, J.M., Alonso-Azcárate, J.: UV to far-IR reflectance spectra of carbonaceous chondrites. I. Implications for remote characterization of dark primitive asteroids targeted by sample-return missions. Mon. Not. R. Astron. Soc. **437**, 227–240 (2014)

Tubiana C., Snodgrass, C., Michelsen, R., Haack, H., Böhnhardt, H., Fitzsimmons, A., Williams, I. P.: 2P/Encke, the Taurid complex NEOs and the Maribo and Sutter's Mill meteorites. Astron. Astrophys. **584** (2015). id.A97

Wasson, J.T.: Large aerial burst: an important class of terrestrial accretionary events. Astrobiology **3**, 163–178 (2003)

Weisberg, M.K., McCoy, T.J., Krot, A.N.: Systematics and evaluation of meteorite classification. In: Lauretta, D.S., McSween, H.Y. (eds.) Meteorites and the Early Solar System II, pp. 19–52. University of Arizona Press, Tucson (2006)

Weissman, P., Hicks, M.D., Abell, P.A., Choi, Y.-J., Lowry, S.C.: Rosetta target asteroid 2867 steins: an unusual E-type asteroid. Meteorit. Planet. Sci. **43**, 905–914 (2008)

Williams, I.P.: Meteoroid streams and the sporadic background, earth. Moon Planets **68**, 1–12 (1995)

Williams, I.P.: Meteoroid streams: successes and problems. J. Int. Meteor Org. **32**, 11–20 (2004a)

Williams, I.P.: The evolution of meteoroid streams. In: Murad, E., Williams, I.P. (eds.) Meteors in the Earth's Atmosphere, pp. 13–32. Cambridge University Press, Cambridge (2004b)

Williams, I.P.: The origin and evolution of meteor showers and meteoroid streams. Astron. Geophys. **52**(2), 20–26 (2011)

Williams, I.P., Jopek, T.J.: The Origin of stream and sporadic meteors, comets or asteroids. In: Jopek, T.J., Rietmeijer, F.J.M., Watanabe, J., Williams, I.P. (eds.) The Meteoroids 2013, pp. 179–192. A. M. University Press, Poznan (2014)

Wolf, S.F., Lipschutz, M.E.: Meteoroid streams: evidence for meteorites recovered on Earth. Earth Moon Planets **68**, 605–637 (1995)

Yau, K., Weissman, P., Yeomans, D.: Meteorite falls in China and some related human casualty events. Meteoritics **29**, 864–871 (1994)

The Chemistry of Solar System Materials: Sun, Planets, Asteroids, Meteorites and Dust

Herbert Palme and Jutta Zipfel

Abstract In this paper we summarize our knowledge of the chemical composition of solar system materials accessible to analysis. In the Sun the three most important rock forming elements Mg, Si and Fe have about the same number of atoms (Mg/Si = 1; Fe/Si = 0.91); the number of Al atoms is a factor of 10 lower (Al/Si = 0.09). Chondritic meteorites have essentially the same chemical signature with some variability, about 20 % for Mg/Si, 50 % for Al/Si and a factor of two for Fe/Si. These variations can be accounted for by variably mixing components that formed by condensation in a cooling gas of solar composition (Mg-silicates, Ca,Al-rich inclusions, NiFe metal). The bulk Earth composition is within this range and may be considered in a broad sense to be chondritic. The bulk compositions of the other terrestrial planets are less well known. They all have a metal core and basaltic surface rocks. Exceptions are Mercury with too much and the Moon with too little iron for a chondritic bulk composition. Asteroids also seem to have chondritic bulk compositions. S-type asteroids have been confirmed to be the parent bodies of ordinary chondrites. Most of the C-type asteroids appear to represent carbonaceous chondrites. The mm to sub-millimeter sized micrometeorites are debris of asteroids and/or comets. They are largely chondritic in composition but the ratio of cometary to asteroidal material is unclear. If there is a significant fraction of cometary material, comets should have chondritic bulk composition, as approximately inferred from the Giotto data.

Interplanetary dust particles (IDP), micrometer to sub-micrometer in size are also largely chondritic. They often contain GEMS (glass with embedded metal and sulfides), nano-meter sized particles which scatter around chondritic bulk compositions and are considered by some authors to be undisturbed interstellar material, the parental material of the solar system.

If material left over from the formation of the Sun was CI-chondritic with respect to rock forming elements, then massive redistribution of high temperature

H. Palme (✉) • J. Zipfel
Forschungsinstitut und Naturmuseum Senckenberg, Senckenberganlage 25, Frankfurt am Main D-60325, Germany
e-mail: Palmeherbert@googlemail.com

J.M. Trigo-Rodríguez et al. (eds.), *Assessment and Mitigation of Asteroid Impact Hazards*, Astrophysics and Space Science Proceedings 46,
DOI 10.1007/978-3-319-46179-3_3

components must have occurred in the early solar nebula to account for the enrichment of refractory elements in the Earth.

Within this framework we are addressing the following fundamental questions: To what extent are the objects of the solar system chemically uniform? What is the relationship between meteorites, asteroids, comets and dust? Are meteorites building blocks of the Earth and other terrestrial planets? How can we account for the enrichment of refractory elements in the Earth.

1 The Composition of the Sun

The history of the solar system begins with the formation of the Sun, about 4.7 billion years ago. The Sun contains 99.86 % of the total mass of the solar system while the rest is accounted for by planets, moons, asteroids, comets, and primitive and processed dust. The abundance of most elements in the Sun can be determined from absorption lines in the spectrum of the solar photosphere, the visible outer 300 km of the Sun. Determining solar abundances from the absorption spectra is a complicated procedure, which requires models of the solar atmosphere and the line formation process (e.g., Asplund et al. 2009). Spectroscopic abundance determinations are not possible for rare gases and elements with lines not accessible or heavily blended for quantitative spectroscopy in the photosphere (e.g., As, Se, Br, Te, I, Cs). Other elements can only be determined in sunspot spectra (e.g., F, Cl, Tl, In) with relatively large uncertainties (Lodders et al. 2009). Elemental abundance of the Sun usually are reported on a logarithmic scale (astronomical scale) as atoms relative to 10^{12} atoms of hydrogen:

$$A_{ast}\,(El) = \log\ [N\,(El)\,/N\,(H)] + 12 \tag{1}$$

where $A_{ast}(El)$ is the abundance of element El on the astronomical scale, with normalization to 10^{12} atoms of H, and N(El)/N(H) is the atomic ratio of element El to H.

Almost all meteorites deviate more or less from solar composition. There are only five out of more than 50 000 meteorites with strictly solar composition for most elements heavier than oxygen and with the exception of rare gases. These meteorites belong to the group of CI chondrites. The two largest CI-meteorites Orgueil (14 kg) and Alais (6 kg) are preserved in museums and material is available for study. In order to compare meteoritic abundances to solar abundances it is necessary to convert solar abundances on the astronomical scale to the meteoritic scale, using the normalization to 10^6 Si atoms. The conversion is given by the following equation:

$$\log\ A_{ast} = \log\ A_{met} + 1.533 \pm 0.03 \tag{2}$$

where A_{met} is the abundance on the meteoritic scale with normalization to 10^6 atoms of Si (Palme et al. 2014).

The 20 most abundant elements in the Sun, arranged with decreasing solar abundance, are shown in Fig. 1 and listed in Table 1. The meteorite data in Fig. 1 of CI chondrites and CV chondrites, a related but compositionally different group of

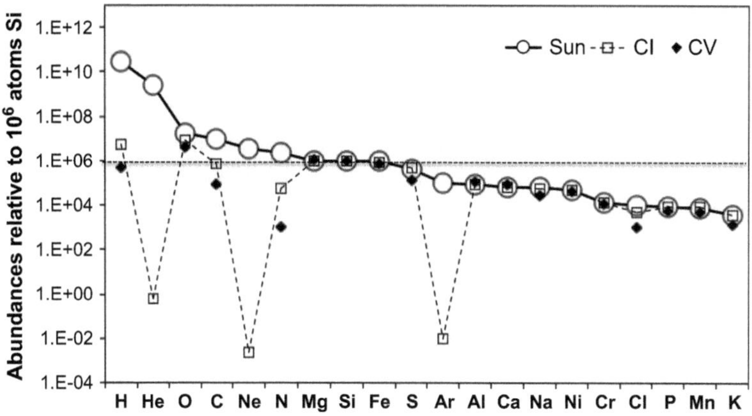

Fig. 1 Comparison of solar abundances and meteoritic abundances of CI and CV chondrites of the 20 most abundant elements in the Sun normalized to 10^6 atoms of Si

Table 1 The 20 most abundant elements in the Sun and CI carbonaceous chondrites

	Elements	Solar abundances	Meteoritic abundances	Meteoritic/solar abundance ratio
1	**H**	3.02×10^{10}	5.13×10^{6}	1.07×10^{-4}
2	**He**	2.54×10^{9}	0.60	2.37×10^{-10}
3	**O**	1.62×10^{7}	7.76×10^{6}	0.48
4	C	9.55×10^{6}	7.76×10^{5}	0.081
5	**Ne**	3.39×10^{6}	2.43×10^{-3}	6.92×10^{-10}
6	**N**	2.19×10^{6}	5.62×10^{4}	0.026
7	**Si**	1.00×10^{6}	1.00×10^{6}	1.00
8	**Mg**	1.05×10^{6}	1.05×10^{6}	1.00
9	**Fe**	9.12×10^{5}	8.71×10^{5}	0.95
10	**S**	4.37×10^{5}	4.47×10^{5}	1.02
11	**Ar**	9.55×10^{4}	9.77×10^{-3}	1.02×10^{-7}
12	**Al**	8.91×10^{4}	8.32×10^{4}	0.93
13	**Ca**	6.46×10^{4}	6.17×10^{4}	0.95
14	**Na**	6.03×10^{4}	5.75×10^{4}	0.95
15	**Ni**	5.13×10^{4}	4.90×10^{3}	0.95
16	**Cr**	1.32×10^{4}	1.35×10^{4}	1.02
17	**Cl**	9.55×10^{3}	5.25×10^{3}	0.55
18	**P**	8.71×10^{3}	8.32×10^{3}	0.95
19	**Mn**	7.08×10^{3}	9.33×10^{3}	1.32
20	**K**	3.89×10^{3}	3.72×10^{3}	0.95

Data are in atoms relative to 10^6 Si atoms. (a) Solar abundances are solar photosphere data (Lodders et al. 2009) recalculated to the meteoritic scale using equation (2). (b) Meteoritic abundances are based on analyses of CI chondrites only (Palme et al. 2014)

meteorites, are shown for comparison. All data are taken from Palme et al. (2014) and are based on a data set published in Lodders et al. (2009) that has been updated by recent results in the literature.

The Sun has higher abundances than meteorites for the first six elements, H to N of the Si-normalized data set in Table 1. In the solar nebula these six elements are predominantly present as gas in very volatile compounds and are therefore only partly incorporated in solids which explains their lower abundance in meteorites. Elements heavier than Mg show excellent agreement between meteorite data and solar photospheric abundances, except for the rare gas Ar. Data for type 3 CV carbonaceous chondrites with its most prominent member Allende fit less well with solar abundances, in particular for moderately volatile elements such as S, Na, and Cl. Also, the two refractory elements Al and Ca plot slightly above the CI chondrite level (Fig. 1).

The agreement between solar and meteoritic abundances for heavy elements (Si to K) in Table 1 is 5 % or better (except Ar, Cl, Mn). Argon as rare gas is not condensed, the photospheric Cl abundance has a large error and Mn is an exception as discussed in Palme et al. (2014). From Fig. 1 and Table 1 it is clear that the three elements Si, Mg and Fe (together with appropriate amounts of oxygen) are the most abundant rock forming elements in the Sun. They make up the major fraction of extraterrestrial solids and, as shown later, terrestrial material. As we will see below the sequence of decreasing heavy element abundances in Fig. 1 and Table 1 is typical of many meteorites and planets.

The Sun is a star of spectral type GV2, and it is a member of a big stellar system, the Milky Way, a typical spiral galaxy that contains about 10^{12} solar masses. Many accurate abundances determinations of similar stars in the neighborhood of the Sun have been performed. The composition of these stars is, if corrected for the galactic evolution of heavy elements, similar to that of our Sun. Thus compositionally the Sun is not an exceptional star (Lodders et al. 2009).

2 Classification of Solar System Materials

Meteorites are the prime source for information about the composition of extraterrestrial materials. Meteorites were released during collisions as fragments from asteroidal bodies or planetesimals in the asteroid belt or from the surface of planets, such as Moon and Mars. In most cases meteorites recovered on the surface of the Earth only spent a few million years as individual bodies, so called meteoroids, in space. Most of their lifetime they were buried in the interior of larger planetesimals completely shielded from cosmic rays. Two different types of such planetesimals can be distinguished, those which were once molten and subsequently differentiated into core, mantle and crust and those which were not heated enough to reach melting temperatures, although they may have experienced variable degrees of thermal metamorphism. The latter are called chondritic meteorites because most of them contain sub-mm to mm sized spherules, the chondrules, which were molten in space before accreting into the meteorite parent body. As we shall see, chondritic meteorites have approximately the same chemical composition for rock forming elements as the average solar system i.e., the Sun. They are therefore often referred to as primitive meteorites.

Melting of chondritic planetesimals leads to two major events: formation of a metal core by gravitational settling of metallic FeNi-alloys, and formation of an Al-rich silicate crust by partial melting of the Mg-rich silicate mantle in a process called differentiation. A variety of meteorite classes are derived from such differentiated planetesimals, iron meteorites from the cores (or segregated metal-ponds), stony irons from the core-mantle boundary and achondrites (basaltic meteorites) from the crust. The composition of these meteorites is very different from the average solar system composition. It is likely but difficult to prove in detail that differentiated meteorites come from once molten planets with chondritic bulk composition.

3 Bulk Chemical Composition of Chondritic Meteorites

The large number of chondritic meteorites with similar numbers of Si, Mg, and Fe atoms indicates that solar system material left over from the formation of the Sun is in terms of heavy element composition not very different from the composition of the Sun. In detail, there are small compositional differences among chondritic meteorites and, as pointed out above, only the CI chondrites match the solar element abundances precisely within analytical uncertainties for Si, Mg, Fe, and many more elements (Table 1; for a complete list see Palme et al. 2014). Other groups of chondritic meteorites, although with abundance patterns roughly similar to the Sun for rock forming elements, deviate somewhat from the solar composition. This is demonstrated in Fig. 2, where the abundances of several elements are plotted for members of the three major chondrite classes, the carbonaceous, ordinary, and enstatite chondrites.

The major chondrite classes shown in Fig. 2 are carbonaceous chondrites with CI, CM, CO, and CV groups, ordinary chondrites with H, L, and LL groups, and enstatite chondrites with EH and EL groups. These are the most populated chondrite groups. The large number of meteorites recovered from Antarctica and from hot deserts have led to a significant increase in new groups of chondritic meteorites not shown in Fig. 2. For a detailed discussion the reader is referred to Krot et al. (2003).

The element ratios plotted in Fig. 2 are representative of different cosmochemical components. Each of these components has a characteristic formation temperature indicated by condensation temperatures of elements and phases involved. Condensation temperatures are calculated by assuming thermodynamic equilibrium between solids and a cooling gas of solar composition (see Lodders 2003, for details). The cosmochemical components are attributed to distinct regimes of condensation temperatures between 1825 and >0 Kelvin (K) at 10^{-4} bar. Accordingly, with decreasing condensation temperature, elements representative of the various cosmochemical components are assigned to six groups, which account for the bulk of chondritic meteorites:

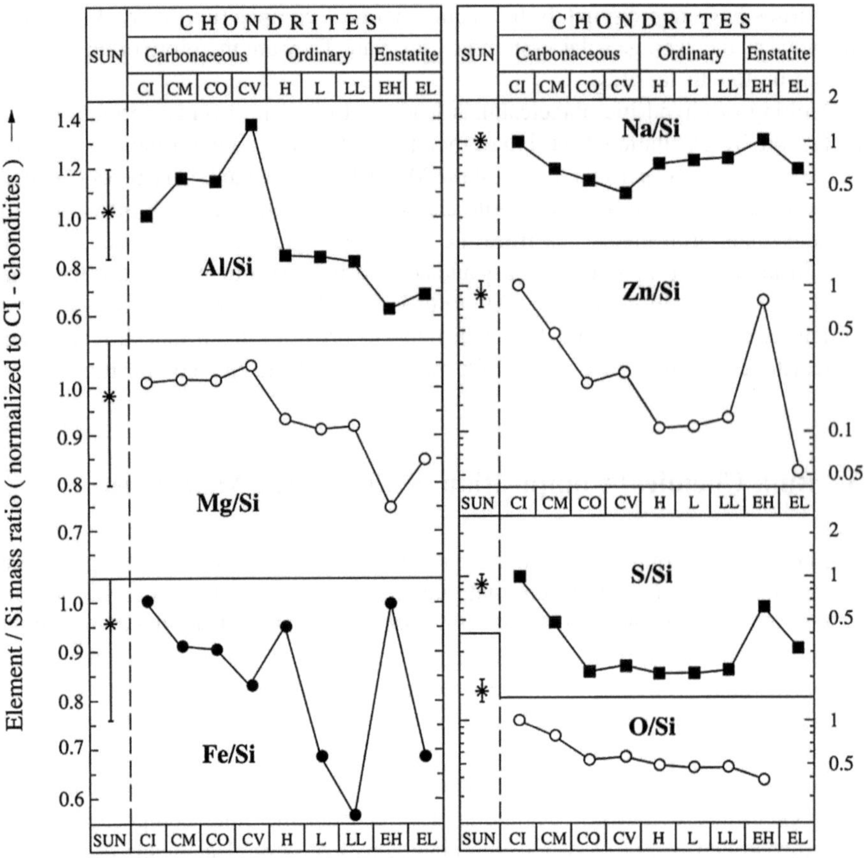

Fig. 2 Element/Si ratios in carbonaceous chondrites, ordinary chondrites, and enstatite chondrites in comparison to the composition of the Sun. Element/Si ratios are weight ratios normalized to the same element pair ratio in CI chondrites. Data are taken from Table 1 and 2. The Figure is from Lodders et al. (2009)

1. Refractory component: The first phases to condense from a cooling gas of solar composition are Ca,Al oxides and silicates rich in trace elements, such as the REE (rare earth elements), Sc, Zr, Nb, Hf, U and Th. These elements are named refractory lithophile elements (RLE), they condense before the condensation of Mg-silicates begins. The refractory siderophile elements (RSE) comprise metals with low vapor pressures, e.g., W, Os, Ir that condense before condensation of the major fractions of Fe and Ni in multicomponent metal alloys. The refractory component makes up about 5 % of the total condensable matter. The Ca,Al-rich inclusions in chondritic meteorites may be regarded as the pure form of the refractory component (Grossman 1980). Variations in Al/Si (Fig. 2) are caused by addition or loss of a refractory component. The CV carbonaceous chondrites have the highest and EL and EH chondrites have the lowest Al/Si ratios. The difference

between the highest and lowest ratio is about 50 %. The refractory element Al is representative of all refractory elements, RLE and RSE. Refractory element ratios among chondritic meteorites are identical in all types of chondrites, at least to within about 5 % (Palme and O'Neill 2014).

2. Mg-silicates: The major fraction of condensable matter is associated with the three most abundant elements heavier than oxygen: Si, Mg and Fe. Mg and Si condense as forsterite (Mg_2SiO_4), which converts to enstatite ($MgSiO_3$) at lower temperatures by reaction with gaseous SiO. Variations in Mg/Si ratios of bulk meteorites are produced by the incorporation of variable amounts of early condensed forsterite. The total range of Mg/Si ratios is about 25 %. Again, the CI ratio fits best with the solar ratio (Fig. 2).
3. Metallic iron condenses as FeNi alloy at about the same temperature as forsterite. The sequence is depending on total pressure. It is important to realize that refractory metals such as Os, Ir, Pt etc. are RSE and condense independent and ahead of the main FeNi-metal phase. Variations in the concentrations of Fe and other siderophile elements in meteorites are produced by the incorporation of variable amounts of metal. Variations in Fe are larger than those of Al and Mg. In Fig. 2 CI chondrites and EH chondrites have about solar Fe/Si ratios. All other chondrites have lower Fe/Si ratios. There are new rare meteorite groups with excess Fe, as for example the CB and CH chondrites (Fe/Mg in CH = 1.7 × CI, CB = 7.8–9.9 × CI) (see Krot et al. 2003).

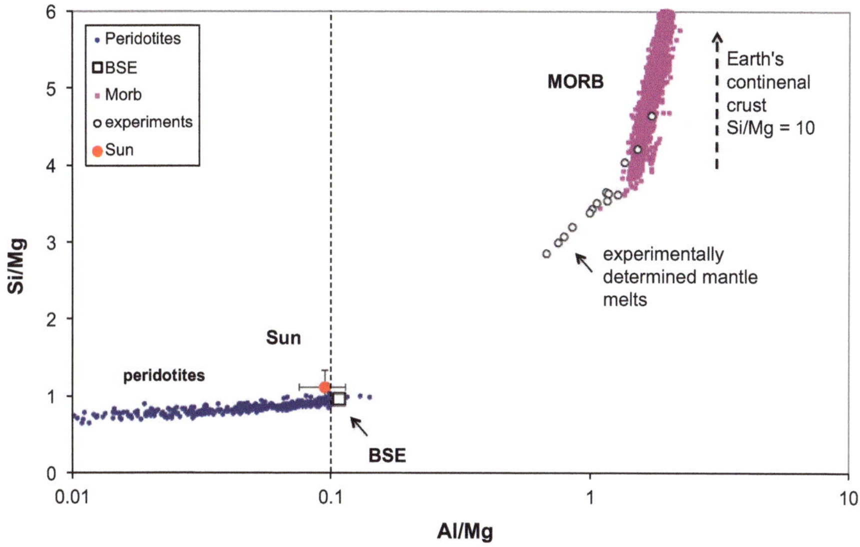

Fig. 3 Plot of Si/Mg vs. Al/Mg weight ratios of Earth's mantle rocks. The bulk silicate Earth composition (BSE) can be deduced from peridotites. The Sun plots slightly above the BSE composition. Data for compositions of partial melt experiments, MORB (analyses of 2912 volcanic glasses), and the continental crust form an array at high Si/Mg and Al/Mg weight ratios. For details and data sources the reader is referred to Palme and O'Neill (2014)

4. Moderately volatile elements have condensation temperatures between those of Mg-silicates and FeS (troilite). The most abundant moderately volatile element is sulfur which starts to condense by reaction of gaseous H_2S with solid Fe metal at 710 K, independent of total pressure. Half of all sulfur is condensed at 664 K. Moderately volatile elements are distributed among sulfides, silicates and metal. The amount and the relative abundances of these elements in meteorites are probably the result of removal of volatiles during condensation (incomplete condensation, see Palme et al. 1988). The elements Na, Zn and S (Fig. 2) belong to the group of moderately volatile elements. Their abundance variations in chondritic meteorites are significantly larger than for Al and Mg. Here the agreement between CI chondrites and the Sun is particularly noteworthy and confirms the unique position of CI-chondrites. It is also important to note that all other groups of chondritic meteorites have lower volatile element contents than CI chondrites. There are no meteorites with excess volatiles (see Palme et al. 1988).
5. Highly volatile elements have condensation temperatures below those of FeS and above water ice. The group of highly volatile elements comprises elements with very different geochemical affinities, such as, for example, the chalcophiles Pb and In and the lithophiles Cs and I. The abundances of the highly volatile elements vary with the petrologic type of the chondrite. The higher the metamorphic grade the lower the highly volatile trace element abundances, presumably because of losses caused by internal heating. In contrast, the contents of the moderately volatile elements S, Zn and Te vary little among the different metamorphic grades of ordinary chondrites. The abundances of highly volatile elements in type 3 ordinary chondrites are similar to those of moderately volatile elements (Keays et al. 1971; Schaefer and Fegley 2010). Processes similar to those invoked for the depletion of moderately volatile elements are responsible for variations in the abundances of these elements. In addition, internal heating of small parent bodies may mobilize highly volatile elements in the chondrite parent body and will ultimately lead to evaporation. There are no highly volatile elements shown in Fig. 2.
6. Ultra volatile elements have condensation temperatures below that of water ice. This group includes H, C N, O, and the noble gases. The O/Si ratios decrease monotonously in the sequence of chondritic meteorites shown in Fig. 2. This can, however, not be interpreted as a depletion trend, as oxygen is not fully condensed, even in CI chondrites (Table 1).

This approach goes back in part to models by Anders (1977) and coworkers who have calculated the bulk chemical composition of meteorite parent bodies and planets from seven cosmochemical components formed by nebular processes, before accretion to larger bodies has occurred.

In summary, Fig. 2 gives some indication for the range of elemental variations in chondritic meteorites. Despite these variations the sequence of decreasing heavy element abundances is the same in all chondritic meteorites and in the Sun with the most abundant elements Si, Mg and Fe, followed by S, Al and Ca. Planetary melting processes such as core formation or partial melting leads to rocks with completely different compositions (see below). We assume that minor compositional variations among chondritic meteorites are produced by nebular processes that occurred before accretion of their parent bodies.

In the next step we explore the bulk composition of larger differentiated solar system objects and compare bulk compositions of planets with the compositions of chondritic meteorites and the Sun. In general, estimating the bulk planet composition from the analyses of differentiated rocks is difficult. The task is easier and leads to more reliable results for the Earth for reasons given below.

4 Earth

There are rocks found on the surface of the Earth with roughly similar numbers of Mg and Si atoms, i.e., Si/Mg atomic ratios that equal 1 corresponding to a Si/Mg weight ratio of 1.15. This is remarkable because Mg is a comparatively rare element in the crust of the Earth. The average crustal Si/Mg weight ratio is about 10. For this, and other reasons, it is thought that many of the high-Mg rocks, called peridotites, come from the mantle of the Earth, either transported to the surface as xenoliths enclosed in basaltic melts or as peridotite massifs that are tectonically emplaced fragments of the lithosphere. In Fig. 3, upper mantle rocks are represented by 365 xenoliths data from localities around the world.

Peridotites range in Al/Mg ratios from below 0.01 to about 0.1. Parallel to the increase in Al/Mg there is a small increase in Si/Mg ratios. Partial melts of the Earth's mantle have high Si/Mg and Al/Mg ratios, because the generation of partial melts such as mid-ocean-ridge basalts (MORB) favor extraction of Al and Si from the source region. The result is a residual peridotitic mantle composition variably depleted in Si, Al, and other incompatible elements, reflecting the loss of partial melts. The inferred BSE (bulk silicate earth) composition at the high end of the peridotite trend indicates the composition of the Earth's mantle after the core had formed and before partial melt extraction. Figure 3 highlights the large compositional differences between the residual mantle (peridotites) and the melts (MORB) generated by partial mantle melting. This difference is the result of eutectic melting. Experimental data for melts produced by melting mantle rocks at appropriate conditions are plotted in Fig. 3.

Knowing the BSE composition, the bulk Earth Fe content can be calculated from mass balance assuming a core size of 32.5 %, a solar Fe/Ni for bulk Earth, and a light element fraction of 10 % in the core. This gives 32 wt. % Fe for bulk Earth and an Fe/Mg weight ratio of 2.14 which is about 10 % above the solar (or CI) ratio. Lowering the estimated light element fraction of the core leads to an increase in the bulk Earth Fe content and would thus further increase the bulk Earth Fe/Mg ratio.

In Table 2 the composition of bulk Earth is compared to selected chondritic meteorites. A crucial point here is the abundance of Si in bulk Earth. Based on evidence from Si-isotopes and metal-silicate partitioning Fitoussi et al. (2009),

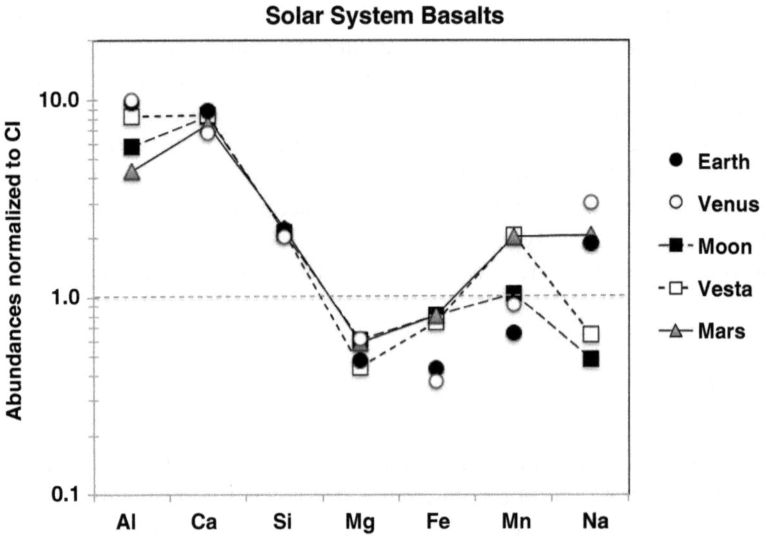

Fig. 4 Comparison of the chemical composition of basalts from Earth, Moon, Venus, Mars and Vesta. Data from Lodders and Fegley (1998)

Table 2 Composition of Chondritic Meteorites and Bulk Earth (wt. ratios)

	Chondritic Meteorites				Bulk Earth	
	CI	CV	H	EH	(1)	(2)
Mg %	9.60	14.5	14	10.6	*14.96*	*14.96*
Al %	0.83	1.62	1.13	0.81	*1.61*	*1.61*
Si %	10.7	15.4	16.9	16.7	*16.60*	*14.32*
Fe %	18.7	22.6	27.5	29.0	*32*	*32*
S %	5.35	2.2	2.0	5.8	*1.7*	*1.7*
Mg/Si	**0.90**	**0.94**	**0.83**	**0.63**	***0.90***	***1.04***
Al/Si	**0.078**	**0.105**	**0.067**	**0.049**	***0.097***	***0.112***
Fe/Si	**1.75**	**1.46**	**1.63**	**1.74**	***1.93***	***2.23***
Fe/Mg	**1.95**	**1.56**	**1.96**	**2.73**	***2.14***	***2.14***

(1) 85 % Fe, 8 % Si in core; (2) 85 % Fe and no Si in core
Sources of data: CC: Wolf and Palme (2001), OC, EC, and S: Wasson and Kallemeyn (1988), Earth: Palme and O'Neill (2014), S in Earth: Dreibus and Palme (1996)

Rubie et al. (2011) and others have concluded that the Earth's core contains some 8 % of Si. Recently, Dauphas et al. (2015) have argued against high Si in the core. In their model the Earth is supposed to continue the positive trend of δ^{30}Si with Mg/Si in chondrites. Such a trend may be explained by local enrichment in early condensing forsterite with fractionated Si-isotopes. In Table 2 we have calculated element/Si ratios for a bulk Earth with 8 % Si in the core and an Earth with no Si in the core. In the second case all three ratios, Mg/Si, Al/Si, and Fe/Si are higher than in chondrites. The deviation from CI-ratios is, nevertheless, within the range of variations seen in chondritic meteorites. To a first order it is therefore justified to say that the Earth has a chondritic bulk composition. The Earth has been considered non-chondritic by Campbell and O'Neill (2012) because of non-chondritic $^{142}Nd/^{144}Nd$ isotope ratios, where ^{142}Nd is the decay product of ^{146}Sm with a half life of 103 Ma. The difference in Nd-isotopes may be the result of early extraction from the mantle of a component enriched in incompatible trace elements such as the light REE. The missing incompatible elements may either be hidden in a reservoir at the core mantle boundary or they were removed from the crust by large impacts (see below).

An additional non-chondritic feature of the Earth is the small negative Tm anomaly relative to the CI composition that is present in terrestrial rocks. This characteristic feature is also seen in ordinary and enstatite chondrites, whereas carbonaceous chondrites have no or a positive Tm anomaly (Dauphas and Pourmand 2015). A positive Tm anomaly is also observed in fine-grained, Al-rich inclusions of CV chondrites. These inclusions have a unique REE pattern, established by condensation from a gas, which had suffered loss of very early condensates, enriched in ultra-refractory elements (the earliest elements that condense, Zr, Hf, Sc). The REE pattern of bulk Allende reflects the presence of fine-grained Al-rich inclusions. The presence of a negative Tm anomaly in the Earth and in ordinary and enstatite chondrites indicates an early fractionation process typical of the material parental to the Earth, OC, and EC. This observation fits with stable isotopes where Earth, OC, and EC show similar fractionations that differ from other groups of carbonaceous chondrites (Warren 2011).

We have earlier mentioned that most chondritic meteorites are depleted in volatile elements. In Table 2 we have listed an estimate for the S content of the Earth, inferred from similar nebular volatilities of Zn and S and the apparent depletion of Zn in the Earth's mantle (Dreibus and Palme 1996). The Earth is lower in S and other volatile elements than most chondritic meteorites. Similarly, the low Na and Mn contents, the chondritic CI-normalized Na/Mn ratio, and the low $^{53}Cr/^{52}Cr$ ratio of Earth's mantle indicate depletion of volatile elements in the Earth making materials by nebular processes shortly after the formation of the solar system (Palme and O'Neill 2014).

According to Table 2 the Fe/Si ratio of bulk Earth is higher than in any of the chondritic meteorites in Fig. 1 and Table 2. As mentioned earlier, recently discovered groups of meteorites widened the known spread of Fe/Si ratios in chondrites significantly. In particular, chondritic meteorites such as the CH and CB groups that have excess FeNi metal, i.e., higher Fe/Si ratios than CI-chondrites, may have received excess Fe by recondensation from a reservoir initially representing a cloud of impact evaporated material (Krot et al. 2003). Thus it seems justified to speculate that similar processes acted on the source region of material that produced the Earth. Another possibility for explaining the high and non-chondritic Fe/Si ratio of bulk Earth was pointed out by O'Neill and Palme (2008). A large impactor hitting Earth after core formation could have removed a substantial fraction of the crust and the underlying mantle. This collisional erosion will lead to a higher Fe/Si ratio of the bulk Earth as the eroded material will preferentially come from the crust and mantle, both low in Fe. This process could also have been responsible for removal of incompatible trace elements, which would have concentrated in the crust of the bulk Earth.

The absolute and relative abundances of refractory elements, the high Mg/Si ratio, the low total Fe content and the strong depletion of volatile elements indicate that the Earth is not exactly CI chondritic in composition. However, by looking at the variability of these indicators in the most common groups of chondritic meteorites one may safely classify the bulk Earth composition as chondritic. In this terminology "chondritic" relates to material that escaped fractionation processes associated with melting. None of the chemical differences among chondritic meteorites can be explained by parent body melting. For example, magmatic processes leading to changes in Mg/Si would be accompanied by fractionations of REE, which is not observed.

One focus of recent cosmochemical work is the study of stable isotopes. In a Δ^{17}O vs. ε^{54}Cr diagram there is a distinct separation of carbonaceous chondrites on one hand and of ordinary and enstatite chondrites on the other hand. The two separated fields of chondritic meteorites are defined not only by stable isotopes but also by characteristic chemical differences (Fig. 2). Warren (2011) has also shown that the same grouping is present in ^{50}Ti and other neutron-rich isotopes.

The Earth plots together with ordinary and particularly close with enstatite chondrites. This has led some authors to postulate that bulk Earth is made of enstatite chondrites (e.g., Javoy et al. 2010). Although stable isotopes favor an enstatite chondrite model for bulk Earth, differences in major elements preclude it as the Earth fits chemically better with carbonaceous chondrites. On the other hand, constraints from stable isotopes do not allow a major fraction of the Earth to be derived from carbonaceous chondrites, which may have formed further away from the Sun than ordinary and enstatite chondrites. The anhydrous nature of the two latter chondrite groups suggests that they formed in the inner solar system, within the formation region of terrestrial planets. Stable isotopes suggest that Earth, Moon and Mars also come from the inner solar system. The basic problem here is that stable isotopes favor a relationship of Earth and ordinary and enstatite chondrites while the chemical composition fits better to carbonaceous chondrites (see Palme and O'Neill 2014).

5 Rocks from Moon, Mars, Venus, Mercury, and Vesta

We have reasonable estimates for the bulk chemical composition of the Earth, because samples of Mg-rich rocks with compositions close to the presumed composition of the Earth's mantle are available for analysis. This is not the case for the other terrestrial planets and the largest differentiated asteroid Vesta. All surface rocks from these planets or planetesimals analyzed in the laboratory or by remote sensing have far too low Mg/Si ratios to qualify as primitive planetary materials. It is therefore difficult to estimate their bulk compositions. Earth, Venus, Mercury,

Table 3 Composition of extraterrestrial basalts

	CV-Chondrite	Venus	Earth	Moon	Mars	Vesta
Al	1.74	8.92	8.08	4.91	3.64	6.90
Ca	1.84	6.22	8.08	7.46	6.86	7.67
Si	16.03	21.93	23.59	22.90	24.00	22.97
Mg	14.83	5.88	4.57	5.83	5.58	4.22
Fe	23.47	7.03	8.11	15.00	15.10	13.90
Mn	0.15	0.18	0.13	0.21	0.40	0.40
Na	0.34	1.50	1.87	0.24	1.03	0.32

Concentrations in percent. Data from Lodders and Fegley (1998)

Mars, Moon and Vesta have an FeNi-metal core and a silicate mantle which to some extent experienced partial melting and/or fractional crystallization.

It is, however, possible to compare the various compositions of the surface rocks of each of these planets. In Table 3 we have compiled data on average compositions of basalts from terrestrial planets and asteroid Vesta. The data plotted in Fig. 4 are normalized to the CI chondritic composition.

The term basalt is used here in a broad sense, essentially meaning pyroxene-plagioclase normative rocks with occasional olivine. Chemically they are high in Al and Si and low in Mg. All rocks shown in Fig. 4 are partial melts of the interior of their parent planets more or less modified by fractional crystallization or assimilation of wall rock material during ascent from the interior to the surface. Alternatively they may be products of fractional crystallization of a larger magma system. Also, the source region of the basalts may be slightly fractionated. All these effects make a comparison difficult, but as we shall see a comparison is reasonable and provides useful insights into the parent planet composition.

In Fig. 4, the Earth is represented by the average composition of mid-ocean ridge basalts (MORB). They are believed to be partial melts from a depleted mantle, which is in major element chemistry very similar to the primitive upper mantle. The similarity of MORB with surface rocks of Venus analysed by three Russian lander missions is obvious (Fig. 4). According to their element pattern and the similarity to terrestrial MORB, the analyzed Venus rocks should be considered basalts (see Fegley 2003, for details). Two chemical characteristics of MORB and Venus basalts are typical of partial mantle melts from a primitive source region: high Al and Ca and enrichment of Si over Mg. The elements Al, Ca, and Si have higher concentrations than the hypothetical mantle composition (CI in Fig. 4), while the concentration of Mg is significantly lower. The low Fe concentrations in MORB indicate a low Fe (the actual species is FeO) mantle, as the Fe partition coefficient between melt and residue approximates 1 and the increase in Fe during partial melting is small. The Fe content of BSE is 6.33 % (Palme and O'Neill 2014) and MORB has an average Fe content of 8.11 % (Table 2). Thus the low Fe content of MORB and the Venus basalts reflect the presence of a metal core, which contains the major amount of Fe, in agreement with high density measurements of Earth and Venus. The same concentration level of Fe in MORB and Venus basalts and the roughly similar densities of Earth and Venus indicate that Venus has about the same bulk Fe content as Earth with an approximately chondritic Fe/Mg ratio and a similar core size.

The Shergotty meteorite in Fig. 4 represents Mars. Shergotty is compositionally similar to MORB, except for higher Fe and Mn contents. The particularly low Al content of Shergotty is typical of many Mars rocks and may reflect the stability of Al-retaining garnet in the deep interior of Mars. If Mars is about chondritic in bulk major element composition it should have a smaller core as more Fe is in silicates. This is in agreement with other geophysical indicators of Mars. The Martian core is believed to be smaller than Earth's core and Mars is higher in volatile elements than Earth and Venus, reflected here in the elevated Mn content. An approximately chondritic bulk composition of Mars is not unreasonable (Wänke 1991).

The lunar basalt 15495 plotted in Fig. 4 is from the Apollo 15 mission. Apollo 12 and many Apollo 15 basalts are relatively primitive with little fractionation of REE and no spectacular enrichment in Ti as in Apollo 11 and Apollo 17 basalts. A general characteristic of lunar basalts is the enrichment in Fe compared to terrestrial basalts. Geophysical data show that the Moon has only a small or no core at all. Thus lunar basalts constrain an upper limit for the mantle Fe content, which is far too low for a chondritic Fe/Mg ratio of the bulk Moon. All Moon formation models have to take into account the low Fe content of the bulk Moon. For some time, models were popular that made the Moon from Earth mantle material. In such a model, the high Fe of the Moon basalts requires that the Earth's mantle had higher Fe at the time of Moon formation. Continuing partitioning of Fe into the core over geologic time was postulated to explain the lower Fe in todays Earth mantle (Ringwood 1979). The giant impact models where most of the mass of Moon was provided by the impactor required core formation in the impactor. The core of the impactor would mix more or less completely with the core of the Earth and the bulk Moon Fe would be inherited from the mantle of the impactor. In present Moon formation models the Moon is again almost entirely made of Earth mantle material, mainly based on constraints by stable isotopes (e.g., Canup 2012).

If the bulk Moon has a chondritic Si/Mg ratio is not known. The Mg content of the Moon is rather uncertain. Highland rocks suggest higher Mg than mare basalts (Warren 2003). Moon, Earth and Venus are depleted in Mn at about the same level as CV chondrites. But compared to Earth the Moon is depleted in Na and other alkali elements (Fig. 4).

The composition of low Ti lunar basalts from Apollo 15 landing site is similar to that of eucrites, basalts from the 500 km diameter asteroid Vesta. The very close agreement is seen in Fig. 4 and shown in Table 3 (see also Ruzicka et al. 2001). Eucrites have high Fe as lunar and Martian basalts, but different from Earth and Venus. An important difference between Moon and Vesta is, however, the lower Mn content of lunar basalts compared to basaltic rocks from Vesta.

There are several estimates on the composition of Vesta based on meteorites from this asteroid. But its Mg/Si or Fe/Si bulk ratios are very uncertain. Some modelers even start by assuming a chondritic bulk asteroid composition. From the general element pattern in Fig. 4 one may assume a similar composition of the mantle source of Vesta as of other planets, such as Earth and Moon. The parent planet of eucrites may have essentially chondritic composition.

It thus appears that the terrestrial planets have to a first order chondritic bulk compositions with roughly chondritic relative abundances of Si, Mg and Fe. It is possible that the range of variations in Mg/Si and Fe/Si ratios among planets is within the range of chondritic meteorites. Exceptions are the Moon with too little total Fe and Mercury with too high total Fe. While the Moon probably has a different origin than the other planets this is less clear for Mercury. Recent data from the Messenger mission showed a surface composition that is fractionated in the same direction, although to a lesser extent, than surface rocks of other terrestrial planets, with low Mg/Si and high Al/Si ratios and very low Fe/Si ratios (Nittler et al. 2011). The bulk Fe content of Mercury is, however, much too high for a chondritic planetesimal. One possibility is that an impact on Mercury after core formation has removed a large fraction of the silicate mantle.

Ratios among refractory elements in bulk planetesimals are assumed to be chondritic. A well-known example is REE patterns in chondritic meteorites which are flat when normalized to CI abundances. Differentiated meteorites have fractionated patterns of REE and other refractory incompatible elements, because of severe fractionations of these elements during magmatic differentiation. In favorable cases it can be tested if the pattern of refractory elements was unfractionated, i.e., chondritic before differentiation. Using the decay of ^{146}Sm to ^{142}Nd with a half-life of 108 Ma it is possible to resolve very early events in the history of planets and planetesimals. Extrapolation of the ^{142}Nd/^{144}Nd ratios to the time of planet formation allows determining the initial Sm/Nd ratio of a planet or planetesimals. For example, the non-chondritic ^{142}Nd/^{144}Nd ratio of terrestrial rocks requires either a superchondritic Sm/Nd ratio in the bulk silicate Earth or a partial melting event in the very early history of the Earth. This early melting event would preferentially remove the more incompatible Nd and leave behind a residue with super-chondritic Sm/Nd ratio, which is actually observed (Boyet and Carlson 2006). In a similar approach Sprung et al. (2013) concluded that Earth and Moon accreted with chondritic Sm/Nd and Lu/Hf ratios, which would probably indicate unfractionated patterns of all refractory elements. These results may, however, be influenced by nucleosynthetic anomalies in Nd and Hf isotopes, which may also influence the isotopic abundance of the radiogenic isotope ^{142}Nd (Burkhardt et al. 2016). As we have seen in there are variations between stable isotope signatures in chondrites and the Earth. Careful considerations of these effects are necessary to correctly interpret the high precision analyses.

6 Asteroids and Dust

Reflectance spectroscopy at various wavelengths has been the traditional tool for studying asteroids for many years (Nelson et al. 1993; DeMeo and Carry 2013 and references). The main asteroid groups are S-types and C-types, where S stands for siliceous and C for carbonaceous. About 15–20 % of all known asteroids are S-types, populating the inner part of the main asteroidal belt. The C-type asteroids make up the majority of asteroids. They dominate the middle and outer main belt. Minor groups include M-types (M for metal) and E-types (E for enstatite).

Direct information of asteroidal material from laboratory studies are available only for few asteroids. By far the most material from a known asteroid is from Vesta, the source of HED meteorites (howardites, eucrites, diogenites). The identification of Vesta as the parent body of HED meteorites has been strengthened by the results of the Dawn mission (McSween et al. 2013). The Japanese spacecraft Hayabusa landed on the Itokawa asteroid and brought back to Earth a few grains, which were studied in the laboratory (Tsuchiyama 2014).

Finally, there are samples from the Almahata Sitta meteorite. The parent asteroid 2008 TC3, with approximately 4 m diameter, was detected and studied in space 19 h before the meteoroid hit the Earth atmosphere and fragmented into thousands of pieces. The recovered samples (>680 individual rocks) comprise the meteorite Almahata Sitta. This is the only meteorite of which its parent asteroid's spectral features are known. The meteorite belongs to the F-class asteroids, a subgroup of the C type asteroids. Almahatta Sitta is classified as a polymict ureilite. Ureilites are a rare group of meteorites that formed as residual rocks in the interior of planetesimals. Almahatta Sitta is polymict, because it contains abundant fragments of enstatite and ordinary chondrites (Horstmann and Bischoff 2014). The spectra of 2008 TC_3 are, however, best explained if the surface of the parent asteroid was dominated by ureilite material and not by S-type asteroidal spectra (Goodrich et al. 2015). This example demonstrates the difficulty in assigning observed reflectance spectra to specific types of meteorites.

Space probes approaching asteroids use X-rays and gamma-rays from asteroids to characterize the surface composition. The Near Earth Asteroid Rendezvous (NEAR)–Shoemaker mission, for example, was designed to measure the chemical composition and mineralogy of the near-Earth S type asteroid 433 Eros. Equipped with a gamma-ray and an X-ray detector the space probe could determine the major element composition of different areas of the asteroid. An enhanced X-ray flux from the Sun during solar flares helped to increase the statistics improving the precision of the data. Most of the data were obtained from a distance of 35–50 km above the surface of the asteroid. Although the uncertainties are comparatively large a basically chondritic composition was obtained. The Mg/Si wt. ratio varied from 0.59 to 0.85 within the range of chondritic ratios (Table 2). The Al/Si and Fe/Si ratios also cover the range of chondritic ratios. The S/Si is below 0.01, about ten times lower than a typical ordinary chondrite ratio. The reason for this extreme depletion of the volatile S is unknown. The approximately chondritic Al/Si ratio is significant because planetary differentiation produces crusts with high Al/Si (see Fig. 4). The low Al/Si ratio of the 433 Eros surface is in accordance with its ordinary chondrite bulk composition (Trombka et al. 2000).

Hayabusa also had an X-ray spectrometer on board. Data obtained during enhanced solar activity gave average mass ratios of Mg/Si of 0.78 ± 0.09 and Al/Si of 0.07 ± 0.03 for the crust of Itokawa consistent with ordinary chondrites (Okada et al. 2006). Although there are no compositional data from other asteroids available it is reasonable to assume that the Eros data are more or less characteristic of S type asteroids, given the similarity in reflectance spectra. Since S type asteroid spectra do not exactly match with spectra of ordinary chondrites obtained in the laboratory it was earlier assumed that S type asteroids are not exactly ordinary chondrites. The data from the returned Itokawa grains clearly showed that Itokawa is an LL-chondrite and that deviation in spectral reflectance of laboratory specimen of ordinary chondrites from asteroids are due to space weathering (Tsuchiyama 2014; Krot 2011 and references). The Itokawa samples are the first samples from an asteroid that were returned to Earth by spacecraft and could be analyzed in the laboratory.

It is thus reasonable to conclude that all S type asteroids of the inner solar system are basically chondritic in their major element composition.

Asteroids with basaltic crusts are very rare in the asteroidal belt. Furthermore, the mass of Vesta and Vestoids is by far dominating the population of basaltic asteroids. Their number is much smaller than one would expect from the number of iron meteorites, cores of destroyed differentiated planetesimals. The corresponding mantles and crusts may have been fragmented to sizes that cannot be observed (Moskovitz et al. 2008). Therefore, nothing can be said about the bulk compositions of these bodies.

The largest group of asteroids, are generally considered to consist of primitive carbonaceous material. They have more or less featureless spectra typical of carbonaceous chondrites (Nelson et al. 1993). One may expect that these objects have basically chondritic composition, although no surface analyses exist.

There is, however, one way to gain some insight into the composition of C-type asteroids. The S- and C-type asteroids are the main contributors to micrometeorites (0.5–2 mm diameter) found in Antarctica and other cold deserts on Earth. Nesvorný et al. (2010) claim that most of the micrometeorites are of cometary origin. Brownlee et al. (1997) have studied the bulk compositions of 500 micrometeorites. Many of the particles were partially or completely melted which did however not affect the non-volatile lithophile elements (Si, Mg, Al, etc.). Based on these elements the authors conclude that the majority of micrometeorites has the composition of CM-chondrites. Genge et al. (2008) also conclude that most of the unmelted micrometeorites have approximately CI or CM compositions. Rudraswami et al. (2015) assume on the basis of oxygen isotopes in relic olivine grains of micrometeorites a relationship to chondrules in carbonaceous chondrites. Particles similar to ordinary and enstatite chondrites are missing. The significance of his observation is unclear. If, as Nesvorný et al. (2010) claim, most micrometeorites are of cometary origin one would be forced to conclude that CI and CM chondrites are derived from comets. The issue is presently unresolved. The problem is to distinguish between an asteroidal and a cometary origin of micrometeorites. It is important to find better criteria for distinguishing between these two alternatives. Stable isotopes could proof very useful.

The much smaller interplanetary dust particles ($<10\ \mu m$) are not affected by atmospheric entry. Many of them have chondritic bulk composition but may be of cometary origin (Keller and Messenger 2005 and literature).

The composition of comet Halley was determined by the Giotto mission. An approximately chondritic composition was found by measuring the dust composition with a time-of-flight mass spectrometer (Jessberger et al. 1988). The Mg/Si and Al/Si ratios are about half of the respective CI-ratios. It is not clear if this is the extension of a chondritic trend, shown in Fig. 2. Alternatively, this may reflect large uncertainties in the mass spectrometric measurements.

A quite different picture emerges from the Stardust mission, which returned tiny dust grains from comet 81P/Wild 2. The expectation was that Wild 2 is dominated by submicron interstellar grains coated with radiation-processed organic mantles. The opposite was found. The Wild 2 grains are dominated by high temperature mineral assemblages and lithic clasts which are familiar from meteorites, i.e., chondrules, chondrule fragments, Ca,Al-rich inclusions and various mineral clasts. These materials must have formed at high temperatures in the inner part of the solar system and were then transported outward to the comet forming region (Brownlee 2014). This indicates a net flux of material from near the Sun to the outer parts of the solar system, a completely unexpected result.

The smallest particles with approximately chondritic bulk compositions are GEMS, an acronym for glass with embedded metal and sulfides. With just 0.1–0.5 μm in diameter, GEMS consist of nanometer-sized sub-grains of FeNi metal and Fe-Ni sulfide in a Mg-Fe-Al amorphous silicate matrix. Keller and Messenger (2011) point out the comparatively large chemical variability of GEMS. On average, the Si content of GEMs is enhanced by about 60 % over Mg, Al, Ca, Fe and S, or alternatively these elements are uniformly depleted relative to Si. Keller and Messenger (2011) postulate an origin in the solar nebula by non-equilibrium condensation and believe that the non-GEMS part of an IDP has the complementary chemical composition, making a chondritic chemistry for the bulk IDP. Bradley (2013) maintains a pre-solar origin of GEMS. The enhancement of Si found by Keller and Messenger (2011) is according to Bradley (2013) the result of the storage and contamination of GEMS-rich IDPs in Si-oil.

7 Summary

The abundances of rock forming elements in the Sun are the same as the composition of solid rocks of all sizes in the present solar system. The bulk composition of the terrestrial planets is essentially solar. This characteristic composition prevails, with more or less variability in km-sized asteroids, meter-sized chondritic meteorites, sub-mm to mm sized micrometeorites, micrometer- sized interplanetary dust particles and sub-micrometer-sized GEMS, which are by some regarded as interstellar dust. The major variables are the ratio of metallic to oxidized Fe and the extremely variable depletions of volatile elements.

The larger the objects the more representative should the mixture be. Thus larger bodies should match better with solar abundances than smaller bodies. This is not the case. The bulk Earth comprising more than 50 % of the mass of the inner solar system has very low volatile element abundances and very high Al/Si ratios, indicating stronger deviations from CI-chondritic chemistry than most chondritic meteorites. The objects of the inner solar system must have experienced an early stage of global fractionation: vaporization, condensation, grain-gas separation and re-accretion have modified the uniform chemical composition of interstellar matter producing refractory enriched and volatile poor reservoirs, where embryos and planetesimals, the starting materials of the Earth, formed.

References

Anders, E.: Chemical composition of the Moon, Earth and eucrite parent body. Phil. Trans. R. Soc. London **A295**, 23–40 (1977)

Asplund, M., Grevesse, N., Sauval, J., Scott, P.: The chemical composition of the Sun. Annu. Rev. Astron. Astrophys. **47**, 481–522 (2009)

Boyet, M., Carlson, R.W.: A new geochemical model for the Earth's mantle inferred from ^{146}Sm-^{142}Nd systematics. Earth Planet. Sci. Lett. **250**, 254–268 (2006)

Bradley, J.P.: How and where did GEMS form? Geochim. Cosmochim. Acta **107**, 336–340 (2013)

Brownlee, D.: The stardust mission: analyzing samples from the edge of the solar system. Annu. Rev. Earth Planet. Sci. **42**, 179–205 (2014)

Brownlee, D.E., Bates, B., Schramm, L.: The elemental composition of stony cosmic spherules. Meteorit. Planet. Sci. **32**, 157–175 (1997)

Burkhardt, C., Borg, L.E., Brennecka, G.A., Shollenberger, Q.R., Dauphas, N., Kleine, T.: Meteoritic Nd isotope constraints on the origin and composition of the earth. 47th Lunar and Planetary Science Conference #**1908** (abstract) (2016)

Campbell, I.H., O'Neill, H.St.C.: Evidence against a chondritic earth. Nature **483**, 553–558 (2012)

Canup, R.M.: Forming a Moon with an Earth-like composition via a giant impact. Science **338**, 1052–1055 (2012)

Dauphas, N., Pourmand, A.: Thulium anomalies and rare earth element patterns in meteorites and Earth: nebular fractionation and the nugget effect. Geochim. Cosmochim. Acta **163**, 234–261 (2015)

Dauphas, N., Poitrasson, F., Burkhardt, C., Kobayashi, H., Kurosowa, K.: Planetary and meteoritic Mg/Si and δ^{30}Si variations inherited from solar nebula chemistry. Earth Planet. Sci. Lett. **427**, 236–248 (2015)

DeMeo, F.E., Carry, B.: The taxonomic distribution of asteroids from multi-filter all-sky photometric surveys. Icarus **226**, 723–741 (2013)

Dreibus, G., Palme, H.: Cosmochemical constraints on the sulfur content in the Earth's core. Geochim. Cosmochim. Acta **60**, 1125–1130 (1996)

Fegley, B.: Venus. In: Holland, H.D., Turekian, K.K. (eds.) Meteorites, Comets and Planets, vol. 1, pp. 487–508. Elsevier-Pergamon, Oxford (2003)

Fitoussi, C., Bourdon, B., Kleine, T., Oberli, F., Reynolds, B.C.: Si isotope systematics of meteorites and terrestrial peridotites: implications for Mg/Si fractionation in the solar nebula and for Si in the Earth's core. Earth Planet. Sci. Lett. **287**, 77–85 (2009)

Genge, M.J., Engrand, C., Gounelle, M., Taylor, S.: The classification of micrometeorites. Meteorit. Planet. Sci. **43**, 497–515 (2008)

Goodrich, C.A., Hartmann, W.K., O'Brien, D.P., Weidenschilling, S.J., Wilson, L., Michel, P., Jutzi, M.: Origin and history of ureilitic material in the solar system: the view from asteroid 2008 TC_3 and the Almahata Sitta meteorite. Meteorit. Planet. Sci. **50**, 782–809 (2015)

Grossman, L.: Refractory inclusions in the Allende meteorite. Annu. Rev. Earth Planet. Sci. **8**, 559–608 (1980)

Horstmann, M., Bischoff, A.: The Almahata Sitta polymict breccia and the late accretion of asteroid 2008 TC3. Chem. Erde **74**, 149–183 (2014)
Javoy, M., Kaminski, E., Guyot, F., Andrault, D., Sanloup, C., Moreira, M., Labrosse, S., Jambon, S., Agrinier, P., Davaille, A., Jaupart, C.: The chemical composition of the Earth: enstatite chondrite models. Earth Planet. Sci. Lett. **293**, 259–268 (2010)
Jessberger, E.K., Christoforidis, A., Kissel, J.: Aspects of the major element composition of Halley's dust. Nature **332**, 691–695 (1988)
Keays, R.R., Ganapathy, R., Anders, E.: Chemical fractionation in meteorites IV Abundances of fourteen trace elements in L4 Chondrites; implications for cosmothermometry. Geochim. Cosmochim. Acta **35**, 837–868 (1971)
Keller, L.P., Messenger, S.: The nature and origin of interplanetary dust: high-temperature components. In: Krot, A.N., Scott, E.R.D., Reipurth, B. (eds.) Chondrites and the Protoplanetary Disk ASP Conference Series, vol. 341, pp. 657–667. Astronomical Society of the Pacific, San Francisco (2005)
Keller, L.P., Messenger, S.: On the origins of GEMS grains. Geochim. Cosmochim. Acta **75**, 5336–5365 (2011)
Krot, A.N.: Bringing part of an asteroid back home. Science **333**, 1098–1099 (2011)
Krot, A.N., Keil, K., Goodrich, C.A., Scott, E.R.D., Weisberg, M.K.: Classification of meteorites. In: Davis, A.M. (ed.) Meteorites, Comets and Planets. Treatise on Geochemistry, vol. 1, pp. 83–128. Elsevier, Oxford (2003). Holland, H.D., Turekian, K.K. (eds.)
Lodders, K.: Solar system abundances and condensation temperatures of the elements. Astrophys. J. **591**, 1220–1247 (2003)
Lodders, K., Fegley, K.: The planetary scientist's companion. Oxford University Press, New York (1998)
Lodders, K., Palme, H., Gail, H.-P.: In: J.E. Trümper (ed.) Abundances in the Solar System, Landolt-Börnstein, New Series, vol. VI/4B, pp. 560–598. Chap. 4.4, Springer, New York (2009)
McSween, H.Y., et al.: Dawn; the Vesta–HED connection; and the geologic context for eucrites, diogenites, and howardites. Meteorit. Planet. Sci. **48**, 2090–2104 (2013). doi:10.1111/maps.12108
Moskovitz, N.A., Jedicke, R., Gaidos, E., Willman, M., Nesvorný, D., Fevig, R., Ivezić, Ž.: The distribution of basaltic asteroids in the main belt. Icarus **198**, 77–90 (2008)
Nelson, M.L., Britt, D.T., Lebofsky, L.A.: Review of asteroid compositions. In: Lewis, J., Matthews, M.S., Guerrieri, M.L. (eds.) Resources of Near-Earth Space, pp. 493–522. University of Arizona Press, Tucson (1993)
Nesvorný, D., Jenniskens, P., Levison, H.F., Bottke, W.F., Vokrouhlický, D., Gounelle, M.: Cometary origin of the Zodiacal cloud and carbonaceous micrometeorites. Implications for hot debris disks. Astrophys. J. **713**, 816–836 (2010)
Nittler, L.R., et al.: The major-element composition of Mercury's surface from MESSENGER X-ray spectrometry. Science **333**, 1847–1850 (2011)
O'Neill, H.S.C., Palme, H.: Collisional erosion and the non-chondritic composition of the terrestrial planets. Phil. Trans. R. Soc. A **366**, 4205–4238 (2008)
Okada, T., et al.: X-ray fluorescence spectrometry of asteroid Itokawa by Hayabusa. Science **312**, 1338–1341 (2006)
Palme, H., O'Neill, H.S.C.: Cosmochemical estimates of mantle composition. In: Holland, H.D., Turekian, K.K. (eds.) Treatise on Geochemistry, vol. 3, 2nd edn, pp. 1–39. Elsevier, Oxford (2014)
Palme, H., Larimer, J.W., Lipschutz, M.E.: Moderately volatile elements. In: Kerridge, J.F., Matthew, M.S. (eds.) Meteorites and the Early Solar System, pp. 436–461. University of Arizona Press, Tucson (1988)
Palme, H., Lodders, K., Jones, A.: Solar system abundances of the elements. In: Holland, H.D., Turekian, K.K. (eds.) Treatise on Geochemistry, vol. 2, 2nd edn, pp. 15–36. Oxford, Elsevier (2014)
Ringwood, A.E.: Origin of the Earth and Moon. Springer, New York (1979)

Rubie, D.C., Frost, D.J., Mann, U., Asahara, Y., Nimmo, F., Tsuno, K., Kegler, P., Holzheid, A., Palme, H.: Heterogeneous accretion, composition and core-mantle differentiation of the Earth. Earth Planet. Sci. Lett. **301**, 31–42 (2011)

Rudraswami, N.G., Shyam Prasad, M., Nagashima, K., Jones, R.H.: Oxygen isotopic composition of relict olivine grains in cosmic spherules: links to chondrules from carbonaceous chondrite. Geochim. Cosmochim. Acta **164**, 53–70 (2015)

Ruzicka, A., Snyder, G.A., Taylor, L.A.: Comparative geochemistry of basalts from the Moon, Earth, HED asteroid, and Mars: implications for the origin of the Moon. Geochim. Cosmochim. Acta **65**, 979–997 (2001)

Schaefer, L., Fegley Jr., B.: Volatile element chemistry during metamorphism of ordinary chondritic material and some of its implications for the composition of asteroids. Icarus **205**, 483–496 (2010)

Sprung, P., Kleine, T., Scherer, E.E.: Isotopic evidence for chondritic Lu/Hf and Sm/Nd of the Moon. Earth Planet. Sci. Lett. **380**, 77–87 (2013)

Trombka, J.I., et al.: The elemental composition of asteroid 433 Eros: results of the NEAR-Shoemaker X-ray spectrometer. Science **289**, 2101–2105 (2000)

Tsuchiyama, A.: Asteroid Itokawa a source of ordinary chondrites and a laboratory for surface processes. Elements **10**, 45–50 (2014)

Wänke, H.: Chemistry, accretion, and evolution of Mars. Space Sci. Rev. **56**, 1–8 (1991)

Warren, P.H.: The Moon. In: Holland, H.D., Turekian, K.K. (eds.) Meteorites, Comets and Planets, vol. 1, pp. 559–600. Oxford, Elsevier-Pergamon (2003)

Warren, P.H.: Stable-isotopic anomalies and the accretionary assemblage of the Earth and Mars: a subordinate role for carbonaceous chondrites. Earth Planet. Sci. Lett. **311**, 93–100 (2011)

Wasson, J.T., Kallemeyn, G.W.: Compositions of chondrites. Phil. Trans. R. Soc. London **A325**, 535–544 (1988)

Wolf, D., Palme, H.: The solar system abundances of P and Ti and the nebular volatility of P. Meteorit. Planet. Sci. **36**, 559–572 (2001)

Shape Models and Physical Properties of Asteroids

T. Santana-Ros, G. Dudziński, and P. Bartczak

Abstract Despite the large amount of high quality data generated in recent space encounters with asteroids, the majority of our knowledge about these objects comes from ground based observations. Asteroids travelling in orbits that are potentially hazardous for the Earth form an especially interesting group to be studied. In order to predict their orbital evolution, it is necessary to investigate their physical properties. This paper briefly describes the data requirements and different techniques used to solve the lightcurve inversion problem. Although photometry is the most abundant type of observational data, models of asteroids can be obtained using various data types and techniques. We describe the potential of radar imaging and stellar occultation timings to be combined with disk-integrated photometry in order to reveal information about physical properties of asteroids.

1 Introduction

Asteroids play an important role in the formation and evolution models of the Solar System and have a direct link to life on Earth. They are connected to the delivery of water and probably also organic material to our planet and therefore are crucial for the development of life. On the other hand, some of them are considered as potentially hazardous for our future.

News about the discovery of a new hazardous asteroid appear regularly in the media. The last popular case was the close fly-by on 30th October 2015 of asteroid 2015 TB145 (the so-called "Halloween asteroid"). However, the news coverage usually tends to be sensationalist rather than scientific, as these asteroids don't represent an imminent risk of impact on Earth. The identification of a potential hazard comes from the forward integration of the asteroid's motion and evolution of its orbit, and the calculation of a probability of an impact (usually less than one in a few thousand chance) with our planet over the next decades. An example of these

T. Santana-Ros (✉) • G. Dudziński • P. Bartczak
Faculty of Physics, Astronomical Observatory Institute, Adam Mickiewicz University, Słoneczna 36, 60-286 Poznań, Poland
e-mail: tonsan@amu.edu.pl; g.dudzinski@amu.edu.pl; przebar@amu.edu.pl

J.M. Trigo-Rodríguez et al. (eds.), *Assessment and Mitigation of Asteroid Impact Hazards*, Astrophysics and Space Science Proceedings 46,
DOI 10.1007/978-3-319-46179-3_4

predictions can be found in the Sentry Risk Table (http://neo.jpl.nasa.gov/risks/) maintained by NASA's Jet Propulsion Laboratory (JPL). In order to obtain the best possible accuracy in these calculations, there are two crucial actions to be taken: (1) a very precise determination of the orbital parameters by astrometric measurements and (2) a study of the physical properties of the body.

The reason for the first action is obvious. Asteroids approaching the Earth have their orbits modified due to gravitational interactions and therefore regular astrometric measurements are required to constrain the orbital parameters. The better the orbit of an asteroid is defined, the greater will be the accuracy of our future predictions. However, for longer term predictions (i.e. several decades), second-order effects such as nongravitational forces plays a key role on the evolution of the orbit. The most important nongravitational perturbation is caused by the Yarkovsky effect (Neiman et al. 1965; Öpik 1951) which is due to radiative recoil of anisotropic thermal emission and causes asteroids to undergo a secular semimajor axis drift. The Yarkovsky acceleration depends on several physical quantities such as spin state, size, mass, shape and thermal properties (Vokrouhlický 1999). On the other hand, the Yarkovsky–O'Keefe–Radzievskii–Paddack (YORP) effect slowly modifies the spin rate of asteroids with irregular shapes, which in turn affects the Yarkovsky acceleration rate. Rubincam (2000) showed that YORP is strongly dependent on an asteroid's shape, size, distance from the Sun, and orientation.

Unlike asteroid's distance from the Sun, which can be trivially calculated with good accuracy when knowing the orbital parameters, deriving other physical properties of the asteroid, like spin state or shape, requires advanced inversion techniques and well-planned observations with a good coverage of various viewing geometries. Relative photometry is, by far, the richest source for deriving asteroid models. However, other observing techniques—such as Doppler-delay radar imaging, adaptive optics or stellar occultations—can provide valuable information of the asteroid's shape and, more importantly, they allow for the model to be scaled.

In this paper we briefly review the importance of investigating the shapes and spin states of asteroids in the context of identifying potentially hazardous bodies. We describe the requirements of photometric data in order to solve the inversion problem, and we discuss different shape solutions. In the last chapter, we review the potential of combining lightcurves with other data types.

2 The Importance of Asteroid Modelling in the Assessment of Asteroid Impact Hazard

The evolutionary processes that asteroids undergo have been traditionally explained by gravitational perturbations (e.g. gravitational pulls in close encounters with other bodies, orbital resonances) and collisions between these small bodies. The classical asteroid evolution model has been useful to explain how asteroid populations have evolved with time. Particularly interesting was efforts to understand the main source of Mars-crossing and Near Earth Object (NEO) populations, which we believe are fed with material from the main belt delivered by the effects of secular resonances (Wetherill 1979; Wisdom 1983). However, classical models are unable to explain some of the physical characteristics observed in the NEO population. For instance, according to the classical model, the only processes able to inject these bodies into orbital resonances are asteroid collisions. Meteoroids delivered through this process should present cosmic-ray exposures (CRE)—a measure of their ages—of the order of million years (Gladman et al. 1997). However, the observed CRE for NEO population are hundreds of times higher.

In turn, these high CRE values can be well explained when introducing nongravitational effects to the evolution model, because such effects can result in a slow delivery of material to orbital resonance zones. Specifically, the Yarkovsky effect induces a tiny force to small bodies by the reradiation of sunlight in the form of thermal energy. This force slowly changes the object's semimajor axis, changing it's orbit inwards (for objects rotating with retrograde sense) or outwards (prograde sense of rotation) with respect to the Sun. Yarkovsky effect is divided into two types of perturbations: (1) a diurnal perturbation due to the body rotation and (2) a seasonal perturbation that depends on the heliocentric longitude of the object. The acceleration da/dt for each perturbation is given by the following equations (see Bottke et al. (2006) for further details):

$$\left(\frac{da}{dt}\right)_{diurnal} = -\frac{8\alpha}{9}\frac{\Phi}{n}F_{\omega}(R, l_v, \Theta)\cos\gamma + \mathcal{O}(e) \tag{1}$$

$$\left(\frac{da}{dt}\right)_{seasonal} = \frac{4\alpha}{9}\frac{\Phi}{n}F_{n}(R, l_v, \Theta)\sin^2\gamma + \mathcal{O}(e) \tag{2}$$

where α is the albedo-factor, Φ is the radiation pressure coefficient and γ is obliquity of the spin axis. The function $F(R, l_v, \Theta)$ depends on the radius of the body R, the penetration depth l_v and the thermal parameter Θ [see the explicit form of this function in Vokrouhlický (1999)]. The total acceleration is the superposition of the diurnal and seasonal terms. Thus, the magnitude of the Yarkovsky effect depends on the object's distance from the Sun, the spin axis orientation, and the body's physical characteristics (i.e., size, shape, thermal properties, and rotation period).

On the other hand, another nongravitational effect called YORP, is capable of modifying the spin rates and axis orientations of asteroids. Reemitted photons apply a recoil force $d\mathbf{f}$ normal to the surface. If the body is not perfectly symmetric, the sum of these forces produces a thermal torque (see Bottke et al. (2006) for further details):

$$\mathbf{T} = \int \mathbf{r} \times d\mathbf{f} \tag{3}$$

where $\mathbf{r}$ is the position vector of a surface element $d\mathbf{S}$.

In this case, the effect strongly depends on the body's shape (i.e. it's irregularities), and to calculate the effect it is necessary to model the body's surface temperature distribution [see for instance Dobrovolskis (1996) or Vokrouhlický and Capek (2002)].

Thus, in order to include these nongravitational effects in the orbital calculations it is necessary to know in detail the physical properties of the asteroid. For NEOs the most commonly used technique to obtain the body's size is radar ranging (see Sect. 4 for further details). Surface thermal properties are related to the roughness of the body surface and its regolith depth. Such properties can be derived, for instance, using infrared interferometric observations. For modelling the Yarkovsky effect, it is essential to know the asteroid's spin state and its axis orientation. A convex representation of the body shape is usually enough to solve the lightcurve inversion problem, what is the main source of asteroid models. However, as the YORP effect is very sensitive to irregularities of the body shape, a high resolution shape model is required to calculate this effect. In this sense, shape models obtained by spacecraft in situ measurements represents the ideal case. Obviously, this kind of observations are limited to a bunch of asteroids which have been visited by spacecrafts, therefore generally we have to rely on remote observations. Radar echo can be useful to retrieve a complex shape model, including concavities. However, shapes obtained with this technique are not always reliable, and care should be taken when deriving results from this technique alone. Moreover, before deriving the shape from the radar echo, it is necessary to know the asteroid's spin axis orientation. Lightcurves are a great source of information regarding the asteroid's rotational state. As the Sun-asteroid-observer geometry changes so does the observed lightcurve. If the observations are gathered at a variety of geometries (see Sect. 3.1 for requirements) it is possible to reconstruct a shape and spin of an asteroid. In the next chapter we describe the data required to solve the inversion problem, as well as the shape representations commonly in use.

3 Models Based on Photometry

Deriving asteroid's spin state and shape is a necessity in order to model the nongravitational forces. To that end, photometry is by far the most fruitful observing technique. The classical photometric observations of asteroids (henceforth "dense lightcurves") that have been collected during the last decades, are the main source of our knowledge about asteroids and their physical parameters. However, gathering enough photometric data to derive a model is an arduous task, which requires good planning and, often, a collaboration between several observers. When the collected data fulfils the requirements, an inversion technique can be applied to obtain a model of an asteroid. Such model includes asteroid's rotational state as well as an approximation of the shape of the body. In this sense, different shape representations can be used (e.g. triaxial ellipsoid, convex or nonconvex figures). In this chapter we summarize the modelling process, from the obtention of data, to finding the solution of the inversion problem.

3.1 Requirements for the Lightcurves

Lightcurves can be obtained by comparing the apparent brightness of an asteroid with that of comparison stars (relative photometry), or with that of photometric standard stars (absolute photometry). It might seem that performing absolute photometric measurements should always be preferred. However, absolute photometry limits observable targets to bright asteroids, due to the use of filters, not to mention the requirement of excellent weather conditions. These constraints are of special importance when observing asteroids with small amplitude range (e.g. below 0.1 mag), as uncertainties of absolute magnitude measurements can be of a comparable range.

The usual lightcurve format for one apparition (the period during which the asteroid is observable from the Earth) is basically a series of photometric measurements collected during several observing nights (e.g. Fig. 1), with a 0.05 magnitude precision at worst. Ideally, a lightcurve should contain at least 50 well placed data points with a precision better than 0.01 magnitude. A general practice is to do continuous asteroid exposures 1–5 min long, depending on the object brightness. The field of view (FOV) should be large enough to include three comparison stars of

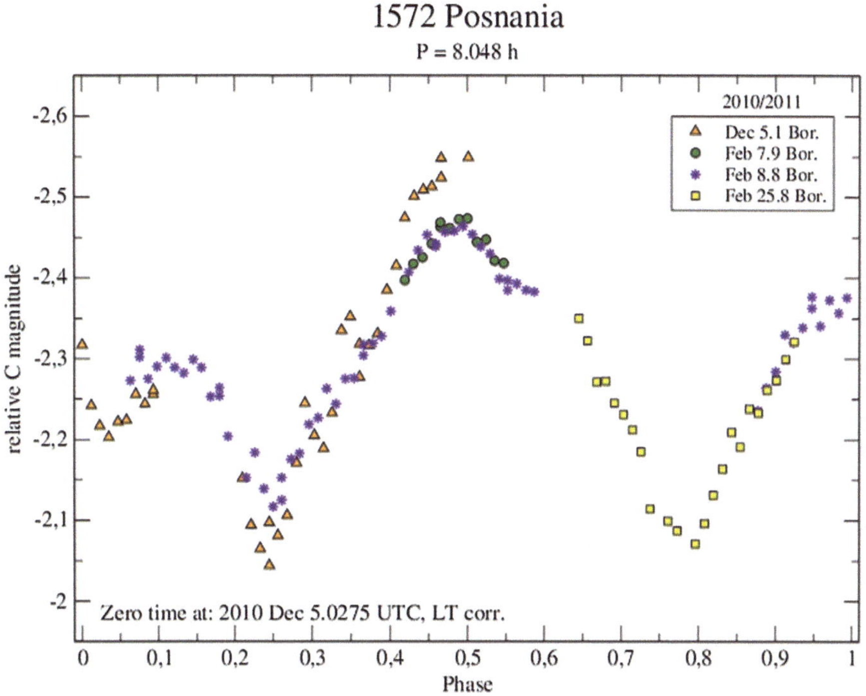

Fig. 1 Composite lightcurve of asteroid (1572) Posnania observed with a 0.4 m Newtonian telescope at the Borowiec observatory

brightness similar to that of an asteroid and, preferably, also of similar colour. When the lightcurve is complete, the rotational period of the asteroid is well covered but the quantity of information on the body shape is limited, as the viewing geometry of the asteroid is almost constant during observations. Consequently, to obtain a unique spin and shape solution, we need a set of dense lightcurves observed at a large span of viewing geometries (i.e. well-spread ecliptic longitudes and a substantial span of phase angles). This observational constraint makes this technique highly time consuming, what is significantly limiting the number of objects for which we have enough dense lightcurves to derive a complex shape of the body.

The quality of photometric observations is related to signal-to-noise ratio (SNR), which is a statistical term that defines the ratio between the useful signal (photons from the object) versus the total signal received (photons from the object, sky background, inherent noise in the chip, etc). The larger this number, the more signal (photons) from the target or comparison star only. An ideal SNR value would be above 100, which means that the noise is about 1 % of the total signal, or in other words, that the measurements are of about 0.01 mag precision. In practice one can still get good results when the SNR drops to 50 or 30. Getting the necessary SNR depends on many factors: size of the telescope, type of CCD camera, whether or not filters are used, the sky background brightness, how fast the asteroid is moving, the quality of dark and flat frames, etc.

3.2 Shape Models

Long term variations in apparent brightness of an asteroid depend mainly on its distant to the Sun and to the observer, and the angle between their pointing vectors (the so-called phase angle). However, any asteroid with non-spherical shape (practically each asteroid) has also shorter cyclical variations due to its rotation. The lightcurve characteristics (e.g. amplitude, period) depends on the asteroid's spin state, but also on its shape. If observed in equatorial view, an elongated body will produce a lightcurve with large amplitude, while a nearly spherical object will present a lightcurve with low amplitude. However, if the observation is taken in a pole-on viewing geometry, its lightcurve will be almost flat, no matter the body's shape.

In order to reproduce such variations, the inversion method has to include a recreation of the real asteroid's shape. A 3-axis ellipsoid shape can be a fairly good approximation to solve the inversion problem for the majority of cases (Connelly and Ostro 1984; Michałowski 1993; Torppa et al. 2008). Such ellipsoid can be defined as the region bound by a surface given by the equation:

$$(x/a)^2 + (y/b)^2 + (z/c)^2 = 1 \tag{4}$$

where a, b and c are the semi-axes and satisfy the condition $a \geq b \geq c$.

Most of the asteroids observed at low phase angles show two maxima and two minima per rotational cycle. Such a lightcurve can be explained considering a ellipsoidal shape rotating with a given sidereal period (P) about its spin axis (which orientation is described by its north pole position in the ecliptic coordinates λ, β) which usually is an axis of the biggest moment of inertia. The shape of the ellipsoid is then defined by two parameters, namely, the ratios of the lengths of the principal axes ($\frac{a}{b}$ and $\frac{b}{c}$). A model based on such a representation of shape is completed with an initial rotation angle ϕ_0 and the sense of rotation of the body (prograde or retrograde, defined by a sign of β). Using these parameters, it is possible to explain the variation in brightness of an asteroid, not only due to a rotation itself, but also due to the changes of the viewing geometry for the Sun–observer–asteroid system. Analytically, the brightness of the asteroid at a given time t, is proportional to the surface area seen from a given reference frame (cross-section of the asteroid presented to the observer). The cross-section can be calculated using the following equation:

$$S = \pi abc \frac{\cos^2 \phi \sin^2 \gamma}{a^2} + \frac{\sin^2 \phi \sin^2 \gamma}{b^2} + \frac{\cos^2 \gamma}{c^2} \tag{5}$$

where ϕ is the asteroid's rotation angle and γ is the aspect angle (the angle between the rotation axis and the asteroid–observer line of sight). As we change the rotation angle ϕ, so does the cross-section observed, thus we obtain a sinusoidal variation of brightness.

Nevertheless, some lightcurve shapes cannot be explained by the use of a simple triaxial ellipsoid model. Asteroids with complex shapes can produce lightcurves with 3 or more maxima per cycle. In the majority of cases these asteroids are modelled using a convex representation of their real shapes (Kaasalainen and Torppa 2001; Kaasalainen et al. 2001), which despite being a first-order approximation of the real shape of the body, have been proven to be good enough to fit the lightcurves and to derive asteroid's main physical parameters. In short, this method attempts to fit a set of parameters namely:

- A convex shape represented as a collection of triangular facets
- Sidereal rotation period
- Pole direction
- Albedo-dependent coefficient for Lommel-Seeliger and Lambert scattering laws

The standard solution of the inversion problem involves minimizing the residuals between disk-integrated photometric data and synthetic brightness generated by the model. The process relies on the Minkovski minimization stability of convex bodies (Lamberg 1993) which makes the method not very sensitive to random noise in data. This inversion technique has been used by several authors during the last decade [e.g. Ďurech et al. (2007), Marciniak et al. (2011), Marciniak et al. (2012)] resulting in around a hundred of convex asteroid models based on dense lightcurves.

However, from direct images of asteroids obtained by radar, adaptive optics or during space missions like NEAR Shoemaker or Hayabusa, we know that the real shapes of asteroids are not convex, but generally are full of concavities. In order to obtain a more accurate (realistic) shape model, alternative methods have been proposed. For instance, Bartczak et al. (2014a) recently developed a new inversion method called SAGE (Shaping Asteroids with Genetic Evolution) capable to derive nonconvex shape models for single and binary asteroids relying on their disk-integrated photometric measurements. In this case, the optimization problem is tackled by a genetic algorithm, which randomly mutates the model parameters and selects the best trial solutions until the evolution stabilizes. These models confirm the pole directions and rotation periods derived with previous methods, and additionally highly detailed description of the asteroids' shape allows more accurate determinations of their physical properties, like the volume and in turn, density.

In all the cases, the inversion techniques generally relies on relative photometry, so the resulting models are also relative in terms of dimensions. In order to scale them, we need an absolute measurement of the asteroid size. This can be obtained from other observation techniques like the time chords recorded during a stellar occultation by an asteroid, or direct imaging techniques, like radar (see Sect. 4).

Several approaches to the multi-data inversion problem have been developed during the last years. For instance, the KOALA [Knitted Occultation, Adaptive optics and Lightcurve Analysis (Carry et al. 2012)] algorithm solutions are based on lightcurves and AO silhouette contours, while ADAM [All-Data Asteroid Modeling (Viikinkoski et al. 2015)] is a collection of functions from which one can tailor an inversion procedure for multiple data sources including direct imaging, radar and interferometry.

For all the methods described, the main constraint for enlarging the number of derived models is the availability of good-quality photometric data fulfilling the requirements described in Sect. 2. The organization of observing campaigns, can potentially generate enough dense photometric data to derive a few tens of new models per year.

On the other hand, during the last years, some observatories around the world have conducted sky surveys mainly focused on detecting new NEAs or to improve their orbital parameter (e.g. USNO in Flagstaff, or Catalina Sky Survey). As a by-product of these astrometric survey programs a vast amount of sparse-in-time photometric measurements for tens of thousands of asteroids have been retrieved. For each object some tens, or often hundreds of discrete observations were collected for different geometries and illuminations. Combining these datasets with dense lightcurves allowed some authors to increase the modelled population of asteroids from 100 (classical photometry) to 400 (combination of classical and sparse photometric data), using a modified version of the convex lightcurve inversion method [e.g. Hanuš et al. (2011), Hanuš et al. (2013)]. The resulting models resemble the ones obtained with dense lightcurves (are equivalent in terms of spin solution) but the shape model is usually a low-resolution, "angular" convex shape, due to the limited quality of the data.

In turn, *Gaia* observations will generate a similar sparse set of photometric measurements during its 5 years operation. But the data improvement will be considerable, both in terms of quantity (observations are expected for ~300,000 asteroids, Mignard et al. 2007) and quality (the photometric accuracy is estimated to be ∼0.01 mag for asteroids up to 18 magnitude, and ∼0.03 mag up to 20 magnitude Cellino et al. 2006, 2009). As a result of this enormous amount of new data, asteroid models for at least 10,000 objects are expected. This means an improvement of two order of magnitudes from our current knowledge level.

However, on average *Gaia* will observe each asteroid 50–70 times during 5 years. Despite of the high data quality, this number of measurements is not enough to constrain a complex model shape by its own. Moreover, processing such enormous amount of data would be highly CPU demanding. For these reasons, the inversion method chosen for inverting *Gaia* photometric data of asteroids is a low CPU demanding method: a triaxial ellipsoid model, which brightness can be analytically calculated as a function of the asteroid–Gaia–Sun position vectors and the Lommel-Seeliger law (Cellino et al. 2015). This method, while simple, has been proved to be effective even when inverting synthetic data generated with nonconvex shapes (Santana-Ros et al. 2015). The results coming out from *Gaia* are expected to have a direct impact on the Solar System formation theories, as a statistically large sample of objects with known properties may reveal physical effects which play an important role for the whole population.

It is worth noting that even such precise data will provide models not completely free from various biases, or selection effects, that favour i.e. elongated targets with extreme obliquities of spin axes (Santana-Ros et al. 2015). It is important for ground-based studies to focus on those targets that will not be fully covered by studies based on data from *Gaia* or other future surveys.

3.3 Models of Binary Asteroids

One particularly interesting case are the asteroids with satellites. Such systems are specially appreciated by the Solar System researchers as they give a unique opportunity to derive the mass of the components directly from the third Kepler's law. For this reason, they are invaluable targets for studies on internal structure and composition.

The synchronous binary systems have been extensively studied and modelled [e.g. in Michałowski et al. (2002), Michałowski et al. (2004) and Kryszczyńska et al. (2009)]. Recently a new algorithm capable to generate model solutions for binary asteroids has been developed using a nonconvex shape representation of the components (a model for 90 Antiope obtained with this technique is shown in Fig. 2). As the model is able to reproduce body concavities, the relative volume obtained for the components is more accurate than for the previous models, which were based in Roche ellipsoids (Descamps et al. 2007), having a direct impact on the density calculation.

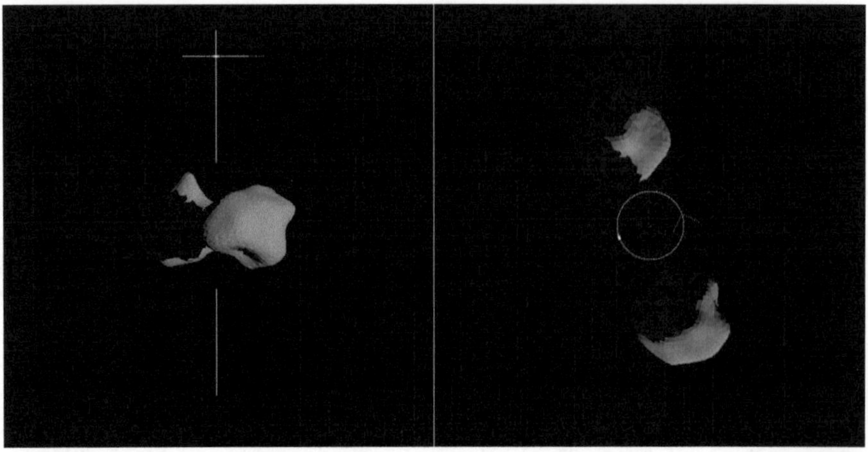

Fig. 2 Two different spatial views of the nonconvex model for 90 Antiope binary system shown at equatorial viewing (on the *left*), and the pole-on view on the *right* [model from Bartczak et al. (2014a)]

We currently know of more than a 100 binary asteroids in the main belt, and about three hundred in total if binary NEOs and TNOs are included. The majority of them have been discovered by recording their mutual events in a classical dense lightcurve. Resolved imaging such as the ones obtained from radar or adaptive optics have allowed to confirm or, in a few cases, discover such objects. The number of asteroids with known satellites it is expected to be increased significantly due to the huge amount of data expected from surveys like *Gaia*. To that end, it is necessary to develop automated strategies to find binary candidates in such large datasets.

It is thought that NEO population can contain a high number of binary (or multiple) asteroids [see for instance Pravec and Harris (2007)]. One possible explanation for a high rate of multiplicity among this population could be the catastrophic disruption of rubble piles due to YORP spin up. Thus, inversion techniques capable of deriving models of binary asteroids can help us to better understand their formation processes and the physics of nongravitational effects.

4 Models from Various Data Types

Models derived from disk-integrated relative photometry can be combined with other data types in order to derive additional physical properties of asteroids. For instance, radar echo is a very powerful technique to study the NEO population, while stellar occultation is an affordable technique to obtain sizes of main belt (or even trans-neptunian!) objects and discover satellites. In this chapter we briefly describe both techniques and we outline a method to combine them with lightcurves. Finally, it is worth noting that this is not intended to be an extensive and comprehensive review of the data types that can complement photometric observations. Other techniques like adaptive optics, spectroscopy, thermal infrared observations or polarimetry can be also combined with photometry to investigate asteroids. Here we provide the description of joining lightcurve data with radar imaging and with stellar occultation timings.

4.1 Photometry and Radar

An Asteroid radar image is a reconstruction of a radio signal sent from the Earth and reflected by body's surface. For this reason, this technique is best working for objects approaching the Earth, such as NEOs. One dimension of such image comes from time delay, as the signal has to travel different distances depending on which a part of asteroid's surface it is reflected from. Second dimension is directly associated with body's rotation. Received echo's frequency is shifted with respect to incident ray due to Doppler effect and depends on radial velocity of a surface element which increases as we move away from asteroid's rotation axis. Range of frequency shift depends on asteroid's rotation period and aspect (angle between rotation axis and direction to the observer). As a result one can produce range-Doppler image where each pixel value corresponds to echo power at certain distance and radial velocity.

Radar techniques can only be used when precise astrometry is available. Importance of knowing body's orbit accurately, especially in case of NEOs, cannot be overestimated.

Radar imaging is a rare situation in astronomy where we conduct an actual experiment by controlling the signal that gets reflected from a target body. It becomes possible to probe asteroids surface features comparable in size with signal wavelength or even have a glance at sub-surface properties. Asteroid's shape is represented on images derived from radar observations in a form of a blend of top and bottom view of asteroid (in respect to line of sight). By examining radar images astronomers can determine large scale surface features such as big concavities or adjudge whether an object has satellites. Body size constrains can also be obtained.

Radar images are a very rich source of information and are used to create accurate three dimensional asteroid models that consist of asteroid's shape, spin axis orientation and rotation period. It can be done by using radar data only or by combining them with photometric data. Shape program (Magri et al. 2007, Greenberg and Margot 2015) is broadly, if not the only, such algorithm in use. It is an iterative method that starts with triaxial ellipsoid and by gradually changing initial shape and spin parameters arrives at a final model. Both radar and optical scattering laws are assumed prior to modelling process and stay fixed.

In each iteration a simulated asteroid image is created and compared with the observations. If radar data is used alongside photometric data, both images are computed separately from the same model. Every point on synthetic lightcurve is a

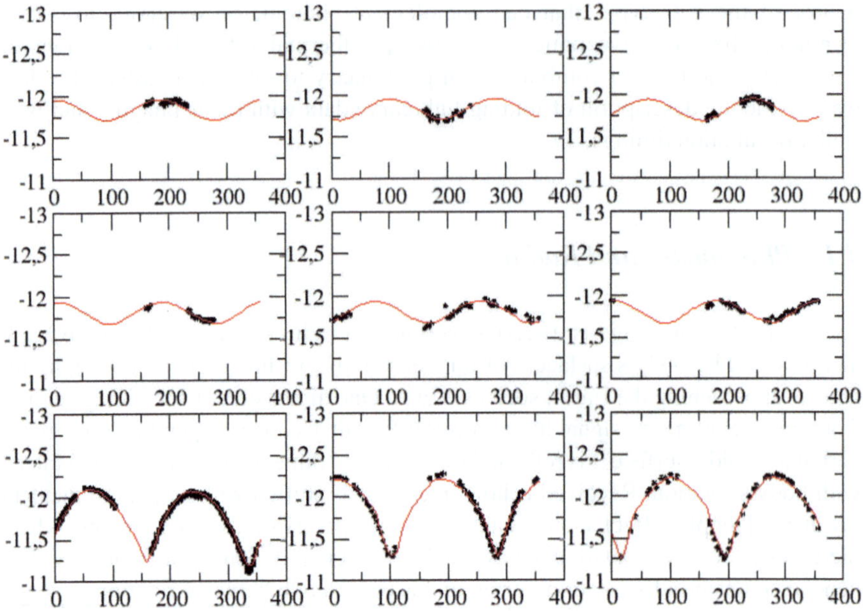

Fig. 3 Selected light curves of 1996 HW1. *Black points* represent observational data while *solid line* represent synthetic lightcurves modelled with SAGE

calculated amount of light reflected by model's surface (Figs. 3 and 4). Similarly, a radar range-Doppler image is created. Then a χ^2 is determined by minimizing the differences between synthetic and observational measurements. Data is weighted depending on the type and quality, and additional penalty functions defined by user are taken into consideration. This approach steers modelling process into global minimum. Final model has to fit both radar and photometric observations.

The shape of an asteroid is derived in three stages and best fitting model from previous stage becomes initial model of the next. The first stage uses only triaxial ellipsoid as a shape model; in the second stage, model is represented by spherical harmonics to be then transformed into polyhedral model described by vertices and triangular surface elements in the last stage. Concavities are allowed only at the last stage as they are difficult to represent using spherical harmonics (especially those of a low degree).

The described method is not fully automatic, meaning that it needs considerable human interference. Modelling is initialized with different values for parameters that are being further changed during the process to help the algorithm to come up with the model that reflects all important features present in observational data. This has it's reflection in the low number of modelled asteroids based on radar observations.

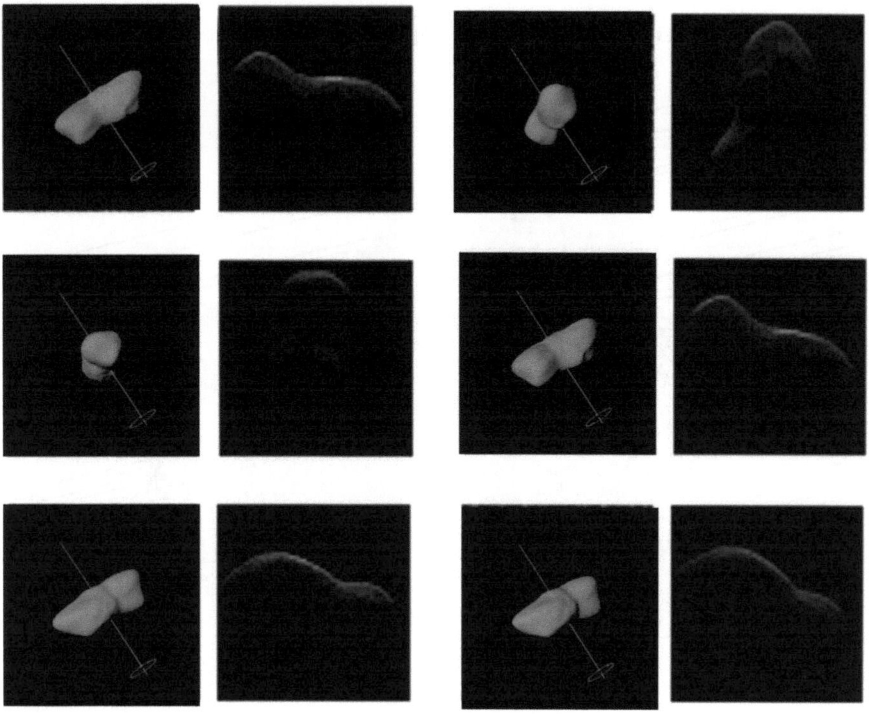

Fig. 4 Comparison of range-Doppler images of 1996 HW1. *First* and *third column*: simulated asteroid model as seen by observer; *second* and *last column*: simulated range- Doppler image based on a SAGE model

4.2 Photometry and Stellar Occultations

Stellar occultation is a interesting technique of asteroid imaging. The idea is to measure asteroid's shadow cast on Earth's surface while it passes in front of a star. Observers are set on shadow path and each of them notes a time of the beginning and end of occultation (e.g. Fig. 5).

In order to predict a shadow path and allow observers to choose observing sites properly, an asteroid's ephemeris as well as an occulted star position have to be very precise. Only if the ephemeris and star position are known with high accuracy will occultation time measurements result in good coverage of the body's shape. Nowadays, this represents a constraint as the availability of precise star astrometry is limited to the Hipparcos catalogue. Publication of *Gaia* catalogue will greatly improve occultation events predictions; accuracy that is now available for main belt asteroids 100 km in diameter, will be achievable for 15 km asteroids (Tanga and Delbo 2007). On the other hand, this technique is not appropriate to NEOs, as their fast apparent movement against the background of distant stars constrict the possibility of recording the event.

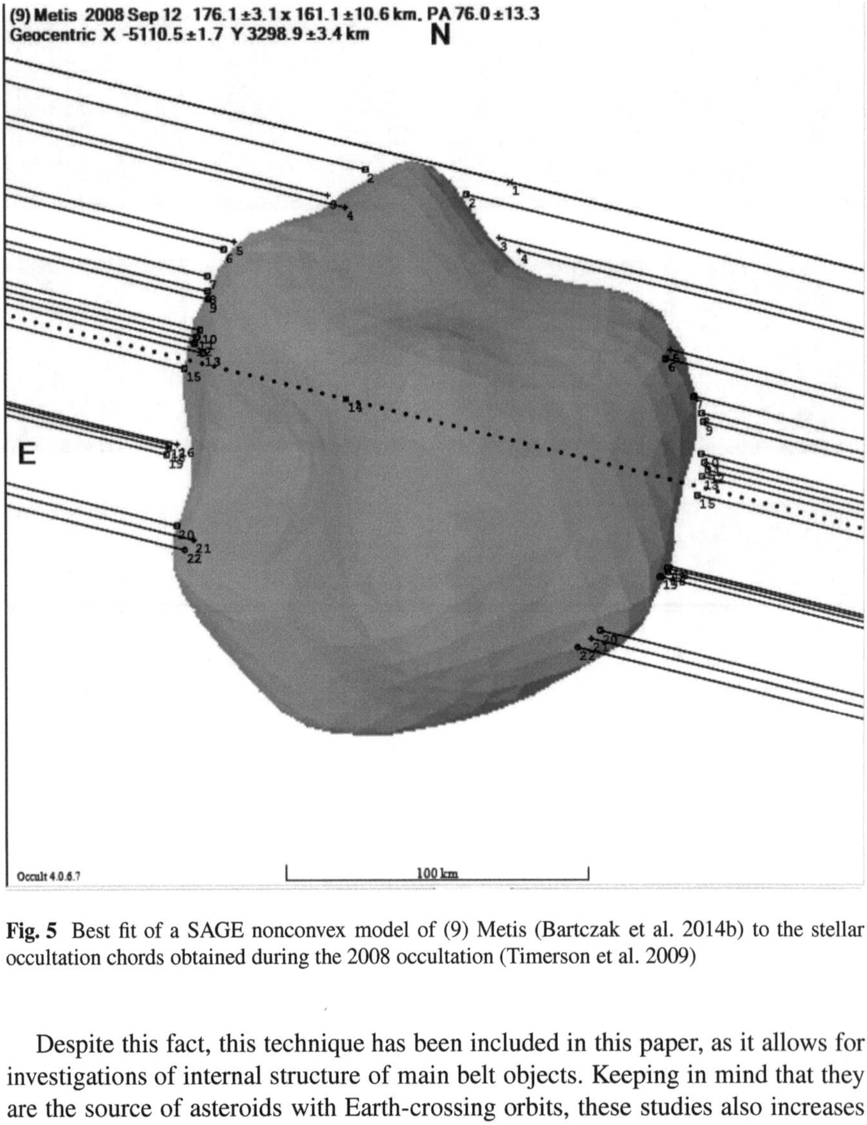

Fig. 5 Best fit of a SAGE nonconvex model of (9) Metis (Bartczak et al. 2014b) to the stellar occultation chords obtained during the 2008 occultation (Timerson et al. 2009)

Despite this fact, this technique has been included in this paper, as it allows for investigations of internal structure of main belt objects. Keeping in mind that they are the source of asteroids with Earth-crossing orbits, these studies also increases our knowledge about NEOs composition.

Choosing observational sites is crucial to successful occultation time measurements. Ideally, observers will cover the path of the shadow evenly and densely. Shadow path prediction however is directly dependent on the knowledge of ephemeris of the observed Solar System body and an estimation of the body's size. With more precise stellar catalogues, like the one ESA *Gaia* mission will produce, more occultation events will be predicted with better precision, even for small bodies.

Occultation observations are mainly carried out by amateur astronomers. Observing groups have to be mobile in order to cover the right area on the ground; fortunately stellar occultation events can be observed with small telescopes if occulted star is a bright one. Systems like GPS are of great help when it comes to establishing the observer's location and time essential to valid measurements.

Every chord (line along the shadow path) with marked beginning and end of the occultation event provides two points on the plane of sky. Given many points (five at minimum, Elliot 1979) an ellipse can be fit to match these points. This is a good estimate of shape for large bodies, e.g. planets and TNOs objects; in case of smaller bodies this method provides rough estimate only as asteroids take a wide range of convex shapes that are irregular in general.

Stellar occultation remains the best available method for determining body's radius its resolution being at the level of kilometres. Given dense chord distribution the resulting body's shape envelope is very precise thus fit for verifying models obtained by inversion techniques against reality.

Models can be enriched with additional information, the size of the body being the most valuable one as it gives the model proper dimensions and allows density and albedo estimations, as was done in case of (90) Antiope [see for instance Bartczak et al. (2014a)]. The described method is also sufficient to determine whether an asteroid is a binary system.

Analysis of lightcurve profiles captured during occultation event (immersion and emersion at the beginning and end of occultation) can tell us about the presence of an atmosphere (Elliot 1979). Moreover, rings around bodies can by detected and studied, like in the case for giant planets in Solar System. It is possible to detect rings around smaller objects (Braga-Ribas et al. 2014) but it requires extremely high precision timings. Still it shows the power of the method, where some features could not be discovered or that precisely measured using other observing techniques.

5 Conclusions

We have shown that asteroid modelling is an effort that needs to be undertaken in order to study the nongravitational forces affecting these small bodies. Disk-integrated photometry is the main technique used to derive spin states and shape models. However, for solving the inversion problem, lightcurves need to fulfil certain requirements in terms of quality and viewing geometries.

A substantial part of this paper has been devoted to describe how to gather such data and what are the inversion techniques commonly applied. The first and simplest solution—which is, however, still useful in some cases due to specific situations, such as the analysis of the huge amount of data generated by the *Gaia* mission—is a triaxial ellipsoid representation of the body shape, for which synthetic brightness can be evaluated analytically, making it ideal for problems with high CPU demand. The so-called lightcurve inversion method, which solution consist of a convex shape model of the asteroid and its spin state, is a worthwhile technique when we are specifically interested in the study of the spin rate and shape outline of the body. However, other techniques producing more detailed shape solutions (i.e. nonconvex shapes) are necessary when investigating further physical properties. In particular, modelling of nongravitational effects acting on asteroids is extremely sensitive to the used shape representation. Some inversion methods are also able to derive shape models of binary asteroids. We have shown the example of SAGE, which is a technique capable of deriving nonconvex models for synchronous binary asteroids from relative photometry only.

It is also possible to combine other observing techniques to derive additional physical properties of asteroids. In this chapter, we have described two techniques—radar imaging and stellar occultation timings—which are mainly used to scale the model in size and, in some cases, derive additional information. This includes the study of fine details in the shape or the discovery of satellites. In addition, we have briefly described some procedures to combine these observations with photometry in the modeling process.

References

Bartczak, P., Michałowski, T., Santana-Ros, T., Dudziński, G.: A new non-convex model of the binary asteroid 90 Antiope obtained with the SAGE modelling technique. Mon. Not. R. Astron. Soc. **443**, 1802–1809 (2014a)

Bartczak, P., Santana-Ros, T., Michałowski, T.: Non-convex shape models of asteroids based on photometric observations. Asteroid, Comets, Meteors 2014. In: Proceedings of the Conference, vol. 29B (2014b)

Bottke, W.F., Vokrouhlický, D., Rubincam, D.P., Nesvorný, D.: The Yarkovsky and YORP effects: implications for asteroid dynamics. Annu. Rev. Earth Planet. Sci. **34**, 157–191 (2006)

Braga-Ribas, F., Sicardy, B., Ortiz, J.L., Snodgrass, C., Roques, F., et al.: A ring system detected around the Centaur (10199) Chariklo. Nature **508**, 72–75 (2014)

Carry, B., et al.: Shape modeling technique KOALA validated by ESA Rosetta at (21) Lutetia. Planet. Space Sci. **66**, 200 (2012)

Cellino, A., et al.: Rotational properties of asteroids from Gaia disk-integrated photometry: a "genetic" algorithm. Adv. Space Res. **38**, 9 (2006)

Cellino, A., et al.: Genetic inversion of sparse disk-integrated photometric data of asteroids: application to Hipparcos data. Astron. Astrophys. **506**, 935 (2009)

Cellino, A., Muinonen, K., Hestroffer, D., Carbognani, A.: Inversion of sparse photometric data of asteroids using triaxial ellipsoid shape models and a Lommel-Seeliger scattering law. Planet. Space Sci. **118**, 221–226 (2015)

Connelly, R., Ostro, S.J.: Ellipsoids and lightcurves. Geom. Dedicata. **17**, 87 (1984)

Descamps, P., et al.: Figure of the double Asteroid 90 Antiope from adaptive optics and lightcurve observations. Icarus **187**, 482 (2007)

Dobrovolskis, A.R.: Inertia of any polyhedron. Icarus **124**, 698–704 (1996)

Ďurech, J., et al.: Physical models of ten asteroids from an observers' collaboration network. Astron. Astrophys. **465**, 331–337 (2007)

Elliot, J.L.: Stellar occultation studies of the solar system. Astron. Astrophys. **17**, 445–475 (1979)

Gladman, B.J., Migliorini, F., Morbidelli, A., Zappalà, V., Michel, P., et al.: Dynamical lifetimes of objects injected into asteroid belt resonances. Science **277**, 197–201 (1997)

Greenberg, A.H., Margot, J.-L.: Improved algorithms for radar-based reconstruction of asteroid shapes. Astron. J. **150**(4), 10 pp. (2015). Article ID. 114

Hanuš, J., et al.: A study of asteroid pole-latitude distribution based on an extended set of shape models derived by the lightcurve inversion method. Astron. Astrophys. **530**, A134 (2011)

Hanuš, J., et al.: Asteroids' physical models from combined dense and sparse photometry and scaling of the YORP effect by the observed obliquity distribution. Astron. Astrophys. **551**, A67 (2013)

Kaasalainen, M., Torppa, J.: Optimization methods for asteroid lightcurve inversion. I. Shape determination. Icarus **153**, 24 (2001)

Kaasalainen, M., Torppa, J., Muinonen, K.: Optimization methods for asteroid lightcurve inversion. II. The complete inverse problem. Icarus **153**, 37–51 (2001)

Kryszczyńska, A., et al.: New binary asteroid 809 Lundia. I. Photometry and modelling. Astron. Astrophys. **501**, 769 (2009)

Lamberg, L.: Ph.D. thesis. University of Helsinki, vol. 315L (1993)

Magri, C., Ostro, S.J., Scheeres, D.J., Nolan, M.C., Giorgini, J.D., et al.: Radar observations and a physical model of Asteroid 1580 Betulia. Icarus **186**, 152–177 (2007)

Marciniak, A., et al.: Photometry and models of selected main belt asteroids. VIII. Low-pole asteroids. Astron. Astrophys. **529**, A107 (2011)

Marciniak, A., Bartczak, P., Santana-Ros, T., Michałowski, T., Antonini, R., et al.: Photometry and models of selected main belt asteroids. IX. Introducing interactive service for asteroid models (ISAM). Astron. Astrophys. **545**, A131 (2012)

Michałowski, T.: Poles, shapes, senses of rotation, and sidereal periods of asteroids. Icarus **106**, 563 (1993)

Michałowski, T., et al.: Eclipsing events in the binary system of the asteroid 90 Antiope. Astron. Astrophys. **396**, 293 (2002)

Michałowski, T., et al.: Eclipsing binary asteroid 90 Antiope. Astron. Astrophys. **423**, 1159 (2004)

Mignard, F., et al.: The Gaia mission: expected applications to asteroid science. Earth Moon Planets **101**, 97 (2007)

Neiman, V.B., Romanov, E.M., Chernov, V.M.: Ivan Osipovich Yarkovsky. Earth Univ. 4, 63–64 [Russ. mag.] (1965)

Öpik, E.J.: Collision probabilities with the planets and the distribution of interplanetary matter. Proc. R. Irish Acad. **54A**, 165–199 (1951)

Pravec, P., Harris, A.W.: Binary asteroid population 1. Angular momentum content. Icarus **190**, 250–259 (2007)

Rubincam, D.P.: Radiative spin-up and spin-down of small asteroids. Icarus **148**, 2–11 (2000)

Santana-Ros, T., Bartczak, P., Michałowski, T., Tanga, P., Cellino, A.: Testing the inversion of asteroids' *Gaia* photometry combined with ground-based observations. Mon. Not. R. Astron. Soc. **450**, 333–341 (2015)

Tanga, P., Delbo, M.: Asteroid occultations today and tomorrow: toward the GAIA era. Astron. Astrophys. **474**, 1015–1022 (2007)

Timerson, B., et al.: A trio of well-observed asteroid occultations in 2008. Minor Planet Bull. **36**, 98–100 (2009)

Torppa, J., et al.: Asteroid shape and spin statistics from convex models. Icarus **198**, 91 (2008)

Viikinkoski, M., Kaasalainen, M., Ďurech, J.: ADAM: a general method for using various data types in asteroid reconstruction. Astron. Astrophys. **576**, A8 (2015)

Vokrouhlický, D.: A complete linear model for the Yarkovsky thermal force on spherical asteroid fragments. Astron. Astrophys. **344**, 362–366 (1999)

Vokrouhlický, D., Capek, D.: YORP-induced long-term evolution of the spin state of small asteroids and meteoroids. I. Rubincam's approximation. Icarus **159**, 449–467 (2002)

Wetherill, G.W.: Steady state populations of Apollo-Amor objects. Icarus **37**, 96–112 (1979)

Wisdom, J.: Chaotic behavior and the origin of the 3/1 Kirkwood gap. Icarus **56**, 51–74 (1983)

Asteroid Mining: Mineral Resources in Undifferentiated Bodies from the Chemical Composition of Carbonaceous Chondrites

Marina Martínez-Jiménez, Carles E. Moyano-Cambero, Josep M. Trigo-Rodríguez, Jacinto Alonso-Azcárate, and Jordi Llorca

Abstract Humanity has been mining Earth deposits for decades in order to extract tiny amounts of economically valuable metals and thereby, producing huge natural devastations of our planet. Recently, asteroids have grabbed our attention since they are fascinating objects carrying the hints of Solar System origin and, at the same time, containing large amounts of valuable resources including platinum group metals (Mining the sky: untold riches form the Asteroids, Comets, and Planets, Reading, 1996), iron, nickel, rare earth elements (REE), and water (Mining the sky: untold riches form the Asteroids, Comets, and Planets, Reading, 1996; The technical and economic feasibility of mining the Near Earth Asteroids. PHD thesis, 1997). At present, 14,036 near-Earth objects (NEOs) are known to travel around an orbit close to the Earth, from which 1684 are considered potentially hazardous asteroids (PHAs). In this scenario, may not be surprising that some private companies start considering asteroid mining. In the present study, we report the bulk rare-earth element (REEs, La-Lu) compositions of 38 carbonaceous chondrites as well as 2 R-chondrites, including 5 falls and 35 finds, by using inductively coupled plasma mass spectrometry (ICP-MS) technique. The CI-chondrite-normalized REE patterns show enormous Ce anomalies and large LREE enrichments never described before, attributed to the small sample size and terrestrial contamination. We have also found the characteristic Tm anomalies described by some authors (Acta 163:234–261, 2015; Geochim. Cosmochim. Acta 176:1–17, 2016) attributed to type II CAIs. We conclude that from the point of view of abundances, REEs are not worth mining yet

M. Martínez-Jiménez (✉) • C.E. Moyano-Cambero • J.M. Trigo-Rodríguez
Institute of Space Sciences (IEEC-CSIC), Meteorites, Minor Bodies and Planetary Science Group, Campus UAB, Carrer de Can Magrans, s/n, 08193 Cerdanyola del Vallès, Barcelona, Spain
e-mail: mmartinez@ice.csic.es

J. Alonso-Azcárate
Fac. de Ciencias Ambientales y Bioquímica, Universidad de Castilla-La Mancha, Avda. Carlos III s/n, Toledo 45071, Spain

J. Llorca
Institut de Tècniques Energètiques i Centre de Recerca de Nanoenginyeria, Universitat Politècnica de Catalunya, Diagonal 647, ETSEIB, Barcelona 08028, Spain

J.M. Trigo-Rodríguez et al. (eds.), *Assessment and Mitigation of Asteroid Impact Hazards*, Astrophysics and Space Science Proceedings 46,
DOI 10.1007/978-3-319-46179-3_5

for PGEs may be reasonable. In any case, the current inequality between supply and demand of rare earths is a real problem that will result in large price instabilities for many sectors of the economy, also having negative effects in new technologies and development. Consequently, we envision that space exploration will be a way to find the new resources required to sustain market economy over longer timescales.

1 Introduction

1.1 Solar System Origin

The galaxy contains molecular clouds, and as its name refers, they are clouds mostly composed by molecular hydrogen, in contrast with the ionized gas that fills the interstellar medium. Consequently, they are cold and dense, and therefore have the potential to form stars and planets under the action of gravity and magnetic fields. When these clouds cross a spiral arm of the galaxy, the star formation is triggered (Fig. 1). Many lines of scientific evidence have demonstrated that the gravitational collapse of the protosolar nebula inside the presolar molecular cloud gave birth to the formation of the Solar System (Lada and Lada 2003; Adams 2010; Bouvier and Wadhwa 2010). Thus, the Sun was not born alone, but in a vicinity full of activity, enhanced by red giants and supernova nucleosynthetic products (Zinner 2003; Trigo-Rodríguez et al. 2009). At the beginning of the collapse, the temperature was high enough to prevent the condensation of minerals from the vapour phase. However, after some time, the major part of the gas was dragged into the Sun starting its nuclear fusion and the solar wind swept most of the gas from the circumstellar medium (Williams and Cieza 2011). Few million years after Sun's formation, the radiation pressure and the appearance of giant planets had reduced the gas density, so the surrounding environment temperature decreased below 2000 K, allowing small mineral particles to condense (Lewis 1974; Cameron and Fegley 1982). Condensates and pre-existing stellar products tend to fall into the ecliptic plane due to rotation, forming the protoplanetary disc (Taylor 2001). The inner disk was dominated by metals, refractory oxides, and silicates, while the outer disk contained complex mixtures of hydrated minerals, organics and ices. Gas drag contributed to the continuous decay of the condensates formed far from the proto-Sun: ice, organic matter and small dust, which were moving inwards (see e.g., Cuzzi et al. 2008; Willacy et al. 2015). At the same time, high temperature minerals formed in the inner region under a strong magnetic field were transported by X-winds, which are high-speed bipolar-collimated jets around young stellar objects powered by enhanced magnetic activities and disk-magnetosphere interactions (Shu et al. 1996; Shu et al. 1997), to the outer regions where they formed comets (Brownlee et al. 2006). With time, the concentration of rotating materials became higher, significantly increasing the density of the disk, and therefore allowing the progressive aggregation of these materials by sticking, which grew bigger and bigger until occasionally achieved kilometre-size diameters

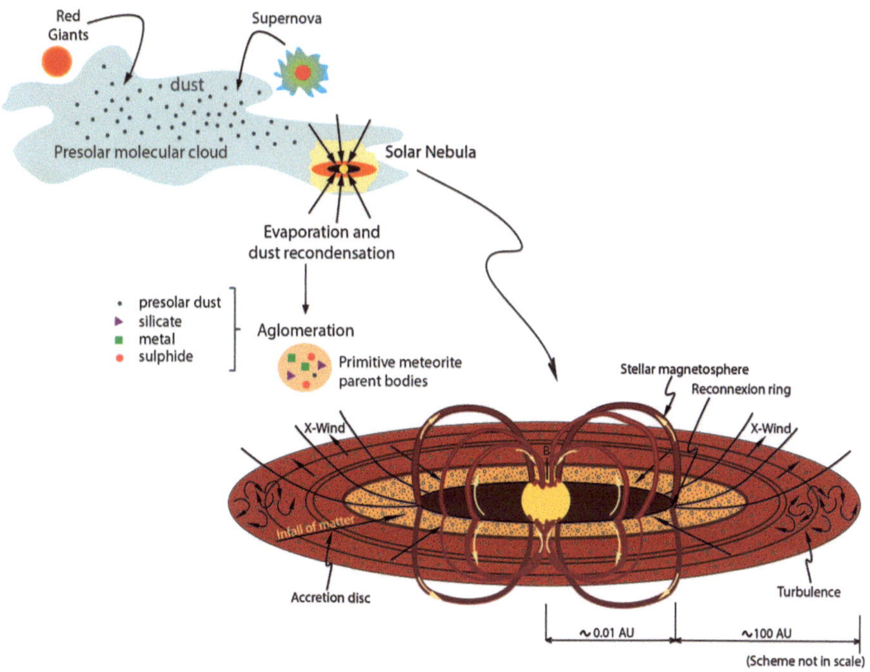

Fig. 1 Schematic view of the gravitational collapse inside the presolar molecular cloud that gave birth to the Solar System, illustrating the magnetospheric accretion model in the protoplanetary disk. Adapted from Lodders and Amari (2005)

(Blum and Wurm 2008). These primordial bodies are called planetesimals, and later participated in Solar System architecture as the building blocks of planetary embryos (Goldreich and Ward 1973; Cuzzi et al. 2008). Calcium-Aluminium-rich inclusions (CAIs) are the oldest materials dated in the Solar System. These refractory inclusions were formed very close to the proto-Sun, in the so-called reconnection ring (Fig. 2) according to the theory from Shu et al. (2001), and contain isotopic anomalies (^{16}O, ^{26}Al, ^{53}Mn, ^{41}Ca, ^{87}Rb) hence indicating that the first solids of the Solar System formed 4567 Ma (hereafter, Ma) ago (Amelin et al. 2002; Bouvier and Wadhwa 2010). Thereby, the protoplanetary disk incorporated lived nuclides in addition to the ones already present that can be used for dating accretionary processes (McKeegan et al. 2006; Nittler and Dauphas 2006). In this scenario, the first million years were characterized by a continuous mixing of the new-forming material, colliding, fragmenting and re-accreting (Figs. 1 and 2) (see e.g. Dominik et al. 2007; Williams and Cieza 2011, and references therein).

Planetesimals evolved differently depending mainly on their accretion time, but also in some extent, on their accreted rock:ice initial ratios, and sizes. The main isotopes contributing to the heat were ^{26}Al and ^{60}Fe (with halftimes of 0.74 and 1.5 Ma, respectively), which decayed during the first few million years (Cameron and Truran 1977; Meyer and Clayton 2000; Takigawa et al. 2008; Wasserburg et al.

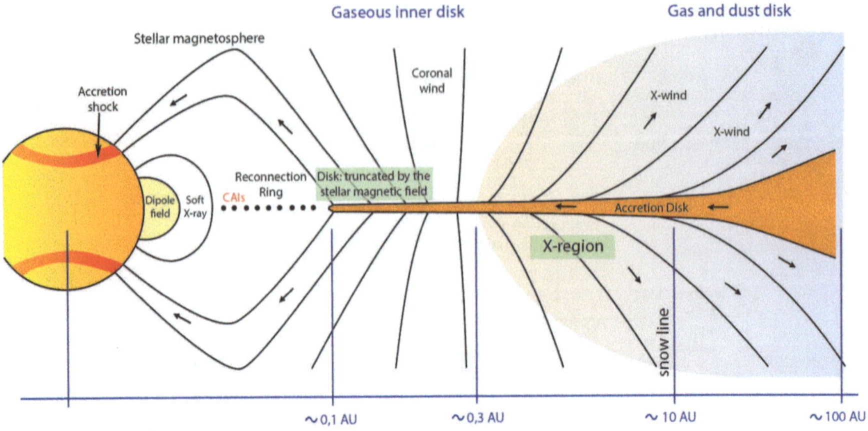

Fig. 2 Section of the protoplanetary disk. The illustration is not at scale

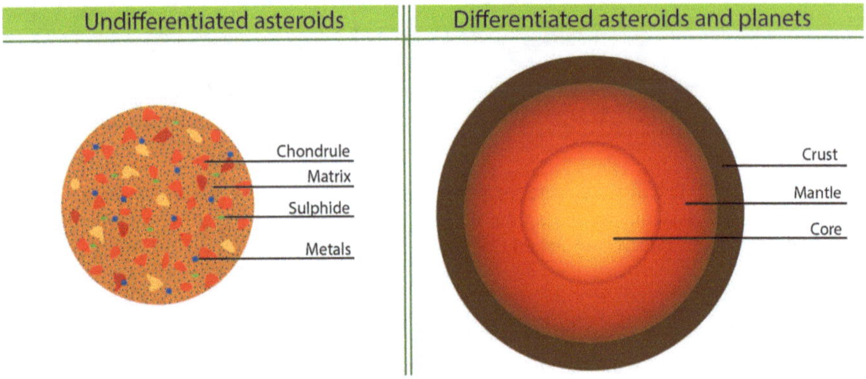

Fig. 3 Simplified illustration showing the differences in the internal structure of undifferentiated asteroids versus differentiated asteroids or planets

2006). Therefore, as ^{26}Al was the major heat source during the first 1.5 Ma, those bodies that formed very early melted the interior and thus differentiated. After some time, when part of ^{26}Al was decayed, the accreted planetesimals that were smaller enough (less than about few hundred km in diameter) remained undifferentiated because the amount of heat was lower and they displayed significant porosity, thus the heat could escape by diffusion through the pores, but their forming materials suffered thermal metamorphism and/or aqueous alteration to some extent (Fig. 3).

Therefore, undifferentiated asteroids are authentic remnants of the first bodies accreted in the solar system (Wetherill 1989; Froeschlé and Greenberg 1989), although often moulded through eons by countless impacts. Collisional processing took place especially during the first formative stages of planets in the Solar System, when their evolution was governed by larger impacts (see e.g. Michel et al. 2015; Beitz et al. 2016). Such evolution has been confirmed by recent studies of the craterisation of Mercury and the Moon (Xiao et al. 2014). In the late solar-system history, one of the largest asteroid disruptions was the L-chondrite parent body breakup (LCPB), in the main belt, occurred ~470 Ma ago during the mid-Ordovician (Schmitz et al. 2003; Schmitz 2013). In the most recent history, the number and size of impacts have been less important and dominated by smaller bodies (Hartmann and Neukum 2001).

1.2 Chemical Fractionations and the Equilibrium Model

Chemical fractionations of elements are observed among chondritic components, which are best understood of quenched equilibrium between phases in a nebula of solar composition (Palme 2001). The equilibrium model assumes that as the initially hot nebula cooled at a given region, the elements in the gas phase with condensation temperatures above the local temperature were incorporated into the planetesimals that were forming. Chondritic meteorites also display volatile depleted species in front of elements with higher condensation temperatures due to the constantly mass loss from the disk, where the gas was being removed. Such model is thought to be consistent with the uniform isotopic composition of chondritic materials, except for oxygen (Palme 2001).

Therefore, elements are classified based on their relative volatility (50 % condensation temperature) of the elements during solar nebula condensation, which is a primary parameter for establishing the chemical composition of chondritic meteorites (Wasson 1985; Palme et al. 2003). In a cooling nebula at 10^{-3} atmospheres, refractory elements condense as Ca, Al, Ti-rich oxides and silicates, along with refractory trace elements like Be, Sr, Ba, Sc, Y, rare-Earth elements (hereafter, REEs), Zr, Hf, V, Nb, Ta, U, Th, Mo W, and Platinum-Group metals (hereafter, PGEs), in a temperature range of 1800–1450 K. At a temperature range of 1450–1350 K, major elements condense as silicates of Mg (basically forsterite and enstatite), and Fe condenses as metallic FeNi along with Co. At 1250–650 K, alkalis and moderately volatile siderophile elements condense, and at 650 K, Fe reacts to form troilite (FeS). Below 650 K, highly volatile elements like the halogens and inert gases condense. As it is seen, elements are also subdivided depending on what they form: siderophiles (condense as metals), litophiles (condense as oxides) and calcophiles (condense as sulphur and/or some other chalcogen other than oxygen). Lodders (2003) provided a complete list of 50 % condensation temperatures of all elements except hydrogen and helium. Trace elements are generally less reliable than major elements, since they condense in solid solution with major phases, and hence, thermodynamic properties are not very well constrained (e.g. Wai and Wasson 1977).

The CI chondrites, a group of carbonaceous chondrites, are the most volatile-rich chondrites, as well as the most aqueously altered group. Their compositions appear to be representative of mean nebular condensable material, and thus, match well with the solar photosphere for all elements with the exception of four (H, C, N and O) (Kallemeyn and Wasson 1981; Wasson 1985; Wasson and Kallemeyn 1988; Lodders 2003).

Fractionations among chondrite groups have been extensively discussed because they provide useful classification information. When normalized to CI-chondrites (solar abundances), they provide insights into evaporation/condensation processes occurred in the nebula (Boynton 1975). Refractory litophile elements (Al, Sc, Ca, Lu, Yb, Eu, Sm, La) in undifferentiated chondrites display almost flat patterns when normalized to CI chondrites (Wasson 1985), with an absolute range of refractory component very limited. Individual REE vary in abundance by a factor of 2.5 with the lowest values in EH and EL chondrites and the highest values in CV chondrites (Evensen et al. 1978; Kallemeyn and Wasson 1981).

1.3 Meteorite Classification

Meteorites are classified based on their mineralogical and petrographic characteristics and on their whole-rock chemical O-isotopic compositions into chondrites, achondrites, iron meteorites, and stony-iron meteorites (Fig. 4). These materials brought to Earth by planetary resonances at work in the MB are naturally sampling tens of parent bodies of different nature. Chondrites are, thus, meteorites coming from undifferentiated bodies and, in particular, carbonaceous chondrites are considered to be the most primitive class, since they come from smaller asteroids formed in the outer parts of the solar system, where they were less affected by solar activity. They contain organic matter in their matrices and also the vast majority are affected in more or less grade by aqueous alteration and/or thermal metamorphism. Only few samples are exempt from alteration, which are reliable samples for studying nebular processes.

1.4 The Rare-Earth Elements Under Cosmochemical Processes

The REEs are the group of elements comprised by the 15 lanthanides (lanthanum (**La**), cerium (**Ce**), praseodymium (**Pr**), neodymium (**Nd**), promethium (**Pm**), samarium (**Sm**), europium (**Eu**), gadolinium (**Gd**), terbium (**Tb**), dysprosium (**Dy**), holmium (**Ho**), erbium (**Er**), thulium (**Tm**), ytterbium (**Yb**), and lutetium (**Lu**)); together with scandium (**Sc**), and yttrium (**Y**) (Fig. 5). They are subdivided into light rare-earth elements (LREE, La-Nd) (in green, Fig. 5), middle rare-earth elements (MREE, Sm-Tb) (in pink, Fig. 5), and heavy rare-earth elements (HREE, Dy-Lu) (in purple, Fig. 5) based on the electron configuration of each REE: LREE have unpaired electrons while HREE, paired electrons. Scandium has other properties that are not similar enough to classify it as either a LREE or HREE. Because of their

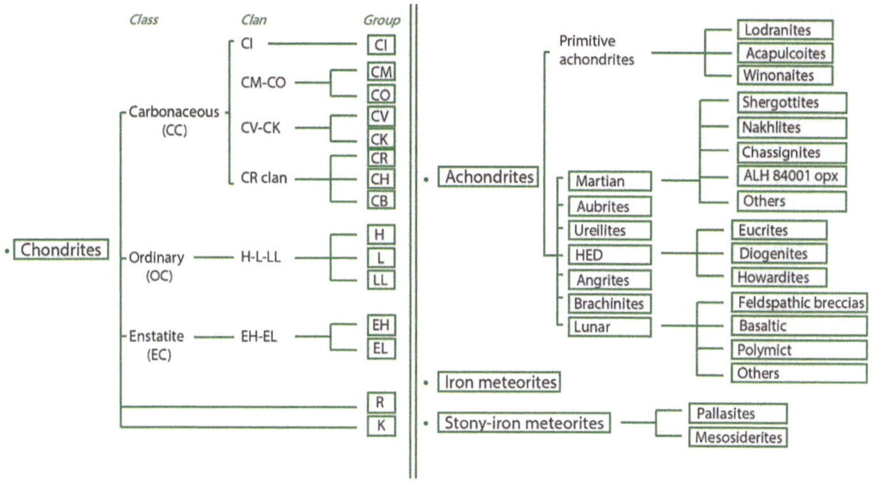

Fig. 4 Diagram showing the systematic meteorite classification, adapted from Weisberg et al. 2006

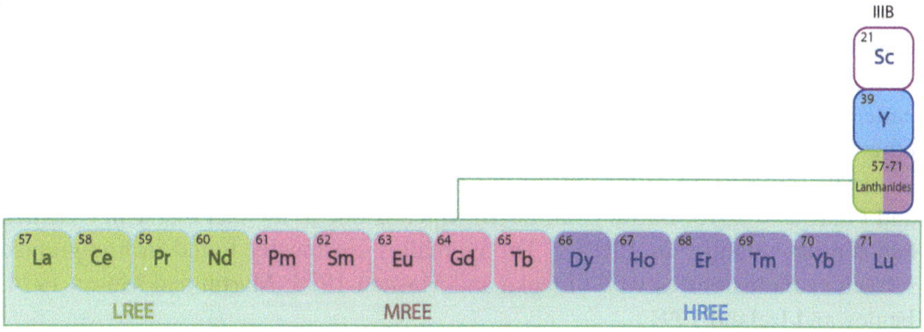

Fig. 5 The rare-Earth elements in the periodic table

similarities in chemical and physical properties, they can easily make substitutions for one another making refinement of pure metal very difficult, and thus, it was not until twentieth century that they were all finally identified (Long et al. 2010).

REEs have similar geochemical behavior since they are all large-ion litophile elements and most of them are trivalent and partition among melts and mineral phases as smooth function of their masses, what is known as lanthanide contraction, whereby ionic radii decreases with increasing atomic masses. This property renders them less susceptible to mutual fractionation than other groups of elements in geochemical processes. There are some exceptions which appear to be anomaly variable among chondrites: Ce and Eu exhibit depletions or enrichments relative to neighbor REEs because they exist as 4+ and 2+ respectively under reducing conditions.

REEs have diagnostic signatures associated with a variety of geochemical and cosmochemical processes which can be either from nebular origin (evaporation and condensation), or from parent-body processes (aqueous alteration, metamorphism, or magmatic differentiation), as well as terrestrial weathering. Usually, to distinguish among these processes is a difficult task. For example, Eu anomalies or fractionation of LREE relative to HREE (Nakamura and Masuda 1973; Nakamura 1974; Evensen et al. 1978) are understood to have a nebular origin (Dauphas and Pourmand 2015). However, anomalies in Ordinary and Enstatite Chondrites show more dispersion as more metamorphosed is the grade (types 4–6) compared to unequilibrated chondrites (types 1–3), suggesting a possible relation with parent-body processing, possibly due to a parent-body redistribution of REEs in various carrier phases during metamorphism (Dauphas and Pourmand 2015).

Grossman (1973) found that CAIs are responsible for the depletion of refractory elements in ordinary and enstatite chondrites. Indeed, CAIs have a clear implication in the pattern of normalized REEs. A common trend observed in about 1/3 of the CAIs analyzed by Fegley and Ireland (1991) is the depletion in the most refractory and volatile REEs (Tanaka and Masuda 1973; Martin and Mason 1974). This characteristic pattern is associated with the condensation sequence of the REEs (Boynton 1975; Davis and Grossman 1979; Kornacki and Fegley 1986). Under reducing conditions, in the region of CAI formation, Eu and Yb are more volatile than LREEs (Dauphas and Pourmand 2015) and hence, they never fully condensed (depletion patterns), but under oxidizing conditions, Ce becomes much more volatile than the other REEs, and Eu and Yb become more refractory (Davis et al. 1982).

On the other hand, some authors have found a positive Tm anomaly (Dauphas and Pourmand 2015; Barrat et al. 2016) which is due to the fact that Tm, a HREE, has similar volatility as LREEs. Tm anomalies are mostly found in fine-grained refractory inclusions as well (Tanaka and Masuda 1973; Martin and Mason 1974; Fegley and Ireland 1991).

1.5 Space Missions to Asteroids

1.5.1 NEAR-Shoemaker

The first space mission performing an *in situ* characterization of the chemical elements present in the surface of an asteroids was the Near-Earth Asteroid Rendezvous—Shoemaker (NEAR-Shoemaker) mission (NASA), which visited the S-type asteroid 433 Eros on December 12, 2001. Eros is the second largest NEA, with a close period orbit of 1 year and a typical Amor orbit, i.e., that crosses Mars' orbit but do not quite reach Earth's. This mission was also the first-ever to orbit an asteroid and the first to touch down its surface. The objectives of the mission were to return data about the bulk physical properties, composition, mineralogy, morphology, internal mass distribution and magnetic field of 433 Eros (Farquhar et al. 1999; Veverka et al. 2000; Zuber et al. 2000).

1.5.2 Hayabusa

Probably the most famous mission to an asteroid so far was the Hayabusa mission, developed by the Japan Aerospace Exploration Agency (JAXA). Its main goal was to return a sample of material from the surface of a small S-type NEA named 25143 Itokawa. It is an Apollo asteroid (crossing Earth's orbit) with a dimension of 550×180 m (Fujiwara et al. 2006). It was launched in 2003 and in 2005 arrived to the asteroid, where it studied the shape, spin, topography, colour, composition, density and evolutionary history. Two Hayabusa touchdowns on the asteroidal surface lifted off regolith grains that were returned to Earth on June 2010. Laboratory studies revealed that the μm-sized grains captured by Hayabusa were indeed asteroid regolith, opening the door for further studies, particularly relevant to better understand space weathering (see e.g. Nakamura et al. 2012).

1.5.3 Hayabusa 2

After the big success of Hayabusa, JAXA decided to promote a second mission to retrieve samples from another asteroid to Earth. Hayabusa 2 plans to visit and return samples from the C-type Apollo asteroid 162173 Ryugu. Hayabusa 2 was launched on December 3, 2014 and it is expected to arrive at its target in July 2018, surveying the asteroid for a year and a half, and departing in December 2019, with a planned return of samples to Earth in December 2020 (Ishiguro et al. 2014).

1.5.4 OSIRIS-REx

Origins-Spectral Interpretation-Resource Identification-Security Regolith Explorer (OSIRIS-REx) will be the first NASA mission to return samples from an asteroid back to Earth. This mission seeks to do a comprehensive map of the asteroid, measure the orbit deviation caused by non-gravitational forces (the Yarkovsky effect), compare observations at the asteroid to ground-based observations, and also doing *in situ* analysis of the samples. The launch will be in September 2016 and will spend 3 years travelling to approach the primitive B-type (a sub category of C-type) Apollo NEA 101955 Bennu. The surface mapping is expected to take 6 months. The spacecraft, then, will pick a location from where its arm will take a sample (Lauretta et al. 2015).

2 Experimental Techniques

Some of the meteorite samples used here were provided by the NASA Johnson Space Center (JSC), responsible for the curation of the Antarctic Search for Meteorites (ANSMET) collection, while others can be found on our Institute of

Space Sciences (ICE) meteorite repository. The specimens were analyzed by using an ICAP 6500 Thermo Electron Inductively Coupled Plasma Atomic Emission Spectrometer (ICP-AES) for major elements, and a Thermo Electron X Series II Inductively Coupled Plasma Mass Spectrometer (ICP-MS) for minor and trace elements. In order to ensure matrix similarity between calibration standards and samples, geological Certified Reference Materials (CRMs) of the United States Geological Survey (AGV-2, BCR-2, GSP-2, SDC-1, BHVO-2 and BIR-1) were used for external calibration. Rhodium was used as internal standard. The solutions of the samples, CRMs and blanks were obtained by alkaline fusion with $LiBO_2$ in Pt-Au crucibles, followed by acid dissolution of the melt: approximately 0.025 g of sample and 0.05 g of flux is fused and the melted glass is poured into a teflon beaker containing 30 mL HNO_3 0.3 N and 1 drop of HF.

3 Results

Lanthanide abundances were analyzed in ICP-MS for 40 chondrites, which are listed in Table 1, along with the results in Table 2, and plotted in Fig. 7. The present CI data used to normalize the results corresponds to that from 6 CI chondrites from McDonough and Sun (1995).

We have to take into account that the bulk analyses of chondrites have been made using a smaller amount of sample (25–50 mg) than commonly used (250 mg). In order to verify the obtained results, we have compared the abundance obtained for 7 elements, which are the common elements found elsewhere in literature (the three main metals Fe, Co, Ni plus the four lanthanides La, Sm, Eu, Yb) in specific carbonaceous chondrites from the CM, CO, CV and CK groups, as well as mean value obtained for several CR chondrites, with the results found in the literature for the same meteorites (Kallemeyn and Wasson 1981; Kallemeyn et al. 1991, 1994, 1996; Wasson et al. 2013; Choe et al. 2010) (Table 3).

Table 1 Meteorite samples analyzed in the present work

Group	Meteorite names
CH	PCA 91467
CI	Orgueil
CK	ALH 85002; LAR 04318; PCA 82500
CM	MIL 07689; SCO 06043; Cold Bokkeveld; EET 96029; LEW 87148; MAC 02606; Murchison; Murray; QUE 99355
CO	DOM 08006; MIL 03377; MIL 05024; ALHA77307; Kainsaz; ALH 82101; ALH 83108; ALHA77003
CR	EET 92159; EET 92042; GRA 95229; LAP 02342; RBT 04133
CV	ALH 84028; Leoville; MET 00430; MIL 07002; QUE 99038; MET 01017
CCs ungr	GRO 95551; EET 83355; EET 96026; GRO 95566; MAC 87300.
R3	MET 01149; PRE 95404

Table 2 Rare-Earth element concentration (μg/g) in carbonaceous chondrites analyzed in the present work

Group	Name	La	Ce	Pr	Nd	Sm	Eu	Gd	Tb	Dy	Ho	Er	Tm	Yb	Lu
Chnd-ung.	GRO 95551	6.18	23.08	1.42	5.68	1.16	0.28	1.01	0.14	0.84	0.16	0.93	0.09	0.51	0.08
C2-ung.	EET 83355	1.75	36.26	0.40	1.72	0.46	0.13	0.63	0.07	0.50	0.11	0.39	0.05	0.31	0.05
C2-ung.	GRO 95566	0.41	1.47	0.12	0.56	0.18	0.06	0.21	0.03	0.23	0.05	0.15	0.03	0.15	0.02
C2-ung.	MAC 87300	2.03	12.04	0.46	1.94	0.45	0.12	0.49	0.07	0.49	0.10	0.43	0.06	0.32	0.05
C4/5	EET 96026	4.73	63.53	1.04	4.19	1.02	0.24	1.30	0.14	0.90	0.18	0.94	0.09	0.55	0.09
CI1	Orgueil	0.16	0.00	0.06	0.36	0.08	0.04	0.13	0.02	0.21	0.04	0.13	0.02	0.14	0.02
CM1	MIL 07689	1.73	4.88	0.40	1.71	0.42	0.12	0.34	0.07	0.45	0.09	0.41	0.05	0.30	0.05
CM1	SCO 06043	2.03	5.62	0.91	4.38	0.51	0.12	0.44	0.08	0.62	0.10	0.40	0.05	0.30	0.05
CM2	Cold Bokk.	0.26	0.46	0.10	0.55	0.18	0.07	0.24	0.04	0.34	0.08	0.22	0.04	0.22	0.03
CM2	EET 96029	0.42	1.15	0.13	0.62	0.20	0.06	0.22	0.04	0.26	0.06	0.16	0.03	0.18	0.02
CM2	LEW 87148	0.80	1.38	0.29	1.50	0.59	0.23	0.72	0.12	0.96	0.22	0.63	0.11	0.60	0.09
CM2	MAC 02606	1.33	3.73	0.31	1.34	0.34	0.09	0.26	0.05	0.34	0.07	0.32	0.04	0.23	0.04
CM2	Murchison	0.39	0.57	0.14	0.79	0.25	0.08	0.26	0.05	0.38	0.08	0.23	0.05	0.24	0.03
CM2	Murray	0.31	0.35	0.10	0.59	0.19	0.08	0.27	0.05	0.47	0.13	0.38	0.04	0.22	0.06
CM2	QUE 99355	2.75	8.17	0.64	2.75	0.71	0.18	0.51	0.10	0.64	0.13	0.62	0.07	0.43	0.07
CR2	EET 92159	0.79	2.30	0.20	0.92	0.24	0.08	0.28	0.05	0.32	0.07	0.26	0.04	0.21	0.03
CR2	EET 92042	1.12	3.37	0.29	1.29	0.36	0.11	0.36	0.06	0.39	0.09	0.32	0.04	0.26	0.04
CR2	GRA 95229	1.47	3.36	0.42	1.92	0.60	0.16	0.45	0.10	0.73	0.15	0.45	0.08	0.44	0.07
CR2	LAP 02342	1.00	2.95	0.27	1.17	0.34	0.10	0.38	0.06	0.44	0.10	0.33	0.05	0.27	0.04
CR2	RBT 04133	2.09	53.67	0.50	2.11	0.55	0.13	1.04	0.11	0.64	0.19	0.97	0.07	0.34	0.12

Table 2 (continued)

Group	Name	La	Ce	Pr	Nd	Sm	Eu	Gd	Tb	Dy	Ho	Er	Tm	Yb	Lu
CO3	DOM 08006	0.47	1.18	0.14	0.69	0.21	0.06	0.24	0.04	0.29	0.06	0.18	0.03	0.19	0.03
CO3	MIL 03377	0.45	1.07	0.14	0.66	0.21	0.06	0.25	0.04	0.31	0.06	0.19	0.03	0.19	0.03
CO3	MIL 05024	0.57	1.31	0.16	0.77	0.25	0.07	0.29	0.04	0.32	0.07	0.20	0.04	0.19	0.03
CO3.0	ALHA77307	6.45	51.74	1.36	5.47	1.31	0.32	1.32	0.16	1.00	0.21	1.08	0.10	0.64	0.10
CO3.2	Kainsaz	0.37	0.86	0.13	0.68	0.24	0.11	0.30	0.05	0.41	0.10	0.26	0.05	0.27	0.04
CO3.4	ALH 82101	7.94	27.22	1.86	7.68	1.59	0.46	1.26	0.20	1.25	0.25	1.30	0.12	0.82	0.12
CO3.5	ALH 83108	0.52	3.91	0.16	0.85	0.25	0.09	0.31	0.06	0.48	0.10	0.24	0.05	0.26	0.04
CO3.6	ALHA77003	2.51	45.60	0.55	2.31	0.60	0.16	0.84	0.09	0.55	0.12	0.52	0.06	0.38	0.06
CV3	ALH 84028	2.28	5.12	0.61	2.58	0.82	0.21	0.61	0.12	0.92	0.19	0.62	0.11	0.57	0.08
CV3	Leoville	0.66	1.26	0.25	1.32	0.47	0.19	0.56	0.10	0.78	0.17	0.48	0.09	0.54	0.08
CV3	MET 00430	3.28	46.49	1.66	7.88	0.73	0.19	0.92	0.13	0.94	0.13	0.68	0.07	0.45	0.07
CV3	MIL 07002	4.01	34.10	0.88	3.70	0.86	0.21	0.88	0.12	0.72	0.15	0.60	0.07	0.44	0.07
CV3	QUE 99038	1.88	4.72	0.45	2.00	0.52	0.13	0.43	0.09	0.59	0.12	0.43	0.06	0.36	0.05
CV3-an.	MET 01017	0.70	1.58	0.21	1.03	0.30	0.10	0.33	0.06	0.39	0.08	0.23	0.04	0.25	0.04
CH3	PCA 91467	5.02	12.16	1.08	4.42	0.89	0.24	0.79	0.15	1.20	0.21	0.70	0.12	0.55	0.09
CK4	ALH 85002	1.66	4.48	0.44	1.94	0.46	0.14	0.38	0.08	0.57	0.12	0.41	0.06	0.35	0.05
CK4	LAR 04318	0.63	4.44	0.19	0.96	0.30	0.09	0.28	0.06	0.44	0.10	0.27	0.05	0.29	0.05
CK4/5	PCA 82500	2.05	6.14	0.62	3.23	1.10	0.15	1.44	0.28	2.09	0.41	1.18	0.19	1.16	0.18
CI[a]	Normalization	0.237	0.613	0.0928	0.457	0.148	0.0563	0.199	0.0361	0.246	0.0546	0.16	0.0246	0.161	0.0246

[a]Value used to normalize the results, which corresponds to the composition of CI chondrites from McDonough and Sun (1995)

Table 3 Comparison of 7 elements in 7 carbonaceous chondrites between literature and ICE-UCLM data (2014)

	Fe (mg/g)	Co (μg/g)	Ni (μg/g)	La (ng/g)	Sm (ng/g)	Eu (ng/g)	Yb (ng/g)
CM							
Murray	213[a]	604[a]	12300[a]	335[a]	210[a]	79[a]	224[a]
	224	589	1061	310	190	80	220
Murchison	209[a]	586[a]	12000[a]	313[a]	202[a]	81[a]	231[a]
	220	590	1034	390	250	80	240
Cold Bokkeveld	197[a]	546[a]	10900[a]	282[a]	175[a]	71[a]	205[a]
	240	545	994	260	180	70	220
CO							
Kainsaz	247[a]	693[a]	14300[a]	399[a]	241[a]	96[a]	275[a]
	207	740	1344	370	240	110	270
CV							
MIL 07002	206[b]	705[b]	13200[b]	410[b]	234[b]	94[b]	321[b]
	205	634	13844	401	860	210	440
CK							
LAR 04318	207[b]	837[b]	19500[b]	390[b]	235[b]	100[b]	222[b]
	217	538	11155	585	270	80	245
CR							
Mean	240[c]	667[c]	13600[c]	342[c]	210[c]	84[c]	236[c]
	222	642	12887	1020	386	110	300
Chondrite ungrouped							
GRO 95566	197[d]	541[d]	11000[d]	321[d]	189[d]	67[d]	164[d]
	194	384	8143	410	180	60	150

[a]Kallemeyn and Wasson (1981)
[b]Wasson et al. (2013)
[c]Kallemeyn et al. (1994)
[d]Choe et al. (2010)

We noticed that our bulk chemistry results are consistent with the available data from literature, which are basically meteorite falls (Tables 3 and 4), considering the intrinsic compositional heterogeneity of different chips from the analysed chondrites. Main differences are in Ni and may be due to the arbitrary distribution of Fe-Ni grains within a sample, which implies a low probability of having them represented in an analysed mass of 25–50 mg. In Fig. 6 we compile the same elements, except for Ni, to exemplify the small differences (within CM group and in the CO Kainsaz).

We have plotted CI-normalized-REE abundances within the different CC groups in Fig. 7.

Table 4 Comparison of REE analyses (in ppm) from Nakamura (1974), Evensen et al. (1978) and the present work for 3 CM falls (Murchison, Murray and Cold Bokkeveld)

	Murchison			Murray		Cold Bokkeveld	
	CM2 Fall			CM2 Fall		CM2 Fall	
	Nakamura 1974	Evensen et al. 1978	Present work	Evensen et al. 1978	Present work	Evensen et al. 1978	Present work
La	0.32	0.25	0.39	0.33	0.31	0.35	0.26
Ce	0.85	0.66	0.57	0.86	0.35	0.92	0.46
Nd	0.65	0.48	0.79	0.66	0.59	0.64	0.55
Sm	0.21	0.15	0.25	0.21	0.19	0.22	0.18
Eu	0.08	0.06	0.08	0.08	0.08	0.08	0.07
Gd	0.28	0.20	0.26	0.27	0.27	0.28	0.24
Dy	0.35	0.26	0.38	0.34	0.47	0.34	0.34
Er	0.23	0.17	0.23	0.22	0.38	0.23	0.22
Yb	0.23	0.20	0.24	0.22	0.22	0.23	0.22
Lu	0.04		0.03		0.06		0.03

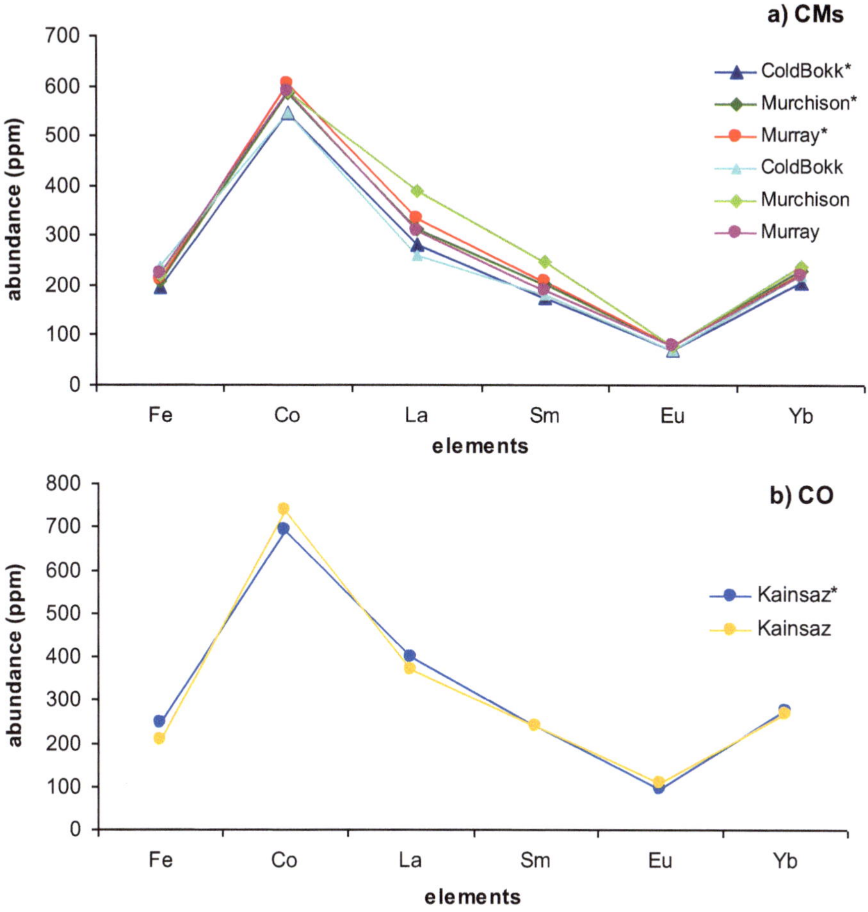

Fig. 6 Comparison for 6 elements in (**a**) 3 CM carbonaceous chondrites and (**b**) 1 CO chondrite between Kallemeyn and Wasson (1981) (*Asterisk*) data and the results in the present work

4 Discussion

4.1 REE Abundances in Undifferentiated Chondrites and Earth Crust

The composition of chondrites (Wasson and Kallemeyn 1988) has been compared with that of the Earth's crust (Rudnick et al. 2005) for REEs, PGEs and other metals in Table 5 in order to have a good approximation about the abundances from literature. As seen in Table 5, there is a limited range of concentrations of REEs and PGEs in chondritic meteorites. Within the elements in the list, only Ni and Cr exhibit much higher abundances in carbonaceous chondrites than in the Earth crust; being these elements concentrated in the nucleus of the planet.

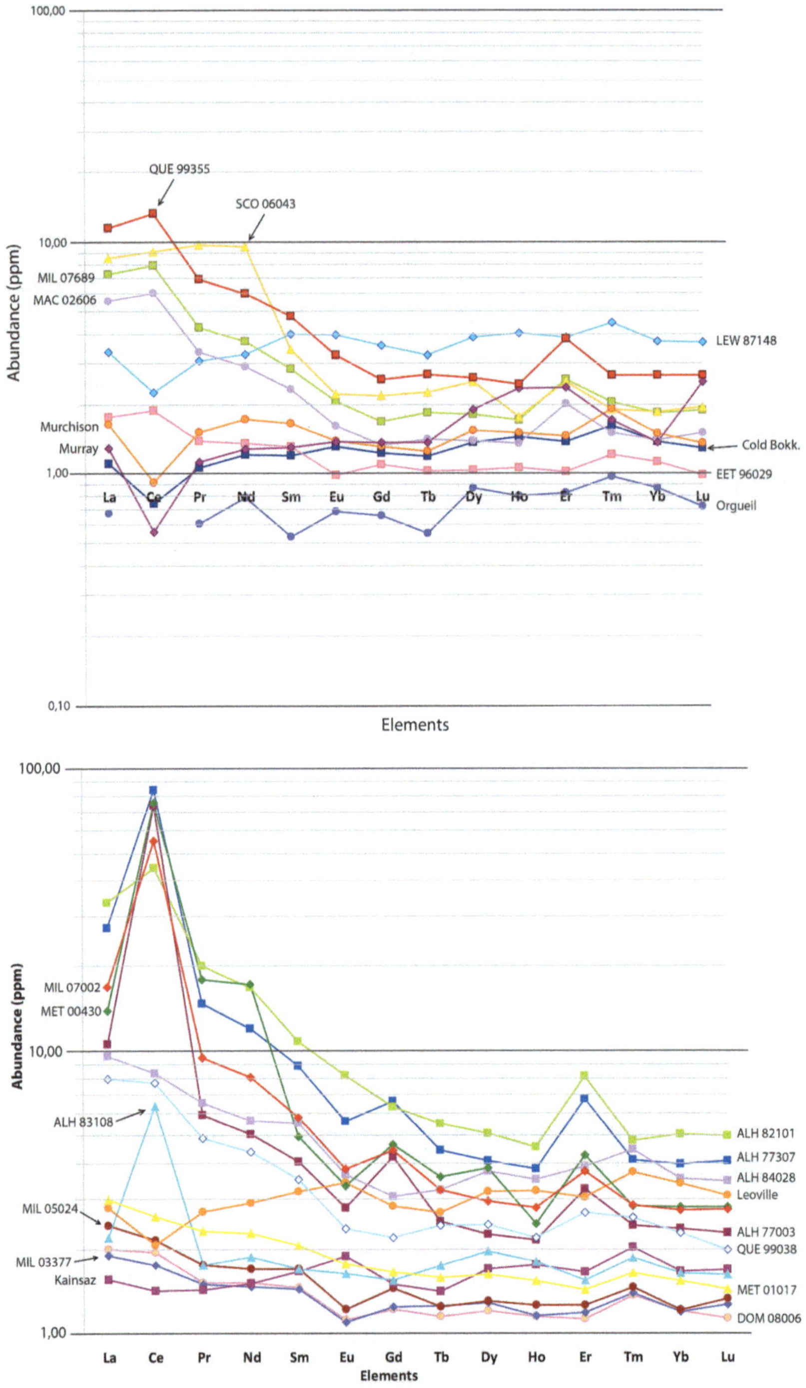

Fig. 7 Lanthanide abundances (in ppm) in 1 CI, 9 CMs, 7 CVs, 5 CRs, 1 CH, 2 CKs, 5 ungrouped CCs, and 2 R-chondrites

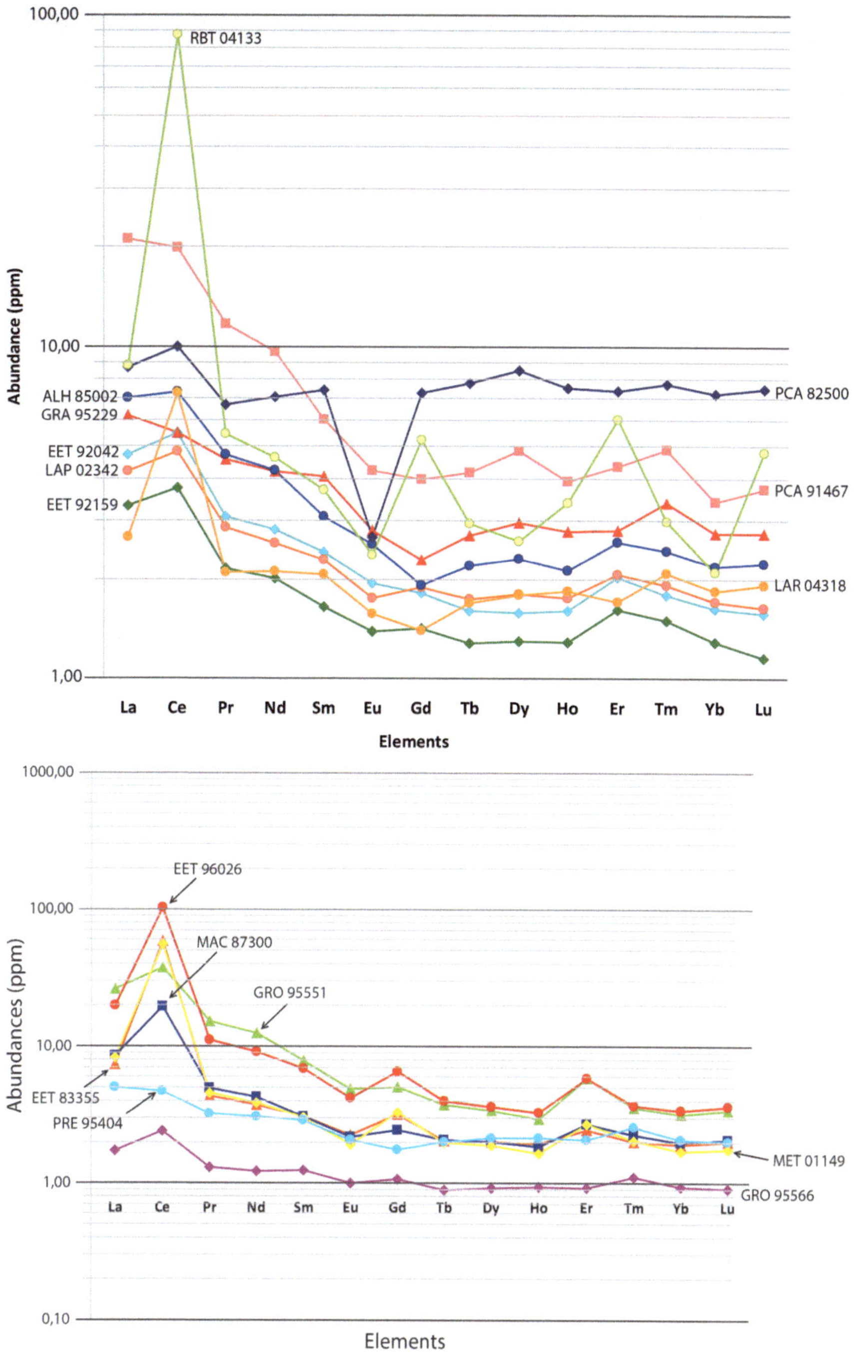

Fig. 7 (continued)

Table 5 Composition of carbonaceous chondrites ([a]Wasson and Kallemeyn 1988) compared to that from the Earth's crust ([b]Rudnick et al. 2005)

Element	CCs[a]	Earth crust[b]
Cr	265–360	135
Co	0.51–0.66	26.6
Ni	1070–1340	59
Cu	0.10–0.12	27
Zn	0.12–0.31	72
Sc	5.80–11.40	21.9
Y	1.44–2.40	19
La	0.24–0.49	20
Ce	0.62–1.29	43
Pr	0.09–0.20	4.9
Nd	0.46–0.99	20
Sm	0.15–0.30	3.9
Eu	0.06–0.11	1.1
Gd	0.20–0.42	3.7
Tb	0.04–0.07	0.6
Dy	0.25–0.48	3.3
Ho	0.06–0.11	0.77
Er	0.16–0.32	2.1
Tm	0.03–0.05	0.28
Yb	0.16–0.32	1.9
Lu	0.03–0.05	0.30
Os	0.49–0.83	0.000041
Ir	0.46–0.76	0.000037
Pt	0.99–1.20	0.0015

How minor are REEs in Earth deposits? They are actually widely distributed throughout Earth's crust, although in very small quantities, and rarely concentrated into mineable ore deposits (Long et al. 2010). That makes them very difficult to mine because it is unusual to find them in concentrations high enough for an efficient economical extraction. Their abundances follow two patterns: firstly, elements with even atomic numbers appear in larger abundances than their odd numbered neighbours, and secondly, LREE (La, Ce, Pr, Nd, Pm, Sm, Eu) are more incompatible and consequently more strongly concentrated in the continental crust. In rock-forming minerals, REE do not occur as metals; instead, they typically occur in carbonates, oxides, phosphates, halides and silicates. Although there are around 200 minerals known to contain REE, only a relatively small number may become commercially significant on Earth, and the vast majority of resources are only associated to bastnäsite, monazite and xenotime, being the first two the primary source of Ce, La and Nd (reported by the British Geological Survey), and the third one dominated by HREE, including Y, Dy, Er, Yb and Ho (Harben, P.W.: Industrial minerals handybook, 4th ed. Rare Earth Minerals and Compounds 2002).

The origin of terrestrial deposits usually has high complexities in forming processes, and it is associated with carbonatites, alkaline igneous rocks, hydrothermal processes, sedimentary processes (placer deposits), or weathering (such as residual clay deposits). Therefore, the diversity of REE-bearing deposits makes extraction methods and processing techniques variable and complex. Usually, in a particular deposit, open-pit mining and the underground mining take place simultaneously. For placers, it depends on whether the deposits are submerged in water or on land. In dry mining placers, scrapers, bulldozers and loaders are needed to collect and transport the ore, which is usually poorly consolidated and shallow (Iluka Mineral Sands Mining. www.iluka.com 2010). When placers are affected by a high quantity of water or even submerged, dredgers are commonly used, which are floating vessels with buckets or a suction device. However, in general, mining operations use labour-intense drilling and blasting techniques, together with explosives, and requires the construction of an underground rail system for material transport.

After mining, the ores have to be processed in order to increase their REE content, rejecting the commercially worthless material (or gangue minerals), through a range of physical and chemical processes that depend on the deposit type. For example, for a hard rock deposit, the following chain of processes is required: crushing and grinding; chemical conditioning producing a slurry; froth flotation, thickening, filtering and drying of it in order to produce 60 % of REO (rare-Earth oxides); and finally, refining, during which chemical processing has to be done to extract the REE itself from REE-bearing minerals. Each step has complicate processes using chemicals and different physical processes (British Geological Survey 2011).

This scenario would change considerably when referring to asteroids but, are carbonaceous chondrite parent-bodies worth mining referring to REEs? As seen in Table 5, they are not but, in any case, a part from REEs, C-type asteroids also contain Platinum Group Elements (PGMs) and non-oxidized metals (Fe, Ni, Cr, Co), which are commonly used in the industry (see e.g. Blair et al. 2000).

4.2 LREE Enrichments and Large Ce Anomalies

By looking at the CI-chondrite-normalized REE patterns from our results, CM chondrites and the CI chondrite Orgueil exhibit the lowest abundances in REEs and also the less scattered values around 1, likely associated with low metamorphosed degrees (types 1–2) (Dauphas and Pourmand 2015). CIs and CMs are the two groups exhibiting the highest porosity (Macke et al. 2011), and those that experienced the most severe aqueous alteration, especially the CI group (Zolensky and McSween 1988; Zolensky et al. 1997). In aqueous alteration systems, REE have very low fluid/rock partition coefficients and hence, their primary abundances are only distributed at relatively high fluid/rock ratios under most weathering, diagenetic, hydrothermal and metamorphic conditions. Therefore, aqueous alteration in the parent bodies could allow the remobilization of REEs. On the other hand, Michard (1989) showed that neutral or alkaline solutions appear to be poorer in REEs than solutions with low-pH at similar conditions. CI chondrites are mainly composed by a very fine-grained material (>99 vol.% of matrix), they are very primitive, and they suffered less compression, hence, they could retain higher amounts of organic matter and ices rich in volatiles (Brearley and Jones 1998; Zolensky and McSween 1988). It is important to keep in mind that the parent body of each group of carbonaceous chondrites could have been broken, providing a possible different path in the Solar System for each asteroidal fragment, which would have suffered additional heat by further collisions, solar approaches, etc. As a consequence, several fragments from the same parent body would show different compositions, explaining the wide range displayed by specimens of the same chondrite group (Trigo-Rodríguez 2015; Beitz et al. 2016).

Hirota et al. (2002) compared carbonaceous chondrites with ordinary chondrites observing that CK chondrites exhibit higher REEs abundances with respect to ordinary chondrites, explained by the higher oxidizing conditions, where Eu was able to exist in its trivalent state, suppressing the redistribution of REEs among minerals that happened in metamorphosed ordinary chondrites during early thermal metamorphic events on the parent body. Nevertheless, some CKs show low abundances of REEs, which is explained by Hirota et al. (2002) as the consequence of intrinsically inherited properties from early solar nebula refractory precursors. They also observed that REE patterns in CK chondrites had minor negative anomalies compared to CV and CO chondrites, up to −15 % on Ce and −25 % on Eu. Regarding ordinary chondrites, Nakamura (1974) concluded that usually the fractionations were relatively small for them, but those for enstatite chondrites were larger. And he also showed that refractory trace elements are depleted in enstatite and ordinary chondrites compared to CI chondrites.

Stracke et al. (2012) found in Allende meteorite that REE concentrations generally varied by more than five times the analytical uncertainty, proving that the refractory element variability is largely caused by the abundance of CAIs. These refractory inclusions host large part of refractory elements, and are much less homogeneous distributed than the phases that host non-refractory elements; and not only vary in abundance but also in composition (Stracke et al. 2012). Also Nakamura (1974) showed that Allende displayed the most fractionated pattern among its analyses.

However, REE abundances previously reported in literature for carbonaceous chondrites (Nakamura 1974; Evensen et al. 1978; Choe et al. 2010; Dauphas and Pourmand 2015) have different signatures than those shown in the present work. When normalized to CI chondrites, our results show large enrichments in LREE and enormous Ce anomalies, and same chondrites that exhibit such Ce enrichments also display small Er enrichments (Fig. 7, Table 6). Indeed, values reported by our analyses seem not real in comparison with literature, but general patterns are coherent without considering the values. For example, Evensen et al. (1978) found that positive Ce anomalies were often in their analyses.

Refractory inclusions were classified by Masson and Martin (1977) and Masson and Taylor (1982) into 6 groups, and all of them were enriched in LREE 20 times relative to CI-chondrite. In Allende meteorite, Stracke et al. (2012) found that the predominant group-type refractory inclusion was group II since they saw important

Table 6 Ce, Er, Tm and Eu anomalies in the carbonaceous chondrites analyzed in the present work

Chondrite group	Positive Ce and Er anomaly	Positive Tm (with negative or no Ce) anomaly	Negative Eu anomaly	Negative Yb anomaly
CM	QUE 99355	LEW 87148		Murray
	MIL 07689	Cold Bokkeveld		
	MAC 02606	Murchison		
CO	ALH77003	Leoville	ALH 77307	
	ALH 82101	Kainsaz	ALH77003	
	ALH 77307	ALH 83108	MIL 05024	
	ALH 83108	MIL 03377		
		DOM 08006		
CV	MIL 07002	MET 01017	MIL 07002	
	MET 00430	ALH 84028	MET 00430	
CK	LAR 04318 (Ce only)	LAR 04318	PCA 82500	
	PCA 82500 (Ce only)			
CR	RBT 04133	GRA 95229	RBT 04133	RBT 04133
			LAP 02324	
CH		PCA 91467		PCA 91467
Chond. ungrouped	EET 96026	GRO 95566	EET 96026	
	EET 83355		EET83355	
	GRO 95551			
R3	MET 01149	PRE 95404	MET 01149	

fractionations and more variable in MREE and HREE than in LREE relative to CI chondrite-like. In our results, fractionations are remarkably visible in all REEs (LREE, MREE and HREE) (e.g. Ce, Eu, Er, Tm, Yb, Table 6), but LREE display more enrichments in general (Fig. 7).

Interestingly, meteorites with large LREE enrichments and huge positive Ce anomalies, also display large (anomaly) bulk Al, Ca and Na abundances. Although REEs are stored in a large part in CAIs, a direct relation with only them is rather difficult, thus we suggest that extensive thermal and/or aqueous alteration (metasomatism) also could lead to extensive fractionation, as well as terrestrial weathering which is explained later. Choe et al. (2010) only found Ce and LREE enrichments in meteorite finds, which is also consistent with our results, suggesting a possible terrestrial contamination related to these anomalies.

On the other hand, as Ce and Eu behave differently depending on the conditions (reducing/oxidizing), would suggest that such anomalies come from nebular processes, but these trends are not reported anywhere else in the literature. Probably, the lower amount of sample in the analyses than commonly used, also have affected the results in some way. It is possible that the admixture of matrix, chondrules and CAIs, as well as metallic grains and sulphides, have not been done equitably and hence, have resulted in depletions and enrichments of different elements from sample to sample, lacking a proper representation of the bulk chondrite.

Among all groups analyzed here, ungrouped chondrites are the least fractionated, which suggests that the components of these chondrites are more homogenized, or that have been formed in a more oxidizing region than other groups. In particular, GRO 95566 displays a flat pattern when normalized to CI chondrites with a small positive Tm anomaly and a small enrichment in LREE. On the other hand, we realize that observed fall chondrites agree better with the literature (Tables 3 and 4), hence, a huge fractionation derived from Antarctic contamination cannot be ruled out. Common weathering processes, as the oxidation of metallic grains and other oxides, are usually accompanied by bulk loss of siderophiles and calcophiles (Huber et al. 2006). Also, cold-desert meteorites are contaminated with elements derived from sea spray (e.g. Ebihara et al. 1989) that may have affected the bulk abundances. Carbonaceous chondrites, for having large proportions of fine-grained material, are easily exposed to these terrestrial changes.

According to Grossman (1973), high temperature condensates would take preferentially Eu by forming solid solutions with host condensates. However, some chondrites from the present work display negative Eu anomalies even though they do have CAIs (Table 6). Hence, we do not attribute this trend to the loss of CAIs.

It is well known that negative Ce and positive Tm anomalies are characteristics of REE abundance pattern of type II CAI (Masson and Taylor 1982; Dauphas and Pourmand 2015). Only few samples display this trend: LEW 87148, Cold Bokkeveld, Murchison and Leoville (Table 6), being three of them CM chondrites. In any case, as already said, the majority of chondrites analyzed in the present work are characterized for having large LREE enrichments and enormous Ce anomalies that have not been described before. Inoue et al. (2004) found something similar in the least-contaminated matrix sections of Allende meteorite, which exhibited

LREE enrichments with positive Ce anomalies, a complementary pattern to that of the chondrules. They concluded that precursor material of matrix in Allende was formed under oxidizing conditions at later stages after removal of chondrules and CAIs, hence associating a nebular origin to this trend. But we did not take only matrix in our samples and pure matrix is difficult to obtain randomly since usually it is contaminated from the other components, hence we rule out this possibility *a priori*, although it cannot be dismissed the possibility of having some cases among them. So, at least part of such anomalies might be associated with the small sample size used, where compositional bias due to sample heterogeneity was not avoided. Hence, as bulk chemical analysis of carbonaceous chondrites depend on the size and distribution of the different constituents (millimeter-size chondrules, CAIs, fine-grained matrix, dark inclusions and lithic fragments), a proper representation of the bulk composition was likely missing in our small sample size.

4.3 Taking Advantage When a NEA Approaches Earth Orbit

A significant number of asteroids are known to exhibit short-period orbits, being forecasted accurately because they cross Earth's orbit every few years (see chapter by Trigo-Rodríguez and Williams 2017). Table 7 shows the current number of PHAs, Near-Earth Asteroids (NEAs), and Near-Earth Objects.

Since the amount of asteroids orbiting near Earth is considerably high, mining them seems to be the best solution to answer the growing demand of precious minerals that will allow the supply of the everyday-growing human necessities.

4.4 Technology for Mining

The technology needed to carry out the mining process has not yet been developed, but possible methods have been suggested, where not overcoming gravity seems to be the major effect to consider. The design requirements for mining and processing equipments will depend on the character of the asteroidal material being used. A device similar to a snow blower, anchored to the surface, could be used to collect loose rubble by using a spinning blade to fling the material through a chute and into a high-strength bag. Many of the mining methods will be similar to those used

Table 7 Current (March 11, 2016) amount of PHA-Km-size, total PHAs, NEA-Km-size, total NEAs and total NEOs (Credit: NASA aeronautics and space administration webpage)

PHA-KM	PHA	NEA-KM	NEA	NEO
154	1686	881	13957	14063

on Earth, and will consist of drilling, blasting, cutting and crushing. Extraction of individual materials, depending upon their properties, will be achieved by either chemical or physical means. Water and volatile compounds (CO_2 and hydrocarbons) can be extracted by heating the solid material (100–400 °C), capturing the vapor and then distilling it. Carbon dioxide is produced by the dissociation of hydrocarbon compounds and by reaction of elemental carbon with oxide phases. On the other hand, electrolysis of molten silicates would produce oxygen, iron and other alloys; and a method called the Mond process could be used to extract nickel (O'Neill et al. 1979). As well as being used for creating industrial wealth on Earth, these raw materials could also be used to actually build structures in space.

An interesting study from Elvis (2014) was done to present an assessment of how many asteroids contain ore, which means commercially profitable material (Sonter et al. 1997). This includes not only a high concentration of a resource but also the consideration of the cost of extraction and its price. In his study, he took into account only NEOs (since MB asteroids are too hard to reach), and the evaluation was done only for two resources: PGMs and water. Therefore, the number of asteroids assumes that only Ni-Fe asteroids are of interest for PGMs, and NEOs have to be larger than 100 m of diameter to constrain other parameters. The result was that the number of ore-bearing asteroids is roughly 10. However, he stated that significant research is needed to improve on asteroidal composition, telescopic discovery and characterization in order to increase this number. The apparently limited supply of potentially profitable NEOs is stark evidence for an acceleration rate both for discovery and characterization surveys.

The first step toward mining asteroids is already launched from the International Space Station: a small spacecraft named Arkyd 3 Reflight (A3R) produced by Planetary Resources, Inc., a company that pretends expand Earth's natural resources. The A3R is a probe to test electronics systems and software during 90 days of operation in low Earth orbit. The Arkyd-6 craft, which will be the next step, gets a little closer to mining readiness, testing out propulsion, communication, avionic, and power systems. Asteroid mining could be one way to cultivate materials in space for construction in low or very-low-gravity environments, where, theoretically, the microgravity environment would make mega-engineering projects easier. Therefore, mining materials from asteroids would eliminate the problem of having to launch them with huge rockets every time to overcome Earth's gravity.

5 Conclusions

Natural resources, which have seemed unlimited in our planet for many years, will eventually run out and thus, by mining asteroids, a significant progress in key technologies for metal extraction, processing equipments, asteroidal retrieval and also deflecting dangerous bodies should be further developed.

This work has been performed in order to consider asteroid mining for REE extraction since the global demand for REEs, PGMs and other metals such as iron, nickel and chromium, has increased as more uses for these elements are found. A major problem arises because the occurrence of these technologically essential elements on Earth's minable deposits is limited, and there is no equality between the REEs supply and demand. However, from the point of abundances, little differences among chondrite parent bodies exist for REEs, and compared to differentiated parent bodies such as the Earth, carbonaceous chondrites have low incompatible element contents (including REEs) and high siderophile element contents, as the lasts are concentrated in the core of differentiated asteroids and planets. Therefore, mining chondritic asteroids makes little sense for REEs but may be reasonable for PGEs, which encourage us to keep studying undifferentiated bodies in the solar system.

In any case, our results for REE abundances may be useful because some meteorites have been analyzed for the first time, and can help to constrain cosmochemical processes in the solar nebula as well as Antarctic alteration hints. As it is known, REE fractionations are correlated with high-temperature components formed in dry conditions. Among all carbonaceous chondrites analyzed in the present work, CI, CMs and ungrouped chondrites are the lest fractionated groups, while CVs, COs, CKs and CRs are the most. The divalent state of Eu, in contrast with the normal trivalent state, is reflected in many samples as negative anomalies, which is common in moderately reducing environments. On the contrary, positive Ce anomalies are associated with oxidizing conditions, in particular, with aqueous solutions. Besides, few samples display positive Tm anomalies described by Dauphas and Pourmand (2015) and Barrat et al. (2016) associated with the incorporation of type II CAI in the sample, since they also display negative Ce anomalies. Our results, thus follow reasonable patterns but display large LREE enrichments and huge Ce anomalies that have not been described before, associated with a small sample size used where variations of refractory element ratios on a bulk meteorite scale has not been considered, and also terrestrial weathering, since most of them are Antarctic finds and surprisingly, chondrite falls match quite good the data from literature (Nakamura 1974; Evensen et al. 1978; Dauphas and Pourmand 2015).

Acknowledgments This work was financially supported by the Spanish Ministry of Science and Innovation projects AYA2011-26522, and AYA2015-67175-P, to which M. Martínez-Jiménez, J.M. Trigo-Rodríguez and C. E. Moyano-Cambero acknowledge financial support. We also thank the NASA Meteorite Working Group, and the Johnson Space Center for the samples provided. This study was done in the frame of a PhD. on Physics at the Autonomous University of Barcelona (UAB) funded as a FPI in AYA2011-26522 (P.I. J.M. Trigo-Rodríguez).

References

Adams, F.C.: The birth environment of the solar system. Annu. Rev. Astron. Astrophys. **48**, 47–85 (2010)

Amelin, Y., Krot, A.N., Hutcheon, I.D., Ulyanov, A.A.: Lead isotopic ages of chondrules and calcium-aluminum-rich inclusions. Science **297**, 1678–1683 (2002)

Barrat, J.A., Dauphas, N., Gillet, P., Bollinger, C., Etoubleau, J., Bischoff, A., Yamaguchi, A.: Evidence from Tm anomalies from non-CI refractory litophile element proportions in terrestrial planets and achondrites. Geochim. Cosmochim. Acta **176**, 1–17 (2016)
Beitz et al. The collisional evolution of undifferentiated asteroids and the formation of chondritic meteoroids. Astrophys. J. In press. (arXiv 1604.02340) (2016)
Blair, B.R.: The role of Near-Earth Asteroids in long-term platinum supply. EB353 Metal Economics, Colorado School of Mines (2000)
Blum, J., Wurm, G.: The growth mechanics of macroscopic bodies in protoplanetary disks. Annu. Rev. Astron. Astrophys. **46**, 21–56 (2008)
Bouvier, A., Wadhwa, M.: The age of the solar system redefined by the oldest Pb-Pb age of a meteoritic inclusion. Nat. Geosci. **3**, 637–641 (2010)
Boynton, W.V.: Fractionation in the solar nebula: condensation of yttrium and the rare earth elements. Geochim. Cosmochim. Acta **39**, 569–584 (1975)
Brearley, A.J., Jones, R.H.: Chondritic meteorites. In: Papike, J.J. (ed.) Reviews in Mineralogy. Planetary materials, vol. 36, pp. 3.1–3.398. Mineralogical Society of America, Washington (1998)
British Geological Survey, Natural Environment Research Council: Rare Earth Elements. Compiled by Walters, A., Lusty, P. In: Minerals UK (www.MineralsUK.com) (2011)
Brownlee, D., Tsou, P., et al.: Comet Wild 2 under a microscope. Science **314**, 1711–1716 (2006)
Cameron, A.G.W., Fegley, M.B.: Nucleation and condensation in the primitive solar nebula. Icarus **52**, 1–13 (1982)
Cameron, A.G.W., Truran, J.W.: The supernova trigger for formation of the solar system. Icarus **30**, 447–461 (1977)
Choe, W.H., Huber, H., Rubin, A.E., Kallemeyn, G.W., Wasson, J.T.: Composition and taxonomy of 15 unusual carbonaceous chondrites. Meteorit. Planet. Sci. **45**, 531–554 (2010)
Cuzzi, J.N., Hogan, R.C., Shariff, K.: Toward planetesimals: dense chondrule clumps in the protoplanetary nebula. Astrophys. J. **687**, 1432–1447 (2008)
Dauphas, N., Pourmand, A.: Thulium anomalies and rare earth element patterns in meteorites and Earth: nebular fractionation and the nugget effect. Geochim. Cosmochim. Acta **163**, 234–261 (2015)
Davis, A.M., Grossman, L.: Condensation and fractionation of rare earths in the solar nebula. Geochim. Cosmochim. Acta **43**, 1611–1632 (1979)
Davis, A.M., Tanaka, T., Grossman, L., Lee, T., Wasserburg, G.J.: Chemical composition of HAL, an isotopically unusual Allende inclusion. Geochim. Cosmochim. Acta **46**, 1627–1651 (1982)
Dominik, C.P., Blum, J., Cuzzi, J.N., Wurm, G.: Growth of dust as the initial step toward planet formation. See Reipurth et al., pp. 783–800 (2007)
Ebihara M., Shinonaga T., Nakahara H., Kondoh A., Miyamoto M., Kojima H.: Depth-profiles of halogen abundance and integrated intensity of hydration band near 3 lm in ALH 77231, Antarctic L6 chondrite. In: Koeberl, C., Cassidy W.A. (eds.) Differences between Antarctic and non-Antarctic meteorites, LPI Technical Report 90-01, pp. 32–37. Lunar and Planetary Institute, Houston (1989)
Elvis, M.: How many ore-bearing asteroids? Planet. Space Sci. **91**, 20–26 (2014)
Evensen, N.M., Hamilton, P.J., O'Nions, R.K.: Rare-Earth abundances in chondritic meteorites. Geochim. Cosmochim. Acta **42**, 1199–1212 (1978)
Farquhar, R.W., et al.: NEAR mission overview and trajectory dessign. J. Astronaut. Sci. **43**, 353–372 (1999)
Fegley, B., Ireland, T.R.: Chemistry of the rare earth elements in the solar nebula. Eur. J. Solid State Inorg. Chem. **28**, 335–346 (1991)
Froeschlé, C., Greenberg, R.: Mean motion resonances. In: Binzel, R.P., Gehrels, T., Matthews, M.S. (eds.) Asteroids II, pp. 827–844. University of Arizona Press, Tucson (1989)
Fujiwara, A., Kawaguchi, J., Yeomans, D.K., Abe, M., Mukai, T., Okada, T., Saito, J., Yano, H., Yoshikawa, M., Scheeres, D.J., Barnouin-Jha, O., Cheng, A.F., Demura, H., Gaskell, R.W., Hirata, N., Ikeda, H., Kominato, T., Miyamoto, H., Nakamura, A.M., Nakamura, R., Sasaki, S., Uesugi, K.: The rubble-pile asteroid Itokawa as observed by Hayabusa. Science **312**, 1330–1334 (2006)

Goldreich, P., Ward, W.R.: The formation of planetesimals. Astrophys. J. **183**, 1051–1062 (1973)
Grossman, L.: Refractory trace elements in Ca-Al-rich inclusions in the Allende meteorite. Geochim. Cosmochim. Acta **37**, 1119–1140 (1973)
Harben, P.W.: Industrial minerals handybook, 4th ed. Rare earth minerals and compounds (2002)
Hartmann, W.K., Neukum, G.: Cratering chronology and the evolution of Mars. Space Sci. Rev. **96**, 165–193 (2001)
Hirota, Y., Tamaki, M., Nakamura, N.: Rare earth element abundances in the CK chondrites including the Kobe meteorite. Geochem. J. **36**, 309–322 (2002)
Huber, H., Rubin, A.E., Kallemeyn, G.W., Wasson, J.T.: Siderophile-element anomalies in CK carbonaceous chondrites: Implications for parent-body aqueous alteration and terrestrial weathering of sulfides. Geochim. Cosmochim. Acta **70**, 4019–4037 (2006)
Iluka Mineral Sands Mining. www.iluka.com (2010)
Inoue, M., Kimura, M., Nakamura, N.: REE abundances in the matrix of Allende (CV) chondrite: Implications for matrix origin. Meteorit. Planet. Sci. **39**, 599–608 (2004)
Ishiguro, M., Kuroda, D., Hasegawa, S., Kim, M.J., Choi, Y.J., Moskivitz, N., Abe, S., Pan, K.S., Takahashi, J., Takagi, Y., Arai, A., et al.: Optical Properties of (162173) 1999 JU3: In preparation for the JAXA *Hayabusa 2* sample return mission. Astrophys. J. **792**, 74–83 (2014)
Kallemeyn, G.W., Wasson, J.T.: The composition classification of chondrites. Part I: the carbonaceous chondrite groups. Geochim. Cosmochim. Acta **45**, 1217–1230 (1981)
Kallemeyn, G.W., Rubin, A.E., Wasson, J.T.: The composition classification of chondrites. Part V: the Karoonda (CK) group of carbonaceous chondrites. Geochim. Cosmochim. Acta **55**, 881–892 (1991)
Kallemeyn, G.W., Rubin, A.E., Wasson, J.T.: The composition classification of chondrites. Part VI: the CR carbonaceous chondrite group. Geochim. Cosmochim. Acta **58**, 2873–2888 (1994)
Kallemeyn, G.W., Rubin, A.E., Wasson, J.T.: The composition classification of chondrites. Part VII: the R chondrite group. Geochim. Cosmochim. Acta **60**, 2243–2256 (1996)
Kornacki, A.S., Fegley Jr., B.: The abundance and relative volatility of refractory trace elements in Allende Ca,Al-rich inclusions: implications for chemical and physical processes in the solar nebula. Earth Planet. Sci. Lett. **79**, 217–234 (1986)
Lada, C.J., Lada, E.A.: Embedded clusters in molecular clouds. Annu. Rev. Astron. Astrophys. **41**, 57–115 (2003)
Lauretta, D.S., Bartels, A.E., Barucci, M.A., Bierhaus, E.B., Binzel, R.P., Bottke, W.F., Campins, H., Chesley, S.R., Clark, B.C., Clark, B.E., Cloutis, E.A., Connolly, H.C., Crombie, M.K., Delbó, M., Dworkin, J.P., Emery, J.P., Glavin, D.P., Hamilton, V.E., Hergenrother, C.W., Johnson, C.L., Keller, L.P., Michel, P., Nolan, M.C., Sandford, S.A., Scheeres, D.J., Simon, A.A., Sutter, B.M., Vokrouhlický, D., Walsh, K.J.: The Osiris-REx target asteroid (101955) Bennu: constraints on its physical, geological, and dynamical nature from astronomical observations. Meteor. Planet. Sci. **50**, 834–849 (2015)
Lewis, J.S.: The temperature gradient in the solar nebula. Science **186**, 440–443 (1974)
Lodders, K.: Solar system abundances and condensation temperatures of the elements. Astrophys. J. **591**, 1220–1247 (2003)
Lodders, K., Amari, S.: Presolar grains from meteorites: remnants from the early times of the solar system. Chemie der Erde **65**, 93–166 (2005)
Long, K.R., Van Gosen, B.S., Foley, N.K., Cordier, D.: The principal rare Earth elements deposits of the United States—a summary of domestic deposits and a global perspective. Scientific Investigations Report 5220, U.S. Department of the Interior, U.S. Geological Survey (2010)
Macke, R.J., Consolmagno, G.J., Britt, D.T.: Density, porosity and magnetic susceptibility of carbonaceous chondrites. Meteor. Planet. Sci. **46**, 1842–1862 (2011)
Martin, P.M., Mason, B.: Major and trace elements in the Allende meteorite. Nature **249**, 333–334 (1974)
Masson, B., Martin, P.M.: Geochemical differences among components of the Allende meteorite Smithson. Contrib. Earth. Sci. **19**, 84–95 (1977)
Masson, B., Taylor, S.R.: Inclusions in the Allende meteorite. Smithson. Contrib. Earth. Sci. **25**, 1–30 (1982)

McDonough, W.F., Sun, S.-s.: The composition of the Earth. Chem. Geol. **120**, 223–253 (1995)
McKeegan, K.D., Aleon, J., Bradley, J., Brownlee, D., Busemann, H., et al.: Isotopic compositions of cometary matter returned by Stardust. Science **314**, 1724–1728 (2006)
Meyer, B.S., Clayton, D.D.: Short-lived radioactivities and the birth of the Sun. Space Sci. Rev. **92**, 133–152 (2000)
Michard, A.: Rare earth element systematics in hydrothermal fluids. Geochim. Cosmochim. Acta **53**, 745–750 (1989)
Michel, P., Richardson, D.C., Durda, D.D., Jutzi, M., Asphaug, E.: Collisional formation and modeling of asteroid families. In: Michel, P., et al. (eds.) Asteroids IV, pp. 341–354. University of Arizona, Tucson (2015)
Nakamura, N.: Determination of REE, Ba, Fe, Mg, Na and K in carbonaceous and ordinary chondrites. Geochim. Cosmochim. Acta **38**, 757–775 (1974)
Nakamura, E., et al.: Itokawa dust particles. A direct link between S-type asteroids and ordinary chondrites. Science 333, 1113–1116 (2011)
Nakamura, N., Masuda, A.: Chondrites with peculiar rare-earth patterns. Earth Planet. Sci. Lett. **19**, 429–437 (1973)
Nakamura, E., Makishima, A., Moriguti, T., Kobayashi, K., Tanaka, R., Kunihiro, T., Tsujimori, T., Sakaguchi, C., Kitagawa, H., Ota, T., Yachi, Y., Yada, T., Abe, M., Fujimura, A., Ueno, M., Mukai, T., Yoshikawa, M., Kawaguchi, J.: Space environment of an asteroid preserved on micrograins returned by the Hayabusa spacecraft. Proc. Natl. Acad. Sci. **109**, E624–E629 (2012)
Nittler, L.R., Dauphas, N.: Meteorites and the chemical evolution of the Milky Way. See Lauretta and McSween, 2006, pp. 147–167 (2006)
O'Neill, G.K. et al.: Space Resources and Space Settlements. Specifically, Gaffey, M.J., Helin, E.F., O'Leary, B.: An assessment of Near-Earth Asteroid resources, pp. 191–204. J. Billingham, W. Gilbreath, B. O'Leary (eds.), NASA SP-428 (1979)
Palme, H.: Chemical and isotopic heterogeneity in protosolar matter. Philos. Trans. R. Soc. Lond. A **359**, 2061–2075 (2001)
Palme, H., Jones, A.: Solar system abundances of the elements. In: Davis, A.M. (ed.) Meteorites, Comets, and Planets. Treatise on Geochemistry, vol. 1, pp. 41–61. Elsevier-Pergamon, Oxford (2003). eds. H.D. Holland and K.K. Turekian
Rudnick, R.L., Gao, S.: Composition of the continental crust. In: TOG, 3.01 (2005)
Schmitz, B.: Extraterrestrial spinels and the astronomical perspective on Earth's geological record and evolution of life. Chemie der Erde **73**, 117–145 (2013)
Schmitz, B., Häggström, T., Tassinari, M.: Sediment-dispersed extraterrestrial chromite traces a major asteroid disruption event. Science **300**, 961–964 (2003)
Shu, F.H., Shang, H., Lee, T.: Toward an astrophysical theory of chondrites. Science **271**, 1545–1552 (1996)
Shu, F.H., Shang, H., Glassgold, A.E., Lee, T.: X-rays and fluctuating X-winds from protostars. Science **277**, 1475–1479 (1997)
Shu, F.H., Shang, H., Gounelle, M., Glassgold, A.E., Lee, T.: The origin of chondrules and refractory inclusions in chondritic meteorites. Astrophys. J. **548**, 1029–1050 (2001)
Sonter, M.J.: The technical and economic feasibility of mining the Near Earth Asteroids. PHD thesis, University of Wollongong (1997)
Stracke, A., Palme, H., Gellissen, M., Münker, C., Kleine, T., Birbaum, K., Günther, D., Bourdon, B., Zipfel, J.: Refractory element fractionation in the Allende meteorite: Implications for solar nebula condensation and the chondritic composition of planetary bodies. Geochim. Cosmochim. Acta **85**, 114–141 (2012)
Takigawa, A., Miki, J., Tachibana, S., Huss, G.R., Tominaga, N., et al.: Injection of short-lived radionuclides into the early solar system from a faint supernova with mixing fallback. Astrophys. J. **688**, 1382–1387 (2008)
Tanaka, T., Masuda, A.: Rare-earth elements in matrix, inclusions, and chondrules of the Allende meteorite. Icarus **19**, 523–530 (1973)

Taylor, S.R.: Composition and chemical evolution of the solar nebula. In: Taylor, S.R. (ed.) Solar System Evolution. A New Perspective, 2nd edn, pp. 73–104. Cambridge University Press, Cambridge (2001)

Trigo-Rodríguez, J.M.: Aqueous alteration in chondritic asteroids and comets from the study of carbonaceous chondrites. In: Lee, M.R., Leroux, H. (eds.) Planetary Mineralogy. EMU notes in mineralogy, vol. 15, pp. 67–87. European Mineralogical Union and the Mineralogical Society, London (2015)

Trigo-Rodríguez, J.M., García-Hernández, D.A., Lugaro, M., Karakas, A.I., van Raai, M., García Lario, P., Manchado, A.: The role of massive AGB stars in the early Solar System composition. Meteor. Planet. Sci. **44**, 627–641 (2009)

Veverka, J., et al.: NEAR at EROS: imaging and spectral results. Science **289**, 2088–2097 (2000)

Wai, C.M., Wasson, J.T.: Nebular condensation of moderately volatile elements and their abundances in ordinary chondrites. Earth Planet. Sci. Lett. **36**, 1–13 (1977)

Wasserburg, G.J., Busso, M., Gallino, R., Nollett, K.M.: Short-lived nuclei in the early solar system: possible AGB sources. Nucl. Phys. A. **777**, 5–69 (2006)

Wasson, J.T.: Meteorites. Their record of early Solar System history. Freeman, New York (1985). Chapter IV

Wasson, J.T., Kallemeyn, G.W.: Composition of chondrites. Philos. Trans. R. Soc. Lond. **A325**, 353–544 (1988)

Wasson, J.T., Isa, J., Rubin, A.E.: Compositional and petrographic similarities of CV and CK chondrites: a single group with variations in textures and volatile concentrations attributable to impact heating, crushing and oxidation. Geochim. Cosmochim. Acta **108**, 45–62 (2013)

Weisberg, M.K., McCoy, T.J., Krot, A.N.: Systematics and evaluation of meteorite classification. In: Lauretta, D.S., McSween Jr., H.Y. (eds.) Meteorites and the Early Solar System II, pp. 19–52. The University of Arizona Press, Tucson (2006)

Wetherill, G.W.: Origin of the asteroid belt. In: Binzel, R.P., Gehrels, T., Matthews, M.S. (eds.) Asteroids II, pp. 661–680. University of Arizona Press, Tucson (1989)

Willacy, K., Alexander, C., Ali-Dib, M., Ceccarelli, C., Charnley, S.B., Doronin, M., Ellinger, Y., Gast, P., Gibb, E., Milam, S.N., Mousis, O., Pauzat, F., Tornow, C., Wirström, E.S., Zicler, E.: The Composition of the protosolar disk and the formation conditions for comets. Space Sci. Rev. **197**, 151–190 (2015)

Williams, J.P., Cieza, L.A.: Protoplanetary disks and their evolution. Annu. Rev. Astron. Asntrphys. **49**, 65 (2011)

Xiao, Z., Strom, R.G., Chapman, C.R., Head, J.W., Klimczak, C., Ostrach, L.R., Helbert, J., D'Incecco, P.: Comparisons of fresh complex impact craters on Mercury and the Moon. Implications for controlling factors in impact excavation processes. Icarus **228**, 260–275 (2014)

Zinner, E.: Presolar grains. In: Davis, A. (ed.) Meteorites, Comets and Planets. Treatise on geochemistry, vol. 1, pp. 17–39. Elsevier, Amsterdam (2003)

Zolensky, M.E., McSween, H.Y.: Aqueous alteration. In: Lauretta, D.S., McSween Jr., H.Y. (eds.) Meteorites and the Early Solar System II, pp. 114–143. The University of Arizona Press, Tucson (1988)

Zolensky, M.E., Mittlefehldt, D.W., Lipschutz, M.E., Wang, M.-S., Clayton, R.N., Mayeda, T.K., Grady, M.M., Pillinger, C.T., Barber, D.: CM chondrites exhibit the complete petrologic range from type 2 to 1. Geochim. Cosmochim. Acta **61**, 5099–5115 (1997)

Zuber, M.T., Smith, D.E., Cheng, A.F., Garvin, J.B., et al.: The shape of 433 Eros from the NEAR-Shoemaker Laser Rangefinder. Science **289**, 2097–2101 (2000)

Atomistic Simulations of Aqueous Alteration Processes of Mafic Silicates in Carbonaceous Chondrites

A. Rimola and Josep M. Trigo-Rodríguez

Abstract There is clear evidence that the parent bodies of some groups of carbonaceous chondrites suffered extensive aqueous alteration. In some pristine chondrites even some minerals can be envisioned to have formed in the proto-planetary disk itself. There is clear evidence that water was available in these materials at early times, and when water alteration occurred element mobilization took place and the primordial chemistry/mineralogy changed. In general, mafic silicates were initially forming chondrules and also matrix fragments, but with time they became transformed into phyllosilicates. With the aim to have atomic-scale in-sights of these water alteration processes, in this chapter we present our first atomistic simulations related to the water alteration of silicates obtained by means of quantum chemical calculations. Results are based on the molecular simulation of the interaction of 12 water molecules with different forsteritic (Mg_2SiO_4) silicate surfaces of different stability and structural state. Relevant structural information (*i.e.*, significant changes due to water interaction) and vibrational properties, including simulation of the infrared spectra, of the water/silicate interface are shown, and energetic features and thermodynamic trends of the water alteration reactions are provided. Finally, we compare such theoretical calculations with the evidence inferred from the mineralogical interpretation of the formation conditions of carbonaceous chondrites. We conclude that parent body aqueous alteration is the preferred scenario to produce the hydration features observed in these primitive meteorites.

A. Rimola (✉)
Departament de Química, Universitat Autònoma de Barcelona, Bellaterra 08193, Spain
e-mail: albert.rimola@uab.cat

J.M. Trigo-Rodríguez
Institute of Space Sciences (IEEC-CSIC), Meteorites, Minor Bodies and Planetary Sciences Group, Campus UAB Bellaterra, Carrer de Can Magrans, s/n, Cerdanyola del Vallès, Barcelona 08193, Spain
e-mail: trigo@ice.csic.es

J.M. Trigo-Rodríguez et al. (eds.), *Assessment and Mitigation of Asteroid Impact Hazards*, Astrophysics and Space Science Proceedings 46,
DOI 10.1007/978-3-319-46179-3_6

1 Introduction

Undifferentiated meteorites usually contain rounded silicate spherules, or chondrules, and are known as chondrites. These meteorites comprise carbonaceous, ordinary, enstatite, Rumuruti (R) classes and anomalous (also known as ungrouped) specimens. All chondrites are subdivided into 15 groups with distinctive chemical and isotopic patterns (Weisberg et al. 2006). Some ordinary and carbonaceous chondrite meteorites are known to have accreted with significant water contents, but here we focus on the aqueous alteration experienced by carbonaceous chondrites (hereafter CCs). Water was probably available in liquid state shortly after their formation, probably in static form in most cases and for millions of years (Zolensky and McSween 1988; Brearley 2003, 2006; Trigo-Rodríguez 2015).

The CI group of carbonaceous chondrites exhibits the most extensive alteration (Tomeoka and Buseck 1988; Endreβ and Bischoff 1996). From the Mn-Cr ages of dolomites we know that parent body accretion took place during the first 10 Ma after accretion (Fujiya et al. 2013). Evidence of progressive aqueous alteration in CM chondrites has been shown by several authors (see e.g. Trigo-Rodríguez et al. 2006; Rubin et al. 2007; Trigo-Rodríguez 2015), pointing towards continuous interaction of water with different minerals, among them the mafic silicates which we model in our calculations. Fine-grained crystalline materials forming the matrix or chondrule mantles in the more-altered CM2 chondrites are rich in phyllosilicates (mainly serpentine with lesser amounts of smectite; Zolensky and McSween 1988; Hua et al. 2002; Chizmadia et al. 2004). This may indicate that the fine-grained crystalline and amorphous anhydrous materials that accreted from the protoplanetary disk were susceptible to aqueous alteration. Their presence of phyllosilicates has been invoked to propose the formation of hydrous minerals in the protoplanetary disk (Ciesla et al. 2003; Ciesla and Lauretta 2005), but phyllosilicates are also found inside altered chondrules, e.g. alteration of the primordial glassy mesostases (Hanowski and Brearley 1997, 2000, 2001). We notice that parent body aqueous alteration microscopic effects are more noticeable at microscopic scale when we use SEM, but mineral features at nano-scale might be produced in the outer protoplanetary disk (Trigo-Rodríguez et al. 2015). Therefore we need to study pristine CCs which have not experienced chemical homogenization after alteration (Trigo-Rodríguez and Blum 2009a). The reason for such homogenization is the mild metamorphism and metasomatism experienced by some of these materials (Zolensky and McSween 1988). In fact, highly unequilibrated carbonaceous chondrites are quite rare or inexistent, particularly for the most hydrated groups, ACFER 094 being a nice but still chemically anomalous (as might be considered a CM-like) exception (Rubin et al. 2007).

The CR chondrites, another group of carbonaceous chondrites, also contain hydrated mineral phases even when they escaped long duration mild thermal metamorphism under oxidizing conditions (Zolensky et al. 1993; Bonal et al. 2013). It seems obvious that the observed differences point towards aqueous alteration at different degrees (Trigo-Rodríguez et al. 2013; Harju et al. 2014; Le Guillou et al. 2015), probably experienced in the distinctive evolution of fragments of a progenitor body (Trigo-Rodríguez 2015). The CR shows a D enrichment (Bonal et al. 2013) that could be the consequence of parent body alteration more than nebular process (Robert 2003). It has been found that the infrared spectra of the fine-grained matrix of CR chondrites point to some similarities with CI and CM chondrites, but with the peculiarity of the absence of the narrow —OH absorption peaking at 3600 cm^{-1}. In general the CR infrared spectra are more similar to the less-altered CM chondrites, such as Murray, Murchison, Mighei, and QUE 97990, than to the more extensively altered CI chondrite Orgueil (Beck et al. 2010). Phyllosilicates produced in oxidizing conditions were also described in some CV chondrites (Tomeoka and Buseck 1990; Brearley 2006). It is accepted that the CI chondrites have undergone the highest degree of aqueous alteration of all carbonaceous chondrite groups (Tomeoka and Buseck 1988), and probably, due to that, can be considered a proxy for the abundances of almost all elements in the early solar system (Anders and Grevese 1989; Lodders 2003).

The plausible alteration conditions for CR, CV and CO chondrite groups were also summarized by Zolensky et al. (1993), which based on the thermodynamic calculations by Krot et al. (1998), indicate that the formation temperatures of these hydrous phases were below 300 °C. Probably the CM group was one of the least affected by mild metamorphism, as revealed by the presence of minerals like e.g. tochilinite, which exhibits an upper limit for thermal stability around ~120 °C (Browning and Bourcier, 1996). In any case, the most altered petrologic type CM1 chondrites could have experienced higher temperatures (Zolensky et al. 1997).

Multiple evidence on the importance and extent of aqueous alteration processes in the parent bodies of CCs lie on the existence of many secondary minerals resulting of parent body hydration. It is obvious that aqueous alteration affects many minerals accreted to form chondritic planetesimals. A compilation of the main mineral phases produced by aqueous alteration are compiled in Table 1 (see also Trigo-Rodríguez 2015). Despite the obvious mineral diversity between groups and the variable circumstances in which aqueous alteration took place, there are some that are quite common. As a first step to form phyllosilicates we present calculations of the initial interaction of water molecules with the crystalline structure of the silicates.

By studying the physico-chemical processes altering CCs we can learn significant features about their parent bodies, which is relevant as we consider them a good proxy for the materials expected to be returned to our laboratories by sample-return missions like *e.g.* Hayabusa 2 or OSIRIS-REx to primitive carbonaceous asteroids which might have experienced significant aqueous processing. Understanding such processes is crucial to interpret the measured abundances of organics and the degree of preservation of presolar grains in these returned materials (Trigo-Rodríguez and Blum 2009a, b).

The CCs evidence indicates that these non-differentiated asteroids are formed by fine-grained minerals embedded in a nanometric matrix that preserves chemical clues of the environment in which they formed. The secondary minerals

Table 1 Aqueous alteration secondary minerals found in CCs (Rubin 1997; Trigo-Rodríguez et al. 2006; Takir et al. 2013; Trigo-Rodríguez 2015)

CI	CM	CR	CV/CK	Ungrouped	Mineral Group
Serpentine	Serpentine Chlorite	Serpentine	Serpentine Chlorite	Serpentine	**Phyllosilicates**
Saponite	Vermiculite	Saponite	Micas	Saponite	
Magnetite	Magnetite	Magnetite	Magnetite	Magnetite	**Oxides**
Calcite	Calcite	Calcite	–	–	**Carbonates**
Dolomite	Dolomite				
Breunnerite	Aragonite				
Siderite					
Apatite	–	–	–	–	**Phosphates**
Merrilite					
Pyrrhotite	Pyrrhotite	Pyrrhotite	Pyrrhotite	–	**Sulphides**
Pentlandite	Pentlandite	Pentlandite	Pentlandite		
Cubanite	Tochlinite				
Sulfur	Awaruite				
–	Brucite	–	–	–	**Hydroxides**
	Tochilinite				
–	Halite	–	–	–	**Halides**

produced alter the reflective properties of these asteroids some of which may become hazardous objects. Massive meteorite falls have been associated with dark, carbonaceous and hydrated asteroid fragments like *e.g.* Murchison or Tagish Lake.

As a first step for gaining deeper insights into the aqueous alteration processes experienced by CCs, we present here atomistic simulations, based on accurate quantum mechanical methods, of the interaction of water layers with different Mg_2SiO_4 silicate surfaces. This kind of simulations have allowed us to obtain useful atomic-scale information such as structural details of the water/silicate interfaces, energetic and thermodynamic data associated with the interaction of water with the silicate surfaces, and a detailed interpretation of the infrared features of the water/silicate surface systems.

2 Methods

All these simulations were carried out using the CRYSTAL14 ab-initio periodic code (Dovesi et al. 2014). This code implements the Hartree-Fock and Kohn-Sham self-consistent field method to solve the Schrödinger equation for periodic systems, and it allows performing geometry optimizations, calculation of one-electron properties, and simulation of vibrational and reflectance spectra of periodic systems. Surface models computed by CRYSTAL are true slab systems, in which the periodicity is along the lattice parameters that define the surface (usually referred to as 2D dimensionality) with an empty space above and below the slab.

The multielectron wave function is described by linear combination of crystalline orbitals, which in turn are expanded in terms of Gaussian-type orbital basis sets.

Calculations for the present work were based on the B3LYP-D2* density functional theory (DFT) method, which includes an empirical a posteriori correction term proposed by Grimme (Grimme 2006) to account for dispersion forces (missed in the pure B3LYP (Becke 1993; Lee et al. 1988) method), but whose initial parametrization (D2) was modified for extended systems (D2*) (Civalleri et al. 2008). The adopted Gaussian functions consisted of a standard all-electron double-ζ 6-31G(d,p) basis set. The shrinking factor of the reciprocal space net defining the mesh of k points in the irreducible Brillouin zone was set to 6, which required diagonalizing the Hamiltonian matrix in 20 *k* points. The accuracy of both Coulomb and exchange series was set to values of overlap integrals of 10^{-7} and 10^{-16}, respectively, which ensure very good numerical accuracy. The self-consistent-field process was stopped when the energy difference between two sub-sequent cycles was smaller than 10^{-7} hartree. Full relaxations of both lattice parameters and internal atomic coordinates by means of analytical energy gradients were carried out.

CRYSTAL14 computes the zero-point energy (ZPE) corrections and the thermodynamic quantities by the standard statistical thermodynamics formulas based on partition functions derived from the harmonic oscillator approximations, which are used to correct the reaction energy values by temperature effects. Vibrational frequencies of the considered systems were computed at the Γ point (point $k = 0$ in the first Brillouin zone, called central zone) within the harmonic approximation by obtaining the eigenvalues from diagonalization of the mass-weighted Hessian matrix. This dynamical matrix was obtained by numerical differentiation (central-difference formula) of the analytical first-energy derivatives, calculated at the geometries obtained by varying, in turn, each of the 3 N equilibrium nuclear coordinates by a small amount $u = 0.003$ Å (Pascale et al. 2004). That is, building up the full mass-weighted Hessian matrix implied performing 3 N + 1 energy plus gradient calculations (N atoms in the unit cell) in the central-difference formula.

3 Results and Discussion

This section is organized as follows. First, analysis of the structure of the different Mg_2SiO_4 surfaces in a bare state and in interaction with the water layers is presented. Then, the simulated infrared spectra of the water/silicate surface systems are shown alongside a discussion of the changes due to water interaction. Finally, energetic results related to the water interaction with the silicate surfaces are reported, which also include some thermodynamic trends as a function of temperature. There is still a debate about the temperature experienced by the parent bodies of CCs, but the formation of hydrous minerals has been described to occur at temperatures below ~100 °C, although some chondrites may have experienced higher temperatures of up to 400 °C (Krot et al. 1998; Brearley 1999, 2006).

3.1 Structure of the Water/Silicate Surface Systems

Silicates are ubiquitous components in carbonaceous chondrites (CCs), and their degree of aqueous alteration in the different groups was first described by Zolensky and McSween (1988). CC groups distinctive mineralogy is indicative that parent body hydration was different depending on the physico-chemical conditions in the interior of each carbonaceous group as we describe in the introduction. Le Guillou et al. (2015) has recently developed experiments that can constrain the environment and reaction kinetics in which aqueous alteration took place in the parent bodies of CCs. Here we introduce a theoretical approach in order to understand the way in which the hydration of crystalline silicates can take place.

Silicates form a class of inorganic materials with a large diversity in chemical composition and structural properties, in which the Si atoms are invariably tetrahedrally coordinated by four O atoms, forming the $[SiO_4]^{4-}$ building block. The negative net electrical charge is compensated by divalent Mg and Fe cations. Si, Mg and Fe are the rock forming elements with the highest cosmic abundance. The different silicate structures are produced by linking the individual $[SiO_4]^{4-}$ units with various levels of complexity. Among the different class of silicates, olivines are actually abundant, not only in CCs, but also in interstellar and circumstellar dust grain particles, interplanetary dust particles, and comets, playing a central role as a solid state matter of Universe. Crystalline olivines present an orthorhombic structure with space group Pbnm and are characterized to have the $[SiO_4]^{4-}$ units linked by the divalent Mg^{2+} and Fe^{2+} cations with general formula $Mg_{2x}Fe_{(2-2x)}SiO_4$ ($x = 0 - 1$), in which the Fe and Mg cations can replace each other in the crystal structure. The two end members are forsterite (Mg_2SiO_4) and fayalite (Fe_2SiO_4).

The bulk crystal structure of forsterite (hereafter referred to as Fo) consists of distorted SiO_4 tetrahedra and MgO_6 octahedra sharing the vertices. Half of the available octahedral voids are occupied by the Mg^{2+} cations. In Fo there are two symmetry independent Mg atoms, named Mg1 and Mg2 (Hazen 1976). The Mg1-centered octahedra share edges forming rods parallel to the crystallographic *c* axis, and the Mg2 octahedra are laterally linked to these rods through the corresponding edges (see Fig. 1a).

The large variability of oxygen sublattice in Fo results in a complex set of crystal planes. For instance, there are already 7 different planes characterized by the smallest Miller indexes: the {100}, {010}, {001}, {110}, {101}, {011}, and {111} crystal forms. Surface modeling is usually performed by cutting out the bulk crystal along the desired direction. For instance, cutting out the Fo bulk crystal structure along the [010] direction (above and below), one gets the (010) surface (perpendicular to the [010] direction) in the form of a slab model of finite thickness (see Fig. 1b). Normally, slab models for silicates are modeled by cutting out the bulk crystal at the Mg—O bonds rather than the Si—O ones to keep the SiO_4 units intact.

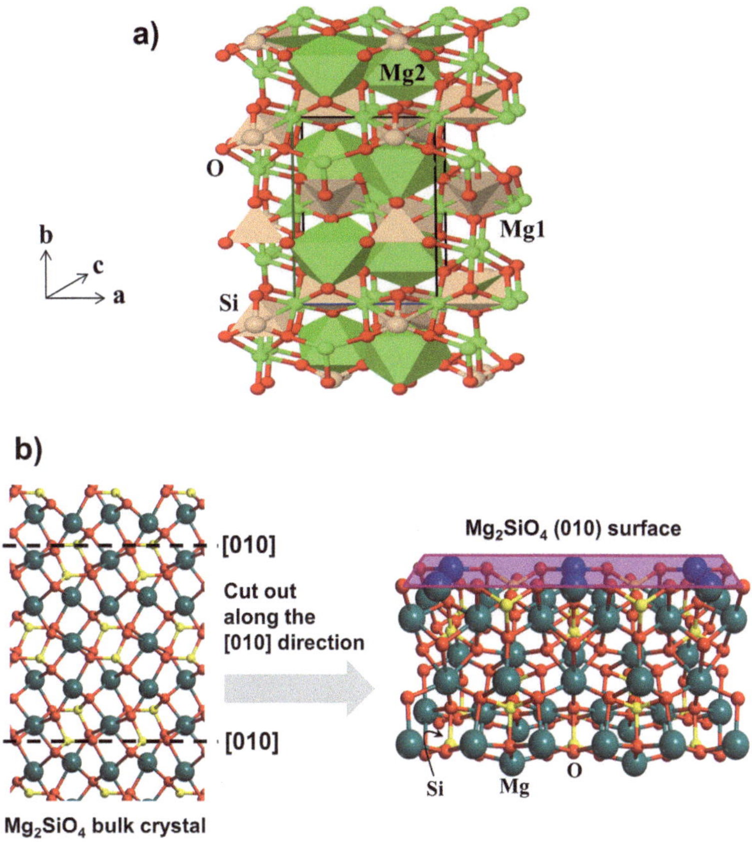

Fig. 1 (**a**) View (along the *c* crystallographic axis) of the crystal structure of forsterite (Mg_2SiO_4). (**b**) Procedure to generate the crystalline periodic slab model of (010) forsterite surface by cutting out the bulk structure along the [010] direction. The actual bare Mg_2SiO_4 (010) surface (*i.e.*, where the water adsorption takes place) is highlighted in *purple*

In this work, four different Fo periodic surface models have been used to study the interaction of water on silicates. Three of them are derived from the bulk crystal structure: the (010), (001) and (110) surfaces. Moreover, because cosmic silicates are abundant in an amorphous state, an amorphous Fo surface model has also been used, which was derived from a bulk amorphous structure of Fo obtained in our laboratory. The structure of the optimized geometries for all these slab models are shown in Fig. 2, both the lateral and top views, including the corresponding unit cells. Detailed analyses of these structures were reported in different papers (Navarro-Ruiz et al. 2014a, b, 2015). All surfaces contain 56 atoms per unit cell and the slab models have thicknesses of about 9.3, 11.4, 8.1, and 10.8 Å for Fo(010), Fo(001), Fo(110), and Fo(amorph), respectively. All these surface models are terminated by undercoordinated Mg^{2+} atoms (labeled by symbols in Fig. 2); *i.e.*,

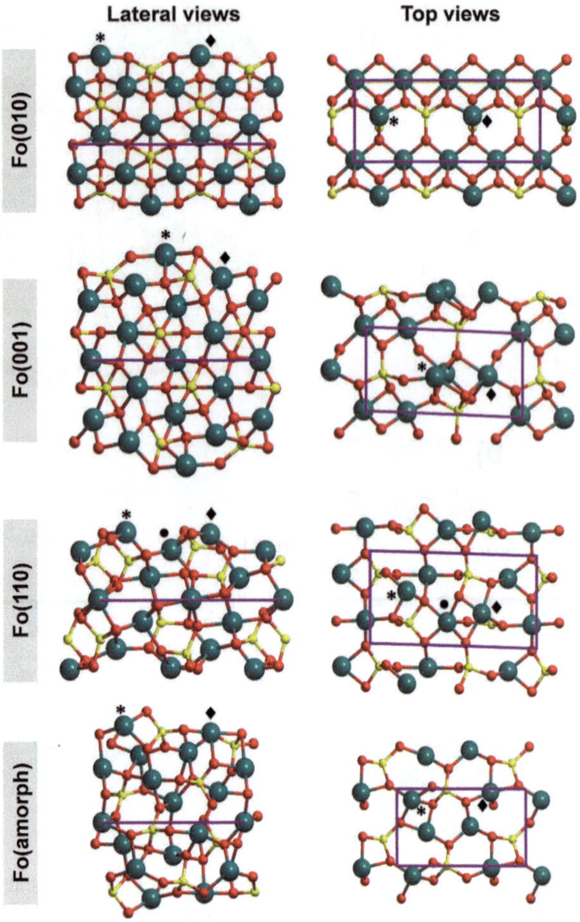

Fig. 2 Lateral and top views of the optimized structures of the different Mg_2SiO_4 slab models used in this work. The unit cells are highlighted in *purple*. Undercoordinated Mg^{2+} cations are labelled by the symbols of *Asterisk*, *filled circle* and *filled diamond*

they are coordinated by fewer O atoms compared to the octahedral coordination for a bulk Mg^{2+} ion. This means that these atoms are prone to interact with electron-donor atoms such as the O atom of H_2O. Moreover, the slab models also have O atoms exposed to outermost positions of the surfaces, so they are able to establish hydrogen bond (H-bond) interactions with the H atoms of H_2O.

It is well-known that the different surfaces of a crystal have different stabilities. In the case of crystalline Fo, theoretical works showed that the (010) surface is the most stable plane in the crystal morphology of Fo, the (110) surface is less stable, while the (001) surface is of intermediate stability (Watson et al. 1997; de Leeuw et al. 2000; Bruno et al. 2014). We calculated the absolute energies (namely, the interaction energies between the nuclei and the electrons of the systems, in Hartree) of our Fo surface models, and with them the relative energies between the surfaces (namely, the energy difference with respect to the most stable surface, in kcal mol^{-1}), which are reported in Table 2. Our results perfectly agree with

Table 2 Absolute energies (E_{abs}) and relative energies (ΔE_{rel}) of the forsterite surface models used in this work

Surface model	E_{abs} (Hartree)	ΔE_{rel} (kcal mol^{-1})
Fo(010)	−7927.4359676317	0.0
Fo(001)	−7927.4117529230	15.2
Fo(110)	−7927.1223118045	196.8
Fo(amorph)	−7927.2883816521	92.6

previous theoretical works as we obtained the following relative stability scale for the crystalline systems (from more to less stable): Fo(010) > Fo(001) > Fo(110). Moreover, the amorphous Fo surface has a stability between Fo(001) and Fo(110). These ladder of surface stability is indicative of the reactive behavior of the Fo surfaces because the less stable the surface is, the more reactive it is.

With these Fo surface models, we studied the interaction of water molecules on them by simulating the adsorption of 12 water molecules. Theoretical studies about water interaction with forsteritic surfaces are very scarce (Prigiobbe et al. 2013; Asaduzzaman et al. 2013; Escamilla-Roa and Moreno 2013; Escamilla-Roa and Sainz-Díaz 2014). The initial guess structures (*i.e.*, those structures built manually and then subjected to geometry optimization) were based on three successive premises: (1) maximization of the interaction between the outermost Mg^{2+} cations and the O atoms of water, (2) maximization of the H-bond interactions between the O surface atoms of Fo and the H protons of water, and (3) maximization of the H-bond interactions between the remaining water molecules. The geometry of the initial guess structures were optimized to reach a stable minimum energy system, which are shown in Fig. 3.

The 12 water molecules interacting on Fo(010) yields the Fo(010)-12W system. In this system, two water molecules interact with each outermost Mg^{2+} cations, in which one of them also H-bonds an O surface atom, two other water molecules interact with the O surface atoms, and the remaining water molecules establish a network of H-bond interactions. The adsorption of the 12 water molecules on Fo(001) gives Fo(001)-12W. The structural features of this system are similar to those of Fo(010)-12W; that is, a first layer of water molecules interacting with the outermost Mg^{2+} cations and O surface atoms, and a second layer of water molecules establishing a H-bond network. Remarkably, neither in Fo(010)-12W nor in Fo(001)-12W dissociation of water due to interaction with the Fo surfaces is observed. In contrast, for water adsorption on Fo(110) (forming the Fo(110)-12W system) significant structural differences are observed. The most important one is that two water molecules interacting with one Mg^{2+} surface cation dissociate, in which one H proton of water is transferred to an O surface atom, thereby resulting in the formation of Mg—OH and Si—OH surface groups. These changes are indicative of water alteration because a spontaneous reactive process occurs when water interacts with this surface. This water alteration can be understood as the first step for the transformation of silicates into phyllosilicates. Phyllosilicates (also called hydrous silicates) can form by reaction of anhydrous silicates with water, and forsterite being the reactant. Formation of chrysotile ($Mg_3Si_2O_5(OH)_4$) can take

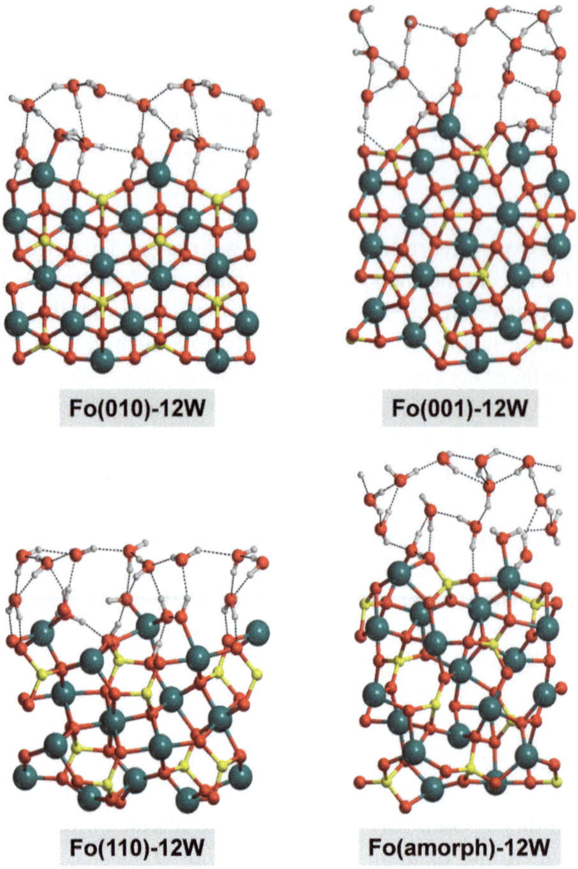

Fig. 3 Lateral views of the B3LYP-D2* optimized structures of 12 water molecules interacting on the Mg_2SiO_4 surface models

place by the reaction of $2Mg_2SiO_4 + 3H_2O \rightarrow Mg_3Si_2O_5(OH)_4 + Mg(OH)_2$. For the 12 water adsorption onto Fo(amorph), similar structural changes are also observed but at a lower degree; only one H_2O molecule transfers its proton to an O surface atom upon Mg^{2+} interaction (see Fo(amorph)-12W in Fig. 3). It is worth mentioning that in both cases, water dissociation only occurs in the first water layer, whereas the second water layer is unalterably engaged by a H-bond network. Interestingly, since direct water interaction with silicates is more likely to occur in the parent bodies of CCs, our results indicate that water alteration could preferentially take place within the chondrites parent bodies, rather than in the cold protoplanetary disk environment (e.g. see proposed scenarios Zolensky and Mc Sween, 1988; Trigo-Rodríguez 2015). This does not mean necessarily that we discard the incorporation of hydrated minerals into primordial aggregates from the protoplanetary disk, but it is likely that the process predated by subsequent parent body alteration at larger micrometric scale.

It is worth mentioning that these structural features of the water/Fo systems perfectly agree with the ladder of stability of the bare Fo surfaces. Indeed, Fo(010) and Fo(001) are the two most stable surfaces (and thus the two less reactive ones) used in this work, so that their interaction with water is not accompanied with water dissociation. In contrast, Fo(110) and Fo(amorph) are the two least stable surfaces (and accordingly the two most reactive ones), so that their interaction with water induces water dissociation to form Mg—OH and Si—OH surface groups. Moreover, for these two surfaces, the degree of water dissociation is consistent with the level of surface instability; that is, as Fo(110) is more unstable than Fo(amorph), Fo(110) presents more water dissociation than Fo(amorph).

3.2 *Infrared Features of the Water/Silicate Surface Systems*

Calculations of the vibrational frequencies are often done in computational chemistry because it allows the simulation of the infrared spectra of chemical systems and the assignment of infrared bands with the corresponding vibrations. In this work we have calculated the vibrational frequencies of all the systems shown in the previous section; namely, the bare Fo surfaces and the surfaces interacting with water. A comparison of the IR spectra and the vibrational features between the bare Fo surfaces with the corresponding hydrated ones is provided, in which the IR differences due to the water interaction is discussed.

The simulated IR spectra are shown in Fig. 4 and Table 3 summarizes the assignments of the IR bands.

The bare Fo surfaces contain 56 atoms per unit cell, which means the presence of 162 vibrational frequencies. Vibrational modes in silicates appear in the 1000–200 cm^{-1} region and are usually classified in two main categories: (1) the internal modes, which include the Si—O stretching (1000–800 cm^{-1}), the O—Si—O bending (700–500 cm^{-1}) vibrations and (2) the external modes, which include rotations and translations of the SiO_4 tetrahedra and translations of the metal cations (450–200 cm^{-1}).

For all the simulated IR spectra of the bare Fo surface models, one can distinguish two separated regions: the bands of the 1100–800 cm^{-1} range corresponding to the internal Si—O stretching modes, and the bands of the 700–100 cm^{-1} range corresponding to the internal O—Si—O bending modes (upper bands) and to the external modes (lower bands). It is worth mentioning that the most energetic Si—O stretching vibrations appear at upper values (above 1000 cm^{-1}) compared to the experimental data (below 1000 cm^{-1}) (Servoin and Piriou 1973; Iishi 1978; Hofmeister 1987; Reynard 1991; Chopelas 1991; Kolesov and Geiger 2004; Suto et al. 2006). This is because experiments recorded the IR spectra of the single crystal of forsterite, whereas in this work we simulate the IR spectra of Fo surface models. A main characteristic of these surfaces is that the outermost Si—O distances are shorter than the internal ones because the metal ions are undercoordinated, and consequently these outer-most Si—O bonds are stronger than the internal ones, hence appearing at higher wavenumbers (Navarro-Ruiz et al. 2014b).

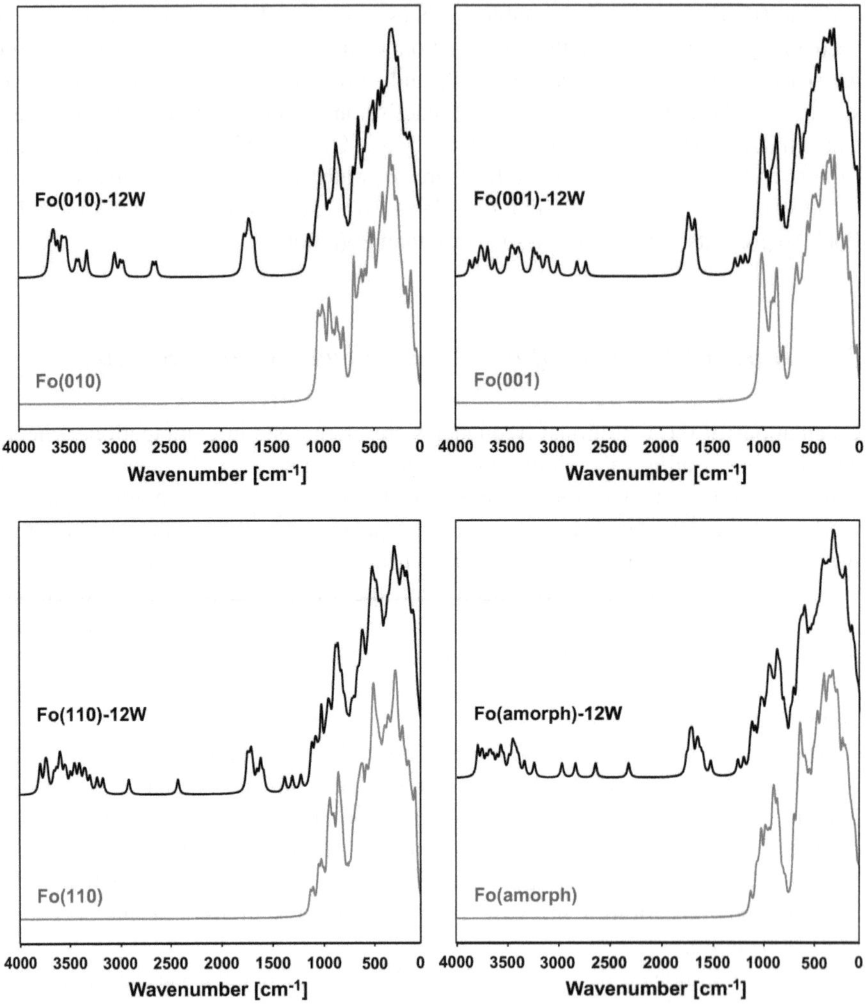

Fig. 4 Simulated infrared spectra of the bare forsterite surfaces (*grey* spectra) and in interaction with water (*black* spectra). Intensities are in arbitrary units

The H_2O/Fo surface systems present, obviously, different IR spectra. As a general feature, these systems have a set of new bands in the 3800–2600 cm^{-1} and 1800–1600 cm^{-1} ranges. They are due to the O—H stretching and H—O—H in-plane bending vibrational modes of water, respectively. Moreover, mixed with the forsterite internal and external modes, bands due to the H—O—H out-of-plane bending vibrations as well as libration and low-frequency-modes of H_2O are also present. Despite these general features, the IR spectra of Fo(110)-12W and Fo(amorph)-12W show an exclusive set of bands, which are not present in Fo(010)-12W and Fo(001)-12W, due to the presence of the SiOH surface groups

Table 3 List of infrared features of the forsterite surface models in a bare state and in interaction with the 12 water molecules

System	Bands (cm^{-1})	Assignment
Fo(010)	1059–794	Si—O stretching
	713–123	Si—O bending + rotations and translations of SiO_4 + translations of Mg
Fo(010)-12W	3687–2635	O—H stretching
	1787–1670	H—O—H in-plane bending
	1152–1074	H—O—H out-of-plane bending
	1063–803	Si—O stretching + H—O—H out-of-plane bending
	802–66	Si—O bending + rotations and translations of SiO_4 + translations of Mg + H_2O libration modes + H_2O low frequency modes
Fo(001)	1050–800	Si—O stretching
	711–114	Si—O bending + rotations and translations of SiO_4 + translations of Mg
Fo(001)-12W	3860–2723	O—H stretching
	1778–1654	H—O—H in-plane bending
	1277–1052	H—O—H out-of-plane bending
	1047–799	Si—O stretching + H—O—H out-of-plane bending
	771–65	Si—O bending + rotations and translations of SiO_4 + translations of Mg + H_2O libration modes + H_2O low frequency modes
Fo(110)	1136–799	Si—O stretching
	773–103	Si—O bending + rotations and translations of SiO_4 + translations of Mg
Fo(110)-12W	3801–3181	O—H stretching
	2925	SiO—H stretching
	2431	SiO—H stretching
	1755–1593	H—O—H in-plane bending
	1389	SiOH bending
	1315	SiOH bending
	1232	SiOH bending
	1135–808	Si—O stretching + H—O—H out-of-plane bending
	802–73	Si—O bending + rotations and translations of SiO_4 + translations of Mg + H_2O libration modes + H_2O low frequency modes
Fo(amorph)	1132–784	Si—O stretching
	705–95	Si—O bending + rotations and translations of SiO_4 + translations of Mg

(continued)

Table 3 (continued)

System	Bands (cm^{-1})	Assignment
Fo(amorph)-12W	3796–3241	O—H stretching
	2969	SiO—H stretching + O—H stretching
	2837	SiO—H stretching + O—H stretching
	2642	SiO—H stretching + O—H stretching
	2321	SiO—H stretching + O—H stretching
	1760–1525	H—O—H in-plane bending
	1137	SiOH bending + H—O—H out-of-plane bending
	1126	SiOH bending + H—O—H out-of-plane bending
	1117	SiOH bending + H—O—H out-of-plane bending
	1112	SiOH bending + H—O—H out-of-plane bending
	1089–770	Si—O stretching + H—O—H out-of-plane bending
	739–68	Si—O bending + rotations and translations of SiO_4 + translations of Mg + H_2O libration modes + H_2O low frequency modes

caused by the spontaneous proton transfer from H_2O to the surfaces. The bands at 2925 and 2431 cm^{-1} in Fo(110)-12W) and at 2969, 2837, 2642 and 2321 cm^{-1} in Fo(amorph)-12W correspond to the stretching vibrational modes of the SiOH groups, and the bands at 1389, 1315 and 1232 cm^{-1} in Fo(110)-12W and at 1137, 1126, 1117 and 1112 cm^{-1} in Fo(amorph)-12W correspond to the bending vibrational modes of the SiOH groups. Note that for Fo(amorph)-12W these bands are coupled with stretching and bending H_2O modes (see Table 3).

Therefore, partial water alteration of mafic silicates in CCs can be identified due to the presence of these bands, which can be detected by means of sensitive IR techniques.

3.3 Energetic Trends of the Water Alteration

In the previous sections, we showed atomic-scale structural details of the interaction of water with forsterite surfaces and we analyzed the vibrational features of these systems. In this section we present results related to the question whether the water alteration processes are favorable or not from an energetic point of view.

To do this, we have calculated the reaction energy of the interaction of 12 water molecules on the forsterite surfaces by extracting the 12 water molecules from the crystal hexagonal water ice (I_h). That is, we calculated the reaction energy of the following process (also shown in Fig. 5a):

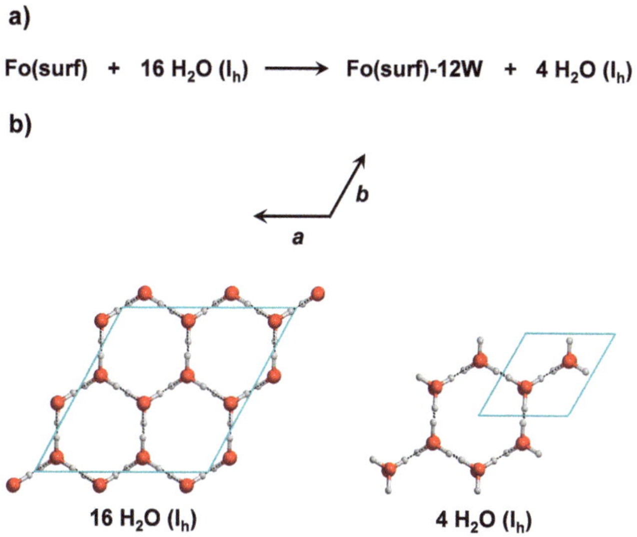

Fig. 5 (**a**) Reaction used to calculate the energetics of the water alteration processes on the silicate surfaces (see text for further details). (**b**) Views (along the *c* axis) of the crystal structure of the hexagonal ice adopting a supercell structure (16 H_2O (I_h)) and a minimal cell structure (4 H_2O (I_h))

$$\mathrm{Fo\,(surf)} + 16\,\mathrm{H_2O}\ \ (\mathrm{I_h}) \rightarrow \mathrm{Fo\,(surf)\,\text{-}12W} + 4\,\mathrm{H_2O}\ \ (\mathrm{I_h})$$

where Fo(surf) corresponds to the forsterite surface models, Fo(surf)-12W to the forsterite surface models with the 12 water molecules, 16 H_2O (I_h) corresponds to the crystalline system of the hexagonal water ice consisting of 16 water molecules per unit cell, and 4 H_2O (I_h) also to the crystal hexagonal water ice but using a unit cell containing 4 water molecules. It is worth mentioning that experimental X-ray diffraction data indicate that the unit cell of the crystal structure of I_h contains 4 water molecules, so that we modeled the "16 H_2O (I_h)" system with a supercell structure, in which the *a* and *b* lattice parameters were enlarged twice (see Fig. 5b).

The calculated reaction energies are reported in Table 3. ΔE_r values refer to those using the absolute potential energy values, and ΔU_r including zero-point energy corrections. Both set of values indicate that the reactions are indeed largely favorable, and accordingly formation of water ice layers on forsterite mineral surfaces is expected to easily takes place. In addition to this, it is also interesting to analyze the calculated values. Accounting for the ΔE_r values, the trend of the reaction energies is, from less to more favorable: Fo(010)-12W < Fo(001)-12W < Fo(amorph)-12W < Fo(110)-12W. This trend is consistent with the stability of the Fo surfaces because Fo(010), which is the most stable surface, presents the less negative reaction energy, followed by Fo(001), Fo(amorph) and Fo(110).

The ΔU_r values of Table 4 are associated with the reactions occurring at 0 K. However, it would be interesting to know if these values can be influenced by

Table 4 Calculated reaction energies of the water alteration processes of the forsterite surfaces, in kcal mol^{-1}

Reaction	ΔE_r	ΔU_r
Fo(010) + 16 $H_2O(I_h)$ → Fo(010)-12W + 4 $H_2O(I_h)$	−58.4	−60.2
Fo(001) + 16 $H_2O(I_h)$ → Fo(001)-12W + 4 $H_2O(I_h)$	−61.9	−64.2
Fo(110) + 16 $H_2O(I_h)$ → Fo(110)-12W + 4 $H_2O(I_h)$	−87.7	−91.0
Fo(amorph) + 16 $H_2O(I_h)$ → Fo(amorph)-12W + 4 $H_2O(I_h)$	−66.7	−64.7

ΔE_r vales are reaction energies calculated with the absolute potential energies, and ΔU_r including zero-point energy corrections

temperature effects. Because of that, the enthalpy and entropy corrections in the 20–400 K range were calculated using the partition functions computed in the frequency calculations, thus obtaining the reaction enthalpies (ΔH), reaction entropies (ΔS) and reaction Gibbs free energies (ΔG) at this temperature range. The thermodynamic trends are shown in Fig. 6. For the four reactions it is observed that: (1) the reaction enthalpies become slightly less negative with increasing temperature; (2) the entropy terms (TΔS) are not significant (at 400 K they are between 1.3 and 5.2 kcal mol^{-1}) and become slightly more positive with temperature; and (3) the reaction free energies are slightly more negative with increasing temperature. Thus, temperature does not exert important effects in the reaction energies and accordingly interaction of water and/or water alteration reactions on the forsterite surface are definitely largely favorable. Moreover, in view that water alteration does not depend on the temperature, it can occur in different environments: such as protoplanetary disk accretion regions, or planetesimal aggregation.

Another issue is the availability and stability of liquid water in such environments that seems to be less favorable than the radiation-protected interiors of the parent bodies of CCs.

4 Conclusions and Future Perspectives

This chapter is a computational study on the interaction of 12 water molecules on model surfaces mimicking the crystalline (010), (001) and (110) planes and an amorphous surface of forsterite (Mg_2SiO_4). This study provides molecular insights on water alteration processes of silicates in carbonaceous chondrites. We present structural details of the water/silicate interface, vibrational features of these systems, and energetic trends associated with the water interaction reactions. With this approach we arrive at the following conclusions:

- From a structural point of view, it is found that the less stable silicate surfaces are those that present more structural changes due to water interaction. Indeed, some water molecules spontaneously dissociate when they interact with the (110) and amorphous surfaces (the less stable ones), thus forming SiOH and Mg-OH

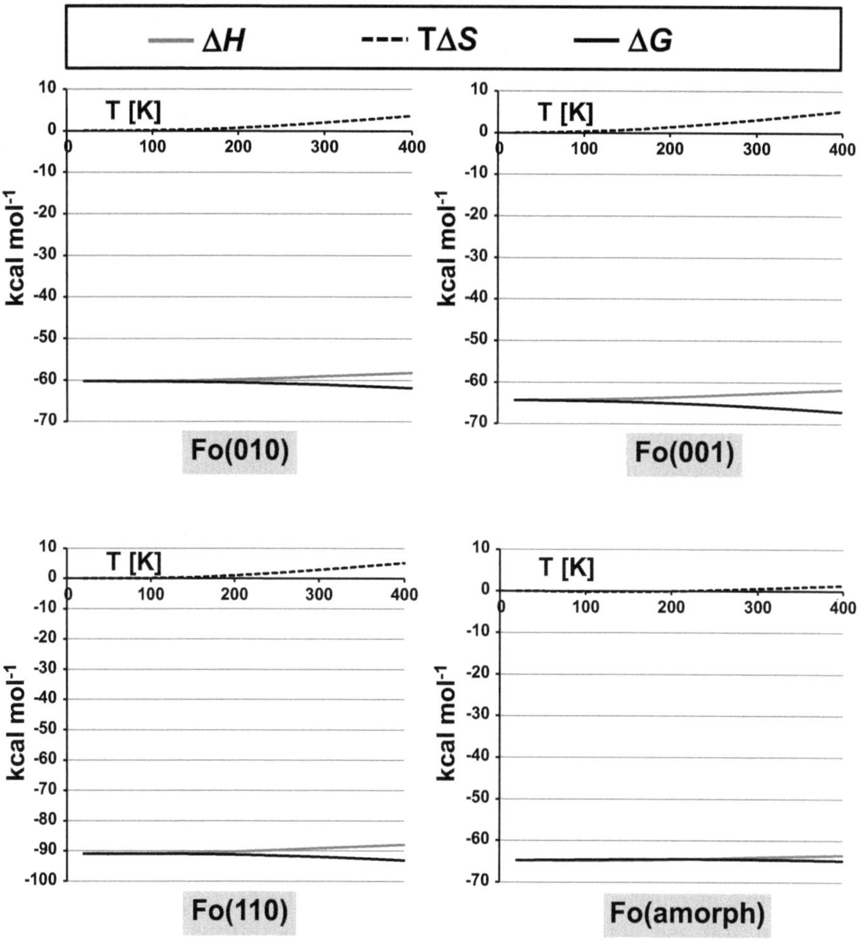

Fig. 6 Thermodynamic trends associated with the water alteration reactions occurring on the forsterite surface models, calculated in the 20–400 K temperature range

surface groups. In contrast, water molecules interacting with the more stable (010) and (001) surfaces do not dissociate and form essentially water layers on them.

- The calculated vibrational features are consistent with these findings; *i.e.*, whereas the simulated IR spectra of water adsorbed on the (010) and (001) forsterite surfaces clearly show an upper region defined by the stretching and bending H_2O vibrations and a lower region by the stretching, bending and rotational SiO_4 modes and metal translations, the IR spectra of water adsorbed on the (110) and amorphous surfaces, in addition to these mentioned bands, a set of new bands associated with the stretching and bending SiOH vibrations are clearly present.

- The reaction energies corresponding to the reaction of 12 water molecules belonging to the crystal of hexagonal water ice with the forsterite surfaces have been calculated, indicating that the reaction energies are for all surfaces large and negative and accordingly very favorable. Moreover, the obtained reaction energies are consistent with the surface stability trend (namely, the more stable the surface, the less negative the reaction energy), in such a way that the reaction energies follow the trend of (from less to more favorable) (010) < (001) < amorphous < (110). Moreover, inclusion of temperature effects in the range of 20 and 400 K does not significantly affect the energetic values since enthalpy and entropic effects are relatively small, and accordingly the calculated reaction Gibbs free energies are likewise very favorable.
- In view of the previously raised conclusions, our results suggest that aqueous alteration found in silicate minerals of primitive meteorites took place preferentially in the parent bodies of CCs, because it is more likely having the required water molecules in direct interaction with the silicate lattices. In view of the current evidence we cannot discard that the much colder protoplanetary outer disk was able to produce the right conditions to produce tiny hydrous minerals seen in the fine-grained structures of the matrix and the chondrule rims, but the extensive aqueous alteration found in several CC groups seems to have originated in their parent bodies.

It should be noted that present results are based on static quantum chemical calculations. That is, although we have employed thermochemical corrections to have insights on the energetic trends of the water alteration, geometry optimizations and IR simulations were performed at 0 K. Accordingly, results are fully representative for these systems at very low temperatures but dynamic effects, which are particularly important at higher temperatures, are not considered. Dynamic effects are expected to be important here since the silicate-water interface present in primitive meteorites can be exposed to a non-negligible range of temperatures (between 100 and 400 °C at most). These dynamic effects can nowadays be computationally simulated by means of molecular dynamics (MD) simulations. These MD simulations are calculations that give a view of the dynamic evolution of the system with time; that is, they reproduce the motion of atoms and molecules with time, and can be performed at a given temperature, which is used as input energy for the motion of the atoms. There are several works in the literature that address the mineral-water interface by means of MD simulations. Some points of these works are of interest for the systems studied here, which are summarized as follows:

1. A very important concept in water-interface systems is the interfacial depth; that is, the region where the water structure is not bulk-like. The structure of liquid water in close proximity to interfaces is different to that of the bulk liquid water. The interfacial depth is a useful parameter to assess if the bulk-like structure of water is largely altered or not in the vicinity of the interface. The interfacial depth

at the air/water (Morita and Hynes 2000; Townsend et al. 1985) and liquid/liquid (Walker and Richmond 2008) interfaces correspond to 1–2 monolayers (∼3–7 Å), in mineral/water systems it is larger (between 2 and 4 monolayers) because the charge of the surface ions induce an electrostatic field that can align water molecules, inducing an orientational order that persists over a certain depth into the bulk water (Rustad et al. 2003; Kerisit and Parker 2004; Philpott et al. 2004; Wang et al. 2006; Kerisit et al. 2008; Marry et al. 2008; Stack 2009; Mignon et al. 2010; Kerisit 2011; Wolthers et al. 2012). Our silicate/water systems present actually two water layers; i.e., one in direct contact with the silicate surfaces and the other interacting with the first one. Of course, the interaction of water is expected to be different with Earth minerals than with meteoritic minerals due to different conditions in both systems. Water interaction with Earth minerals occurs under abundant liquid water. In the case of meteorites water is probably reduced to tens of molecular layers. In this work, although we provide a detailed and accurate description of the actual contact between water and the silicate surfaces, the number of water layers is rather limited (just two) and accordingly more water molecules must be included to cover the interfacial depth in order to have a more general description of the silicate/water interfaces.

2. MD simulations allow us to analyze in detail and at an atomic-scale the driving forces restructuring the water molecules present at the interfacial depth. Such a restructuring depends on the nature of interacting partners at the interface. As mentioned in point (1), MD simulations for cation-bearing minerals indicated that the structural reorganization is due to the electrostatic interactions between the cations and the water molecules. Moreover, the cation-water interactions also influence the H-bonding network of the interfacial depth since water molecules attached to the metal ions present an enhanced H-bond donor behavior. This can partly be observed in our calculations, but MD simulations provide a dynamical view of the H–bonding restructuration by forming and destroying H-bonds with time. Interestingly, this H-bonding restructuring is not exclusive to cation-bearing minerals but also in cation-free surfaces with capabilities to establish H-bonds with water. This is the case of silica (SiO_2) surfaces. Silica surfaces present silanol (SiOH) groups and their hydrophilic/hydrophobic character depends on the surface SiOH concentration. MD simulations have allowed identifying a cooperative effect between SiOH and H_2O, which has a significant role in the interfacial H-bonding restructuration (Musso et al. 2012). This cooperation could also operate between the water molecules and the surface OH groups present in some of our silicate/water systems. MD simulations are required to shed some light into this point.
3. MD simulations would also be useful to characterize the non-dissociative adsorption of water on the silicates surfaces. There are several works that use MD simulations to assess the adsorption state of water on inorganic surfaces. One of the well-studied systems is TiO_2/water (Lindan et al. 1996, 1998; Langel 2002; Zhang and Lindan 2003; Tilocca and Selloni 2003; Kornherr et al. 2004; Beck et al. 2005; Koparde and Cummings 2007; Zhang et al. 2008; Mattioli et al. 2008; He et al. 2009; Cheng and Sprik 2010; Liu et al. 2010; Sumita et al.

2010; Kavathekar et al. 2011; Sebbari et al. 2011; Chen et al. 2013). Depending on the TiO_2 polymorph (rutile, anatase, or titanite) and on the crystallographic plane, water molecules can adsorb either molecularly or in a dissociative way. Some of these dissociative adsorptions were only identified by performing MD simulations: the trajectories indicated that dissociation involved a set of proton jumps between several water molecules, the final jump leading to a proton transfer to the surface. A similar mechanism was also observed in the dissociative water adsorption on bioglasses (Tilocca and Cormack 2009). In other cases, however, MD simulations were useful to determine a non-dissociative water adsorption, as it is the case of FeS_2/water. From an initially dissociated adsorbed water molecule, MD simulations indicated that the recovering of the H to form water is the most favorable dynamical path (Stirling et al. 2003). The added value of the MD simulations is that, since we can take temperature as an energy input, the bonds (in particular O—H bonds) are excited thermally and within a few vibrations the protons easily overcome the energy barriers associated with the H jumps. In relation to the present work, MD simulations are necessary to fully characterize the presence of OH groups at the silicate/water interface. On one hand, it is possible that those systems in which no OH formation is observed, water dissociation can take place due to dynamical and temperature effects introduced along the simulation (similarly to the TiO_2/water systems). On the other hand, MD simulations on those systems with OH formation can show a reversion towards water formation (similarly to the FeS_2/water systems). We, however, don't expect this latter situation since OH formation took spontaneously during the geometry optimization process, which is indicative of large stability.

For all these points exposed above, a major goal and a principal aim which is ongoing in our laboratories is to improve and have more accurate description of the water/silicate interfaces, including the role of the environment, by taking dynamical and temperature effects into account in a more rigorous way by means of MD simulations.

Finally, we would like to mention that the present study is a proof that quantum mechanical calculations can provide relevant chemical information on fundamental processes involving aqueous alteration and, for extension, in the chemistry occurring in or on meteorites, such as the interaction of water and organic compounds, or chemical reactions related to the formation of molecules present in meteorites samples. From this better understanding of these reactions new clues can be obtained about the formation environments of aqueous alteration. A proper description of all these processes from different approaches, including the atomistic one, is a major challenge to be achieved in the following years.

Acknowledgments Financial support from MICINN (projects CTQ2015-62635-ERC and CTQ2014-60119-P) and DIUE (project 2014SGR482) is gratefully acknowledged. JMTR also acknowledges MICINN project AYA 2015-67175-P. A.R. is indebted to Programa Banco de San-tander for a UAB distinguished postdoctoral research contract. A.R. kindly acknowledges BSC-MN for the generous allowance of computing time through the 'QCM-2013-2–0006: Formation of Molecu-lar Hydrogen on Surfaces of Cosmic Dust', 'QCM-2013-3–0015:

Adsorption of Atomic Hydrogen on Defective Non-Stoichiometric Surfaces of Cosmic Dust' and 'QCM-2014–3–0032: Quantum Effects in the Diffusion of Atomic Hydrogen on Interstellar Silicate Surfaces' projects.

References

Anders, E., Grevese, N.: Abundances of the elements: meteoritic and solar. Geochim. Cosmochim. Acta **53**, 197–214 (1989)

Asaduzzaman, A.M., Laref, S., Deymier, P.A., Runge, K., Cheng, H.-P., Muralidharan, K., Drake, M.J.: A first-principles characterization of water adsorption on forsterite grains. Phil. Trans. R. Soc. A **371**, 20110582 (2013)

Beck, T.J., Klust, A., Batzill, M., Diebold, U., Di Valentin, C., Tilocca, A., Selloni, A.: Mixed dissociated/molecular monolayer of water on the TiO2(0 1 1)-(2 × 1) surface. Surf. Sci. Lett. **591**, L267–L272 (2005)

Beck, P., Quirico, E., Montes-Hernandez, G., Bonal, L., Bollard, J., Orthous-Daunay, F.-R., Howard, K.T., Schmitt, B., Brissaud, O., Deschamps, F., Wunder, B., Guillot, S.: Hydrous mineralogy of CM and CI chondrites from infrared spectroscopy and their relationship with low albedo asteroids. Geochim. Cosmochim. Acta **74**, 4881–4892 (2010)

Becke, A.D.: Density-functional Thermochemistry. III. The role of exact exchange. J. Chem. Phys. **98**, 5648–5652 (1993)

Bonal, L., Alexander, C.M.O.'.D., Huss, G.R., Nagashima, K., Quirico, E., Beck, P.: Hydrogen isotopic composition of the water in CR chondrites. Geochim. Cosmochim. Acta **106**, 111–133 (2013)

Brearley, A.J.: Origin of graphitic carbon and pentlandite inclusions in matrix olivines in the Allende meteorite. Science **285**, 1380–1382 (1999)

Brearley, A.J.: Nebular vs. parent body processing. In: Davis, A.M. (ed.) Meteorites, Comets, and Planets, vol. 1, pp. 247–268. Elsevier–Pergamon, Oxford (2003) (eds.H. D. Holland and K. K. Turekan)

Brearley, A.J.: The action of water. In: Lauretta, D.S., McSween Jr., H.Y. (eds.) Meteorites and the Early Solar System II, pp. 584–624. University of Arizona Press, Tucson (2006)

Browning, L., Bourcier, W.: Fluid conditions during the alteration of CM chondrites. Meteorit. Planet. Sci. **31**, A22 (1996)

Bruno, M., Massaro, F.R., Prencipe, M., Demichelis, R., De La Pierre, M., Nestola, F.: Ab Initio calculations of the main crystal surfaces of forsterite (Mg2SiO4): a preliminary study to understand the nature of geochemical processes at the olivine interface. J. Phys. Chem. C **118**, 2498–2506 (2014)

Chen, J., Li, Y.-F., Sit, P., Selloni, A.: Chemical dynamics of the first proton-coupled electron transfer of water oxidation on TiO_2 anatase. J. Am. Chem. Soc. **135**, 18774–18777 (2013)

Cheng, J., Sprik, M.: Acidity of the aqueous rutile TiO_2 (110) surface from density functional theory based molecular dynamics. J. Chem. Theory Comput. **6**, 880–889 (2010)

Chizmadia L.J., Brearley A.J.: Aqueous alteration of carbonaceous chondrites: New insights from comparative studies of two unbrecciated CM2 chondrites: Y-791198 and ALH81002. Lunar Planet. Sci. **35**, abstract#1753. Lunar and Planetary Institute, Houston (2004)

Chopelas, A.: Single crystal Raman spectra of forsterite, fayalite, and monticellite. Am. Mineral. **76**, 1101–1109 (1991)

Ciesla, F., Lauretta, D.: Radial migration and dehydration of phyllosilicates in the solar nebula. Earth Planet. Sci. Lett. **231**, 1–8 (2005)

Ciesla, F.J., Lauretta, D.S., Cohen, B.A., Hood, L.L.: A nebular origin for chondritic fine-grained phyllosilicates. Science **299**, 549–552 (2003)

Civalleri, B., Zicovich-Wilson, C.M., Valenzano, L., Ugliengo, P.: B3LYP augmented with an empirical dispersion term (B3LYP-D*) as applied to molecular crystals. CrystEngComm **10**, 405–410 (2008)

de Leeuw, N.H., Parker, S.C., Catlow, C.R.A., Price, G.D.: Modelling the effect of water on the surface structure and stability of forsterite. Phys. Chem. Miner. **27**, 332–341 (2000)

Dovesi, R., Orlando, R., Erba, A., Zicovich-Wilson, C.M., Civalleri, B., Casassa, S., Maschio, L., Ferrabone, M., De La Pierre, M., D'Arco, P., Noël, Y., Causà, M., Rérat, M., Kirtman, B.: CRYSTAL14: A program for the ab initio investigation of crystalline solids. Int. J. Quantum Chem. **114**, 1287–1317 (2014)

Endreβ, M., Bischoff, A.: Carbonates in CI chondrites: clues to parent body evolution. Geochim. Cosmochim. Acta **60**, 489–507 (1996)

Escamilla-Roa, E., Moreno, F.: Adsorption of glycine on cometary dust grains:II—effect of amorphous water ice. Planet. Space Sci. **75**, 1–10 (2013)

Escamilla-Roa, E., Sainz-Díaz, C.I.: Amorphous ammonia–water ice deposited onto silicate grain: effect on growth of mantles ice on interstellar and interplanetary dust. J. Phys. Chem. C **118**, 26080–26090 (2014)

Fujiya, W., Sugiura, N., Sano, Y., Hiyagon, H.: Mn-Cr ages of dolomites in CI chondrites and the Tagish Lake ungrouped carbonaceous chondrite. Earth Planet. Sci. Lett. **362**, 130–142 (2013)

Grimme, S.: Semiempirical GGA-type density functional constructed with a long-range dispersion correction. J. Comput. Chem. **27**, 1787–1799 (2006)

Hanowski, N.P., Brearley, A.J.: Chondrule serpentines as indicators of aqueous alteration in CM carbonaceous chondrites (abstract). Lunar Planet. Sci. **28**, 501–502 (1997)

Hanowski, N.P., Brearley, A.J.: Iron-rich aureoles in the CM carbonaceous chondrites Murray, Murchison, and Allan Hills 81002: Evidence for in situ aqueous alteration. Meteorit. Planet. Sci. **35**, 1291–1308 (2000)

Hanowski, N.P., Brearley, A.J.: Aqueous alteration of chondrules in the CM carbonaceous chondrite, Allan Hills 81002: Implications for parent body alteration. Geochim. Cosmochim. Acta **65**, 495–518 (2001)

Harju, E.R., Rubin, A.E., Ahn, I., Choi, B.-G., Ziegler, K., Wasson, J.T.: Progressive aqueous alteration of CR carbonaceous chondrites. Geochim. Cosmochim. Acta **139**, 267–292 (2014)

Hazen, R.M.: Effects of temperature and pressure on crystal-structure of forsterite. Am. Mineral. **61**, 1280–1293 (1976)

He, Y., Tilocca, A., Dulub, O., Selloni, A., Diebold, U.: Local ordering and electronic signatures of submonolayer water on anatase TiO_2(101). Nat. Mater. **8**, 585–589 (2009)

Hofmeister, A.: Single-crystal absorption and reflection infrared spectroscopy of forsterite and fayalite. Phys. Chem. Miner. **14**, 499–513 (1987)

Hua, X., Wang, J., Buseck, P.R.: Fine-grained rims in the Allan Hills 81002 and Lewis Cliff 90500 CM2 meteorites: their origin and modification. Meteorit. Planet. Sci. **37**, 229–244 (2002)

Iishi, K.: Lattice dynamics of forsterite. Am. Mineral. **63**, 1198–1208 (1978)

Kavathekar, R.S., Dev, P., English, N.J., MacElroy, J.M.D.: Molecular dynamics study of water in contact with the TiO_2 rutile-110, 100, 101, 001 and anatase-101, 001 surface. Mol. Phys. **13**, 1649–1656 (2011)

Kerisit, S.: Water structure at hematite–water interfaces. Geochim. Cosmochim. Acta **75**, 2043–2061 (2011)

Kerisit, S., Parker, S.C.: Free energy of adsorption of water and metal ions on the {1014} calcite surface. J. Am. Chem. Soc. **126**, 10152–10161 (2004)

Kerisit, S., Liu, C., Ilton, E.S.: Molecular dynamics simulations of the orthoclase (001)- and (010)-water interfaces. Geochim. Cosmochim. Acta **72**, 1481–1497 (2008)

Kolesov, B.A., Geiger, C.A.: A Raman spectroscopic study of Fe–Mg olivines. Phys. Chem. Miner. **31**, 142–154 (2004)

Koparde, V.N., Cummings, P.T.: Molecular dynamics study of water adsorption on TiO2 nanoparticles. J. Phys. Chem. C **111**, 6920–6926 (2007)

Kornherr, A., Vogtenhuber, D., Ruckenbauer, M., Podloucky, R., Zifferer, G.: Multilayer adsorption of water at a rutile TiO_2 (110) surface: towards a realistic modeling by molecular dynamics. J. Chem. Phys. **121**, 3722–3726 (2004)
Krot, A.N., Petaev, M.I., Scott, E.R.D., Choi, B.-G., Zolensky, M.E., Keil, K.: Progressive alteration in CV3 chondrites: more evidence for asteroidal alteration. Meteorit. Planet. Sci. **33**, 1065–1085 (1998)
Langel, W.: Car–Parrinello simulation of H_2O dissociation on rutile. Surf. Sci. **496**, 141–150 (2002)
Le Guillou, C., Changela, H.G., Brearley, A.J.: Widespread oxidized and hydrated amorphous silicates in CR chondrites matrices: implications for alteration conditions and H_2 degassing of asteroids. Earth Planet. Sci. Lett. **420**, 162–173 (2015)
Lee, C., Yang, W., Parr, R.G.: Development of the Colle-Salvetti correlation-energy formula into a functional of the electron density. Phys. Rev. B **37**, 785–789 (1988)
Lindan, P.J.D., Harrison, N.M., Holender, J.M., Gillan, M.J.: First-principles molecular dynamics simulation of water dissociation on TiO2 (110). Chem. Phys. Lett. **261**, 246–252 (1996)
Lindan, P.J.D., Harrison, N.M., Gillan, M.J.: Mixed dissociative and molecular adsorption of water on the rutile (110) surface. Phys. Rev. Lett. **80**, 762 (1998)
Liu, L.-M., Zhang, C., Thornton, G., Michaelides, A.: Structure and dynamics of liquid water on rutile TiO2(110). Phys. Rev. B **82**, 161415 (2010)
Lodders, K.: Solar system abundances and condensation temperatures of the elements. Astrophys. J. **591**, 1220–1247 (2003)
Marry, V., Rotenberg, B., Turqab, P.: Structure and dynamics of water at a clay surface from molecular dynamics simulation. Phys. Chem. Chem. Phys. **10**, 4802–4813 (2008)
Mattioli, G., Filippone, F., Caminiti, R., Bonapasta, A.A.: Short hydrogen bonds at the water/TiO2 (anatase) interface. J. Phys. Chem. C **112**, 13579–13586 (2008)
Mignon, P., Ugliengo, P., Sodupe, M., Hernandez, E.R.: Ab initio molecular dynamics study of the hydration of Li^+, Na^+ and K^+ in a montmorillonite model. Influence of isomorphic substitution. Phys. Chem. Chem. Phys. **12**, 688–697 (2010)
Morita, A., Hynes, J.T.: A theoretical analysis of the sum frequency generation spectrum of the water surface. Chem. Phys. **258**, 371–390 (2000)
Musso, F., Mignon, P., Ugliengo, P., Sodupe, M.: Cooperative effects at water–crystalline silica interfaces strengthen surface silanol hydrogen bonding. An ab initio molecular dynamics study. Phys. Chem. Chem. Phys. **14**, 10507–10514 (2012)
Navarro-Ruiz, J., Sodupe, M., Ugliengo, P., Rimola, A.: Interstellar H adsorption and H_2 formation on the crystalline (010) forsterite surface: a B3LYP-D2* periodic study. Phys. Chem. Chem. Phys. **16**, 17447–17457 (2014a)
Navarro-Ruiz, J., Ugliengo, P., Rimola, A., Sodupe, M.: B3LYP periodic study of the physicochemical properties of the Nonpolar (010) Mg-Pure and Fe-containing olivine surfaces. J. Phys. Chem. A **118**, 5866–5875 (2014b)
Navarro-Ruiz, J., Martínez-González, J.A., Sodupe, M., Ugliengo, P., Rimola, A.: Relevance of silicate surface morphology in interstellar H_2 formation. Insights from quantum chemical calculations. Mon. Not. R. Astron. Soc. **453**, 914–924 (2015)
Pascale, F., Zicovich-Wilson, C.M., López Gejo, F., Civalleri, B., Orlando, R., Dovesi, R.: The calculation of the vibrational frequencies of crystalline compounds and its implementation in the CRYSTAL code. J. Comput. Chem. **25**, 888–897 (2004)
Philpott, M.R., Goliney, I.Y., Lin, T.T.: Molecular dynamics simulation of water in a contact with an iron pyrite FeS_2 surface. J. Chem. Phys. **120**, 1943–1950 (2004)
Prigiobbe, V., Suarez Negreira, A., Wilcox, J.: Interaction between olivine and water based on density functional theory calculations. J. Phys. Chem. C **117**, 21203–21216 (2013)
Reynard, B.: Single-crystal infrared reflectivity of Pure Mg2SiO2 forsterite and (Mg0.86, Fe0.14)2SiO4 olivine. Phys. Chem. Miner. **18**, 19–25 (1991)
Robert, F.: The D/H ratio in chondrites. Space Sci. Rev. **106**, 87–101 (2003)
Rubin, A.E.: Mineralogy of meteorite groups. Meteorit. Planet. Sci. **32**, 231–247 (1997)

Rubin, A.E., Trigo-Rodríguez, J.M., Huber, H., Wasson, J.T.: Progressive aqueous alteration of CM carbonaceous chondrites. Geochim. Cosmochim. Acta **71**, 2361–2382 (2007)

Rustad, J.R., Felmy, A.R., Bylaska, E.J.: Molecular simulation of the magnetite-water interface. Geochim. Cosmochim. Acta **67**, 1001–1016 (2003)

Sebbari, K., Domain, C., Roques, J., Perron, H., Simoni, E., Catalette, H.: Investigation of hydrogen bonds and temperature effects on the water monolayer adsorption on rutile TiO_2 (110) by first-principles molecular dynamics simulations. Surf. Sci. **605**, 1275–1280 (2011)

Servoin, J.L., Piriou, B.: Infrared reflectivity and Raman scattering of Mg_2SiO_4 single crystal. Phys. Status Solid. B **55**, 677–686 (1973)

Stack, A.G.: Molecular dynamics simulations of solvation and kink site formation at the {001} barite-water interface. J. Phys. Chem. C **113**, 2104–2110 (2009)

Stirling, A., Bernasconi, M., Parrinello, M.: Ab initio simulation of water interaction with the (100) surface of pyrite. J. Chem. Phys. **118**, 8917–8926 (2003)

Sumita, M., Hu, C., Tateyama, Y.: Interface water on TiO_2 anatase (101) and (001) surfaces: first-principles study with TiO_2 slabs dipped in bulk water. J. Phys. Chem. C **114**, 18529–18537 (2010)

Suto, H., Sogawa, H., Tachibana, S., Koike, C., Karoji, H., Tsuchiyama, A., Chihara, H., Mizutani, K., Akedo, J., Ogiso, K., Fukui, T., Ohara, S.: Low-temperature single crystal reflection spectra of forsterite. Mon. Not. R. Astron. Soc. **370**, 1599–1606 (2006)

Takir, D., Emery, J.P., McSween, H.Y., Hibbits, C.A., Clark, R.N., Pearson, N., Wang, A.: Nature and degree of aqueous alteration in CM and CI carbonaceous chondrites. Meteor. Planet. Sci. **48**, 1618–1637 (2013)

Tilocca, A., Cormack, A.N.: Modeling the water-bioglass interface by ab initio molecular dynamics simulations. ACS Appl. Mater. Interfaces **1**, 1324–1333 (2009)

Tilocca, A., Selloni, A.: Reaction pathway and free energy barrier for defect-induced water dissociation on the (101) surface of TiO_2-anatase. J. Chem. Phys. **119**, 7445–7450 (2003)

Tomeoka, K., Buseck, P.: Matrix mineralogy of the Orgueil CI carbonaceous chondrite. Geochim. Cosmochim. Acta **52**, 1627–1640 (1988)

Tomeoka, K., Buseck, P.R.: Phyllosilicates in the Mokoia CV carbonaceous chondrite: evidence for aqueous alteration in an oxidizing condition. Geochim. Cosmochim. Acta **54**, 1787–1796 (1990)

Townsend, R.M., Jan, G., Stuart, A.R.: Structure of the liquid vapor interface of water. J. Chem. Phys. **82**, 4391–4392 (1985)

Trigo-Rodríguez, J.M.: Aqueous alteration in chondritic asteroids and comets from the study of carbonaceous chondrites. In: Lee, M.R., Leroux, H. (eds.) Planetary mineralogy. EMU notes in mineralogy, vol. 15, pp. 67–87. European Mineralogical Union and the Mineralogical Society, London (2015)

Trigo-Rodríguez, J.M., Blum, J.: The effect of aqueous alteration and metamorphism in the survival of presolar silicate grains in chondrites. Publ. Astron. Soc. Aust. **26**, 289–296 (2009a)

Trigo-Rodríguez, J.M., Blum, J.: Tensile strength as an indicator of the degree of primitiveness of undifferentiated bodies. Planet. Space Sci. **57**, 243–249 (2009b)

Trigo-Rodríguez, J.M., Moyano-Cambero, C.E., Mestres, N., Fraxedas, J., Zolensky, M.E., Nakamura, T., Martins, Z.: Evidence for Extended Aqueous Alteration in CR Carbonaceous Chondrites. 44th LPSC, The Woodlands, Texas. LPI Contribution No. 1719, p. 1929 (2013)

Trigo-Rodriguez, J.M., Alonso-Azcárate, J., Abad, M.M., Lee, M.R.: Ultra High Resolution Transmission Electron Microscopy of Matrix Mineral Grains in CM Chondrites: Preaccretionary or Parent Body Aqueous Processing? 46th LPSC, The Woodlands, Texas. LPI Contribution No. 1832, p. 1198 (2015)

Trigo-Rodríguez, J.M., Rubin, A.E., Wasson, J.T.: Non-nebular origin of dark mantles around chondrules and inclusions in CM chondrites. Geochim. Cosmochim. Acta **70**, 1271–1290 (2006)

Walker, D.S., Richmond, G.L.: Interfacial depth profiling of the orientation and bonding of water molecules across liquid-liquid interfaces. J. Phys. Chem. C **112**, 201–209 (2008)

Wang, J., Kalinichev, A.G., Kirkpatrick, R.J.: Effects of substrate structure and composition on the structure, dynamics, and energetics of water at mineral surfaces: a molecular dynamics modeling study. Geochim. Cosmochim. Acta **70**, 562–582 (2006)

Watson, G.W., Oliver, P.M., Parker, S.C.: Computer simulation of the structure and stability of forsterite surfaces. Phys. Chem. Miner. **25**, 70–78 (1997)

Weisberg, M.K., McCoy, T.J., Krot, A.N.: Systematic and evaluation of meteorite classification. In: Lauretta, D.S., McSween, H.Y. (eds.) Meteorites and the Early Solar System II, pp. 19–52. The University of Arizona Press, Tucson (2006)

Wolthers, M., Tommaso, D.D., Dub, Z., de Leeuw, N.H.: Calcite surface structure and reactivity: molecular dynamics simulations and macroscopic surface modelling of the calcite–water interface. Phys. Chem. Chem. Phys. **14**, 15145–15157 (2012)

Zhang, C., Lindan, P.J.D.: Multilayer water adsorption on rutile TiO2(110): a first-principles study. J. Chem. Phys. **118**, 4620–4630 (2003)

Zhang, W., Yang, J., Luo, Y., Monti, S., Carravetta, V.: Quantum molecular dynamics study of water on TiO2 (110) surface. J. Chem. Phys. **129**, 064703 (2008)

Zolensky, M., Barret, R., Browing, L.: Mineralogy and composition of matrix chondrule rims in carbonaceous chondrites. Geochim. Cosmochim. Acta **57**, 3123–3148 (1993)

Zolensky, M.E., McSween Jr., H.: Aqueous alteration. In: Kerridge, J.F., Matthews, M.S. (eds.) Meteorites and the Early Solar System, pp. 114–143. The University of Arizona Press, Tucson (1988)

Zolensky, M.E., Mittlefehldt, D.W., Lipschutz, M.E., Wang, M.-S., Clayton, R.N., Mayeda, T.K., Grady, M.M., Pillinger, C., Barber, D.: CM chondrites exhibit the complete petrologic range from type 2 to 1. Geochim. Cosmochim. Acta **61**, 5099–5115 (1997)

Measuring the Terminal Heights of Bolides to Understand the Atmospheric Flight of Large Asteroidal Fragments

Manuel Moreno-Ibáñez, Maria Gritsevich, and Josep M. Trigo-Rodríguez

Abstract The extent of penetration into the Earth's atmosphere of a meteoroid is defined by the point where its kinetic energy is no longer sufficient to produce luminosity. For most of the cases this is the point where the meteoroid disintegrates in the atmosphere due to ablation process and dynamic pressure during flight. However, some of these bodies have particular physical properties (bigger size, higher bulk strength, etc.) or favorable flight conditions (lower entry velocity or/and a convenient trajectory slope, etc.) that allow them to become a meteorite-dropper and reach the ground. In both cases, we define the end of the luminous path of the trajectory as the terminal height or end height. Thus, the end point shows the amount of deceleration till the final braking. We thus assume that the ability of a fireball to produce meteorites is directly related to its terminal height. Previous studies have discussed the likely relationship between fireball atmospheric flight properties and the terminal height. Most of these studies require

M. Moreno-Ibáñez (✉)
Institute of Space Sciences (IEEC-CSIC), Meteorites, Minor Bodies and Planetary Sciences Group, Campus UAB, c/Can Magrans s/n, 08193 Cerdanyola del Vallés (Barcelona), Catalonia, Spain

Department of Geodesy and Geodynamics, Finnish Geospatial Research Institute (FGI), Geodeetinrinne 2, Masala FI-02431, Finland
email: mmoreno@ice.csic.es

M. Gritsevich
Department of Physics, University of Helsinki, Gustaf Hällströmin katu 2a, P.O. Box 64, FI-00014 Helsinki, Finland

Department of Geodesy and Geodynamics, Finnish Geospatial Research Institute (FGI), Geodeetinrinne 2, Masala, FI-02431, Finland

Institute of Physics and Technology, Ural Federal University, Ekaterinburg, 620002, Russia

Department of Computational Physics, Dorodnicyn Computing Centre, Russian Academy of Sciences, Vavilova str. 40, Moscow, 119333, Russia

J.M. Trigo-Rodríguez
Institute of Space Sciences (IEEC-CSIC), Meteorites, Minor Bodies and Planetary Sciences Group, Campus UAB, c/Can Magrans s/n, 08193 Cerdanyola del Vallés (Barcelona), Catalonia, Spain

J.M. Trigo-Rodríguez et al. (eds.), *Assessment and Mitigation of Asteroid Impact Hazards*, Astrophysics and Space Science Proceedings 46,
DOI 10.1007/978-3-319-46179-3_7

the knowledge of a set of properties and physical variables which cannot be determined with sufficient accuracy from ground-based observations. The recently validated dimensionless methodology offers a new approach to this problem. All the unknowns can be reduced to only two parameters which are easily derived from observations. Despite the calculation of the analytic solution of the equations of motion is not trivial, some simplifications are admitted. Here, we describe the best performance range and the errors associated with these simplifications. We discuss how terminal heights depend on two or three variables that are easily retrieved from the recordings, provided at least three trajectory (h, v) points. Additionally, we review the importance of terminal heights, and the way they have been estimated in previous studies. Finally we discuss a new approach for calculating terminal heights.

1 Introduction

This chapter is dedicated to estimating the terminal heights of fireballs using a dimensionless methodology. The key ideas of this methodology have been described in another chapter (Gritsevich et al. 2017), so we will focus on the simplifications and variables that are able to provide us with a detailed estimation of terminal heights.

We will first introduce the terminology that we are going to follow, defining the terms to be used in this chapter: meteoroid, meteor, fireball and meteorite. Briefly, a meteoroid is, in most cases, a detached part of an asteroid or a comet. The size of a meteoroid may range from few tens of microns to tens of meters in diameter. Meteoroids originating from Mars or Moon are also possible, but they represent a much smaller fraction. When a meteoroid enters into the Earth's atmosphere, it produces a bright luminous path produced by its ablation. The light emitting object is called a meteor. Due to its high entry velocity (up to 73 km/s), the meteor experiences intense aerodynamic forces that produce an abrupt braking. One of the immediate effects is the intense interaction of the meteoroid surface material with atmosphere which causes ionization and subsequent emission of light. This effect is quite helpful to track the meteor on the sky (which is colloquially known as shooting star). Depending on several factors (mass, trajectory slope, size, velocity, etc.) a meteoroid could penetrate into the atmosphere. The atmospheric density of particles is higher and the temperature around the meteor increases which, eventually, melts the external layers and provoke the meteor to lose mass due to the interaction with the surrounding flow of air particles. This is the physical process called ablation. Due to the meteoroid ablation and the aerodynamic pressure during its flight trough the atmosphere most meteoroids eventually disintegrate in the atmosphere. We refer to these deep-penetrating luminous meteors as either fireballs or bolides. The amount of light emitted due to the ablation enhances their visibility from the Earth's surface. Typically, the brightness magnitude of a fireball reaches or overcomes that of Venus (−4). Very bright fireballs are able to get to even lower altitudes, reaching a brightness over −16, being observable at distances of more than 700 km (see e.g.

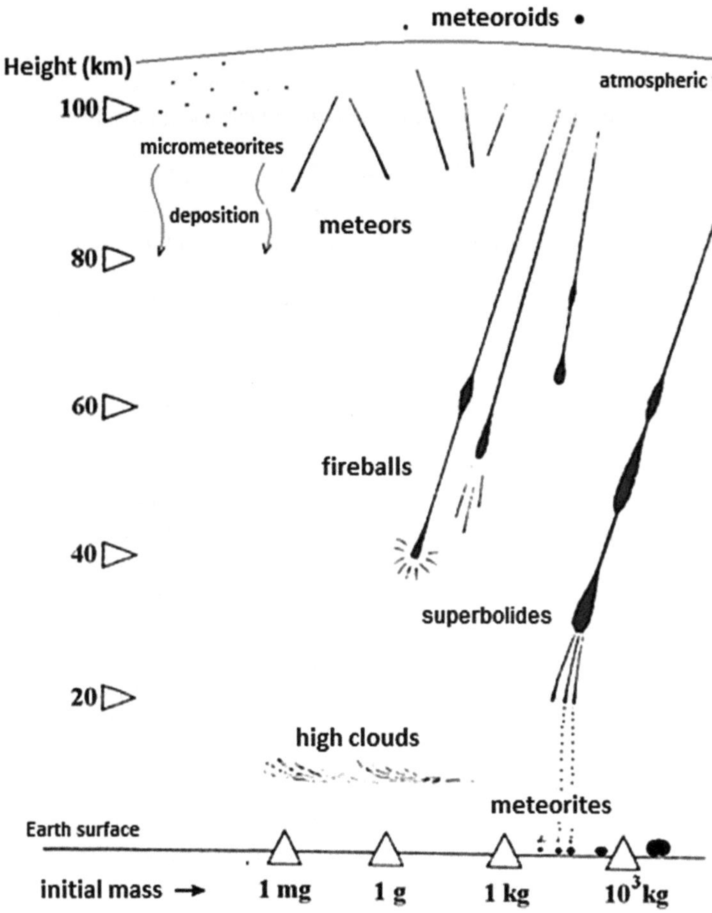

Fig. 1 Graphical description of meteoroids, meteors, fireballs, superbolides, meteorites and micrometeorites. Adapted from Rendtel et al. (1995)

Trigo-Rodríguez et al. 2009). Such bright fireballs are named superbolides; they are usually meteorite-droppers, surviving the ablation processes, reaching the ground as meteorites. Thus, a meteorite is produced by a meteoroid that survived partially its atmospheric flight and reached the ground. In general, it is estimated that less than a 3 % of the incoming (preatmospheric) mass can survive as meteorites (Ceplecha et al. 1998).

Besides, small grains orbiting in space may interact with the atmosphere at low velocities and may survive to atmospheric deceleration with partial or no melting at all. These particles are deposited on the ground as micrometeorites. All these descriptions are illustrated in Fig. 1.

Recovered meteorites are generally classified based on their composition (see e.g. Weisberg et al. 2006). This classification is globally accepted and it associates each meteorite with a particular class according to cosmochemical and mineralogical patterns. Similarly, some classifications for meteors and fireballs were introduced in the scientific literature (see e.g. Ceplecha 1967, Gritsevich et al. 2012). Since meteors and fireballs may disintegrate in the atmosphere, any possible classification may rely on some physical variable of the atmospheric flight. Ceplecha (1967, 1968, 1988) initially discussed that a combination of the beginning height of the luminous trajectory (including the dependence on velocity) with a parameter describing the product of heat conductivity, density and specific heat would lead to classification of sporadic meteors, especially those photographed with a Super-Schmidt camera. Four groups were defined, A, B C and D, each related to a different range of fireball properties and heights. It was also suggested that members of each group show similar orbital dynamics. Alternatively, the work of Ceplecha and McCrosky (1976) suggested that meteors could be classified based on their atmospheric dynamical behavior. As we will see later on, they used the terminal height as a criterion to distinguish between different populations of meteorites. This criterion allowed them to determine the grade of ablation experienced by the body during its atmospheric flight. Four different groups were described: I, II, IIIA and IIIB. There is no reason for extending our review here because an extensive summary of meteor science and more characteristics of these classifications can be found in Ceplecha et al. (1998).

Needless to say that these studies are of great relevance for planetary defense purposes. Meter-sized meteoroids encountering the Earth produce meteorites that turn into hazardous projectiles like e.g. the recent Carancas or Chelyabinsk meteorite falls (Tancredi et al. 2009, Borovička et al. 2013). These falls are able to release a large amount of energy either via a final impact onto the Earth surface, creating a crater, triggering an earthquake or even a tsunami; or if they do not reach the surface, the energy transferred to the atmosphere may produce dangerous airblasts. However, there is not an easy way to carry out an accurate analysis of these phenomena. Lots of considerations shall be taken into account, besides the number of cases and our previous experience on this subject is quite poor due to the small number of hazardous events documented. In 1908, over Tunguska River, a violent event associated with a possible meteorite fall took place. No meteorite was recovered but the shock wave associated with the likely final explosion of the fireball devastated an area of 2150 km^2 approximately, of which, 100 km^2 resulted with burnt trees (Popova et al. 2013, Vasilyev 1998). Local inhabitants felt an Earthquake ranging from 4.5 to 5 on the Richter scale. The energy released by this event has been estimated to 10–50 Mt (Chyba et al. 1993, Collins et al. 2005). Considering different entry conditions and meteor origins Chyba et al. (1993) suggested that for a carbonaceous body with an entry angle ranging from horizontal to 45°, the energetic explosion should have occurred at 14 km above earth's surface. There was no crater and no meteorite was recovered, so it was suggested that the energy was completely released to the atmosphere leading to a massive airblast. For many years, this event was taken as an isolated event which is not fully understood. More recently, in February 2013, over the Russian city of Chelyabinsk another spectacular event took place. This time numerous civilian cameras recorded the phenomenon, thus providing the first well documented hazardous event. It was an approximately 19 m body that entered the atmosphere at 19.03 km/s with a grazing angle of 18.5° (Borovička et al. 2013). The energy released was estimated to be around

500 kt, causing glass damage in nearby towns (Brown et al. 2013). The Chelyabinsk meteoroid suffered intense fragmentation between heights of 45 and 30 km, and only an 8 m hole on ice was found (Borovička et al. 2013).

These two cases are examples of the hazardous potential of tens of meters sized objects. Unfortunately, the small size of these bodies makes it complicated to detect and timely identify those that may lead to future encounters, and therefore, to properly quantify the risk of future events. Conversely, we could, for example, predict the amount of energy released in the atmosphere through a dedicated study on terminal heights.

The sky is constantly observed by photographic and video cameras, devoted to record any fireball event taking place. Most of them are arranged locally under the same image acquisition and reduction software, and the same institution requirements. This set up is commonly known as a fireball network (hereafter FN). Although fireball networks are currently widespread, it was not until 1936 that the first organized couple of cameras were co-pointed to the sky ruled by F.L. Whipple at the Harvard Observatory, the Harvard Meteor Project (Jacchia and Whipple 1956). From that moment and on, researchers realized the relevancy of these observations. Their recordings are fundamental for any meteor research. For example, in this chapter we will later make use of the fireball data provided by the Meteorite Observation and Recover Project in Canada, MORP (1970–1985).

In line with its contemporary FNs, the MORP (Halliday et al. 1978) was created with the idea of gaining knowledge on the origin and properties of fireballs and meteorites. It consisted of 12 observatories located mainly in the south of Canada. The control headquarters were located at the campus of the University of Saskatchewan in Saskatoon. Each of the observatories had five cameras each of which covered 54° of azimuth near the horizon. Neighbor stations covered the part of the unrecorded azimuth area. The cameras used photographic films (see more details in Halliday et al. 1978) which gives an idea of the relative complexity compared to the current digitalization era.

MORP was able to register more than 1010 fireballs, including a meteorite-dropper. MORP 285, known as Innisfree, was recorded on the 6th of February of 1977, and was recovered 12 days later (Halliday et al. 1977, 1978, 1981). Since the fall was accurately observed by two stations, its orbital trajectory could be calculated. This is quite remarkable since the number of recovered meteorites for which we have been able to calculate their orbit is quite low (for a review see: Trigo-Rodríguez et al. 2015).

Current instrumentation combines the use of charged couple devices (CCDs) in photographic and video cameras along with more sophisticated software and optics. Photo and video images are often supported by spectrograph devices coupled to observation cameras which reveal significant information about the chemical composition of meteors. Video frames and pictures provide details on the meteor flight trajectory. In order to correctly deal with pictures, cameras are provided with a chopping shutter, which allows to sequentially splitting the meteor trajectory into shorter sections during the exposition time. Using astrometry techniques (see Ceplecha 1987), the exact position and time of these sections (the procedure is similar for video frames) in the sky is calculated; this is, at each moment we can accurately determine the altitude, latitude and longitude of the meteor. Thus, velocity and height values for the whole meteor trajectory are derived. It is also possible to obtain the light curve of the meteor atmospheric flight which provides alternative ways of estimating the kinetic energy released and any fragmentation occurred during the atmospheric entry. A good determination of the trajectory of the fireball is crucial to extract further flight characteristics and fireball properties using theoretical models.

Once a fireball is registered, the data extracted from these ground-based observations should undergo an analytical study. Thus, the reliability of the mathematical model used is essential. The *Single Body Theory* (Hoppe 1937), a.k.a. *classical model*, has been widely used to describe the dynamical laws of the atmospheric flight. Briefly, this theory considers the following coefficients as constant: drag, luminosity, heat transfer and ionization. The atmosphere is assumed to be isothermal and the fragmentation of the fireball cannot be modeled (except for separation of small particles). The major handicap of this theory is its accuracy. Normally, for the calculation of the value of certain variable we need to set beforehand the values of other variables. These values are not always known, so it is common practice to use the generally accepted mean values. Each case requires careful analysis and high accuracy cannot be assured.

Alternatively, the dimensionless methodology presented by Gritsevich et al. (2016, 2017), overcomes this problem and offers a new point of view. It has been recently applied to the determination of the terminal heights and we will discuss in following lines its validity and any further improvements.

We would like to remark that we will assume no sudden fragmentation in the analysis as described below. Only the main fragment is studied and all changes to its mass along the trajectory are approximated using a physically based approach. The inclusion of discrete fragmentation is only possible when dealing with very well-documented cases (with well-observed fragment trajectories) and it automatically increases the complexity of the study with an unknown number of free parameters, but a good handling of the problem shall improve the performance of the results. For instance, Revelle (2007), based on his previous work with Ceplecha (Ceplecha and Revelle 2005), explored the possibility of improving the *Single Body Theory* estimation of the terminal mass of any bolide. By including a fragmentation model called TPFM (triggered progressive fragmentation model), Revelle suggested that important atmospheric flight values like the end mass or the ablation should have a limited upper value defined by those of the *Single Body Theory*, being, in fact, smaller. The TPFM is mainly based on introducing the variation of the ratio defined by the cross-section area (participating in the deceleration of the body) and the mass of the body during the flight time: m(t)/A(t). In order to get a first approach to the particularities of each event, Revelle (2007) assumed two main subcases. On the one hand, the fragments of the bolide continue flying along with the main fragment; in this case the cross-sectional area increases (more drag) and the general mass has not changed. On the other case, the mass detached from (mainly the back face of) the remaining main fragment move away from it quite fast, the main fragment does not show a variation in its flight configuration and drag is still the same; now, the mass is reduced and the cross-section area remains the same. Though this methodology partially leads to good results, there is still more work to be done (for example other thermal effects should be considered). It is clear that the real advantage of the TPFM is including the meteor fragmentation in a consistent way, which could be crucial in some events.

The topic of this chapter is devoted to the terminal heights which are one of the most relevant and characteristic parameters of the atmospheric flight of fireballs. It corresponds to the final point of the luminous part of the trajectory. This is the point where a fireball disintegrates or, for meteorite-droppers, the last point where luminosity is present. Meteoroids typically disintegrate at a pretty well defined atmospheric height depending on their particular tensile strength (Trigo-Rodríguez and Llorca 2006, 2007). The terminal height also describes the amount of deceleration experienced by the meteor, which in turn means the degree of penetration into the atmosphere.

The study of the flight conditions and fireball properties leads to a better understanding of their terminal heights. And vice versa, we could gain better insight into the composition of the fireball, relate it to a parental body (i.e. a particular asteroid or comet), and, when the trajectory is accurately described, obtain the orbital elements that describe its space motion, analyzing their terminal heights.

Finally, by accurately calculating and observing the terminal height for any fireball, mathematical models can be adjusted and other atmospheric flight properties can be obtained. Both the calculated and the recorded values could differ sometimes, for several reasons. On the one hand, the calculated terminal heights depend on various parameters (entry mass, bulk density, shape coefficient, ablation coefficient, etc.). If the real values of these parameters have not been derived from observations, then, they are commonly assumed to be close to the accepted mean values. On the other hand, the recording of fireballs relies on the spatial resolution and instrumental limitations. Weather conditions do also affect the ability to record the whole flight trajectory.

2 Terminal Heights in the Literature

The chances to gather and extract a lot of information from FN have increased along the years. Previous studies did take advantage of this situation to gain knowledge in meteor science. As we have mentioned, the *Single Body Theory* was the most detailed mathematic model used to deal with calculations. Clear example of this is the work of Ceplecha and McCrosky (1976) on the Prairie Network (PN). Operated in USA between 1963 and 1975, the PN registered more than 2700 fireballs, one of them being the Lost City meteorite (McCrosky et al. 1971). Being conscious of the importance of this database, Ceplecha and McCrosky (1976) undertook a deep analysis aiming to distinguish between ordinary and carbonaceous chondrites within recorded fireballs. Their research was based on the idea that carbonaceous chondrites are more fragile than ordinary chondrites. This means that the atmospheric flight of the carbonaceous chondrites should be shorter, mainly due to their higher ablation. In other words, the terminal heights shall be higher for carbonaceous than for ordinary chondrites. Generally, this should be true, but every meteor has its own peculiarities that modify its trajectory: the trajectory angle, the entry velocity, the shape, etc. Therefore, terminal heights cannot be considered as the sole classifying criterion. In order to account for these relevant parameters in any further classification based on terminal heights, Ceplecha and McCrosky (1976) suggested a new parameter, PE. This parameter is defined by the addition of the logarithms of the entry mass, velocity and zenith distance. The terminal height is included through the air density at that point. Note that all the parameters involved in defining PE express the atmospheric dynamic behavior of the meteor:

$$PE = log\rho_E + Alogm_\infty + BlogV_\infty + Clog\,(cosZ_R) \quad (1)$$

Where the air density at terminal height is ρ_E [g/cm^3], the preatmospheric mass, m_∞ [g], preatmospheric velocity, V_∞ [km/s], and the zenith distance of the meteor radiant, Z_R [degrees]. Constants A, B and C are adjusted for all the meteor trajectories.

Ceplecha and McCrosky (1976) also realized that the observation of each fireball provided extra information that is not included via the PE parameter. For instance, the ablation coefficient (σ [s^{2}cm^{-2}]) and the geometrical coefficient depending on the shape of the object, the drag coefficient and the bulk density, called K [cm^{2}g$^{-2/3}$], were also determined from the trajectory when the observations and the theoretical equation were compared. The information of these two parameters was available for ninety meteors meteors of the PN at each trajectory point. The authors used this valuable information to set a second criterion which considered the sum of the average values of σ and K for the entire meteor trajectory. This SD criterion describes globally the physical changes that the meteor suffers during its atmospheric flight (change in mass, surface, ablation, etc.):

$$SD = \langle logK \rangle + \langle log\sigma \rangle \quad (2)$$

Despite the PE criterion was thought to be used as a unique classifying parameter, the combination with SD criterion could provide more accurate results and unambiguously characterize a fireball.

Ceplecha and McCrosky (1976) showed that (1) and (2) are theoretically related. This is important in three senses. First, (1) was initially intended to be empirical, but it has been proved that the *Single Body Theory* could explain it. Second, it is interesting to note that K is a function of the shape of the object and its density so it indirectly depends on the fireball fragmentation; therefore, one of the *Single Body Theory* weaknesses can be partially overcome including the SD criterion as well.

Finally, it should be mentioned that since SD depends on the second derivative of observed measurements, it is less affected by observational errors than PE. Consequently, both parameters are supplementary.

The ablation and geometric coefficients in (2) along with the atmospheric density at the terminal height in (1), describe the meteor atmospheric flight dynamics.

On the contrary, the classification suggested in Ceplecha (1967, 1968, 1988) relies on the beginning height and pre-atmospheric orbit. Ceplecha and McCrosky (1976) discussed a possible relationship between both classifications but the only objects that could be compared in both studies are those with Taurid-like orbits. This leads to a small sample of nine bodies available, thus no strong conclusions could be derived. They also stated that results using the ablation and geometric coefficients values recorded at the last luminous point of the meteor trajectory instead of the average values seem to provide better results.

This new way of providing fireball classification complements the existing meteorite's composition based classification, and it is an alternative to the already mentioned classification suggested by Ceplecha (1967, 1968, 1988).

Few years later, Wetherill and Revelle (1981) published a work where they also considered the possibility of distinguishing the ordinary chondrites present in the PN data. The authors suggested that a large number of ordinary chondrite falls may not have been found due to both, their small terminal masses and their small sizes. The authors also discussed that previously accepted low bulk density values affected studies of meteorite falls. The authors claimed that, based on previous work from Revelle and Rajan (1979) and Revelle (1980), masses derived photometrically could be ten times higher than real masses, whereas dynamic masses could be two times smaller (just as a short review, the photometric mass is that derived from the light curve of the meteor, we assume the energy emitted represents up to a certain point the amount of kinetic energy of the body; as for the dynamic mass, it is the mass obtained from the deceleration equations of motion). Differences also were spotted regarding the cross-sectional area (about twice higher than for assumed spherical shape), which is related to any possible fragmentation occurred during the atmospheric flight (fragments flying very close to the main fragment increase the effective cross sectional area). Owing to this, chances that small ordinary chondrites could produce meteorite falls increased. However, not every ordinary chondrite meteor may survive its atmospheric flight, as it also depends on the initial mass and entrance geometry. Besides, meteorites with different compositions were present in the PN data set, so we cannot exclude that the terminal heights of these meteorites could be similar to those of ordinary chondrites. However, Wetherill and Revelle (1981) assumed that chances of this non ordinary chondrites presenting such terminal heights (within the PN data) were less than 16 %.

In order to filter the dominant ordinary chondrites from other types, Wetherill and Revelle (1981) stated that any other ordinary chondrite present in the PN should show, scaled up to certain point, the same behavior as the Lost City meteorite. They expressed it mathematically through four criteria. A dedicated review of their work is recommended to any interested reader. We only remark here one of these criteria. Their third criteria stated:

> End height agrees with the Single-Body Theory theoretical value, calculated using dynamic mass, as well as with that of Lost City to within ± 1.5 km, when scaled for mass, velocity, and entry angle in accordance with classical meteor theory.

Once again, the terminal height appears as fundamental parameter of the analysis. Wetherill and Revelle (1981) considered the agreement between the observed and theoretical terminal heights as an indicator of a good adjusted meteorite atmospheric trajectory, this is, they showed similar deceleration, drag coefficient, ablation, etc. A normalization of this terminal height allowed it to be compared to the Lost City corresponding value, assuming a deviation of ± 1.5 km due to the errors in the calculation of the dynamic mass. As we have mentioned, they considered the dynamic mass to be more accurate than the photometric mass used in Ceplecha and McCrosky (1976). Anyway, they were aware of the fact that the derivation of both masses is affected by different errors and assumptions. All in all, as they stated, despite of the different methodologies applied, the final amount of ordinary chondrite fireballs identified within the PN is similar to the previous work of Ceplecha and McCrosky.

3 New way of Calculating the Terminal Height

We have reviewed through a couple of very relevant bibliographic works the importance of the fireball terminal height to understand its atmospheric flight and to derive further properties. Up to now, the efforts to determine this value analytically relied on the classical theory. However, the large number of unknowns involved in the classic theory affects the accuracy of the results. Generally, the use of mean values for these variables is accepted. Nonetheless, there are cases where these values are far from realistic. The introduction of scaling laws and dimensionless variables helps to overcome these inaccuracies. The next lines will show how this new modelling can be applied to the analytical determination of the terminal height. Particularly we are going to focus on the simplifications of the exact solution achieved using this methodology.

Let us consider the dimensionless approach to describe the atmospheric trajectory developed by Stulov et al. (1995), Stulov (1997) and Gritsevich (2007). Since this methodology has been explained in Gritsevich et al. (2016, 2017) we will not delve into every detail but explain its basics. The dynamical behavior of a meteoroid that enters the Earth's atmosphere can be described using the Newton equations of motion. We are mainly interested in the variation with time of the velocity (deceleration) along its path, the height and the mass variation. If we project the meteoroid movement along its trajectory and consider the mass variation equation, we can easily derive the following expressions:

$$M\frac{dV}{dt} = -\frac{1}{2}c_d\rho_a V^2 S \tag{3}$$

$$\frac{dh}{dt} = -V\cdot \sin\gamma \tag{4}$$

$$H^* \frac{dM}{dt} = -\frac{1}{2} c_h \rho_a V^3 S \tag{5}$$

Were M is the mass, V the velocity, γ the slope between the trajectory and the horizon at each time, t is the time, h the height above the Earth's surface, S is the area of the middle section of the body, ρ_a is the density of the atmosphere, c_d is the drag coefficient, c_h is the heat exchange coefficient and H^* is the effective destruction enthalpy. Provided the high entry velocities, the effect of the drag is much higher than the gravity acceleration and this is usually not considered, that explains why we have not included it in (3).

Due to the large number of variables extra equations are required. Normally we accept the atmosphere as isothermal, which leads to an exponential equation for the atmospheric density $\rho/\rho_0 = exp(-h/h_0)$, where ρ_0 is the atmospheric density at sea level and $h_0 = 7.16 \times 10^3$ m is the scale height (note that Lyytinen and Gritsevich (2016) describe how to use more elaborate atmospheric models on the case-by-case basis). In addition to this, Levin (1956, 1961) suggested that the variation of the middle section and the mass of the body are related owing to $S/S_e = (M/M_e)^{\mu}$, where μ is a constant that indicates the spin velocity of the body during the flight (see also Gritsevich and Koschny 2011, and Bouquet et al. 2014) , and the e subscript refers to the values of the variables when the body enters the atmosphere.

In order to study the variation of M and V with height, we combine (3)–(5) and the extra expressions. However, we introduce dimensionless variables ($M = M_e m$, $V = V_e v$, $h = h_0 y$, $S = S_e s$ and $\rho_a = \rho_0 \rho$) and solve the resulting equations with the conditions $y = \infty$ and $v = 1$ (for details see Gritsevich et al. 2017):

$$m = \exp\left[-\left(1 - v^2\right)\beta / \left(1 - \mu\right)\right] \tag{6}$$

$$y = ln2\alpha + \beta - ln\Delta, \quad \Delta = \overline{Ei}(\beta) - \overline{Ei}\left(\beta v^2\right) \tag{7}$$

Where $\overline{Ei}(x) = \int_{-\infty}^{x} \frac{e^t}{t} dt$.

As a consequence of including dimensionless variables two new parameters appear in (6) and (7). The parameter α is a ballistic coefficient which characterizes the drag intensity (8); and β is called the mass loss parameter which characterizes the ablation of the meteor body (9), it is proportional to the fraction of kinetic energy of the unit mass of the body that is transferred to the body in the form of heat divided by the effective destruction enthalpy.

$$\alpha = \frac{1}{2} c_d \frac{\rho_0 h_0 S_e}{M_e sin\gamma} \tag{8}$$

$$\beta = (1 - \mu) \frac{c_h V_e^2}{2 c_d H^*} \tag{9}$$

The exact solution of the problem (6), (7) admits some simplifications. For fast meteors that show little deceleration during the luminous path, we can approximate $v = V/V_e = 1$. In these cases, $\beta \gg 1$ given the high evaporation process that takes place. Stulov et al. (1995) suggested an alternative asymptotic solution of the system (6) and (7) for these cases:

$$v = 1, \quad m^{1-\mu} = 1 - 2\alpha\beta e^{-y}, \quad ln2\alpha\beta < y < \infty \tag{10}$$

However, (10) does not consider the final deceleration in the vicinity of $m = 0$. This would provide unrealistic results in some cases and it does not account for the drag process until that point. An appropriate way of solving this disadvantage consists of combining (10) with (6) which is suitable for arbitrary β values (Gritsevich 2008b):

$$v = \left(\frac{ln\,(1 - 2\alpha\beta e^{-y})}{\beta} + 1\right)^{1/2}, \quad ln2\alpha\beta < y < \infty \tag{11}$$

These two equations (10) and (11) provide the first simplified solutions of the exact analytical solution expressed in (6), (7). Analytically, the terminal height is the last point of the registered luminous path of the meteor atmospheric trajectory (when ablation processes are over). This point is reached when $m = 0$ in (10) (given that v remains constant) and it is calculated via $v_t = V_t/V_e$ for (11). Note that at this point, that the terminal velocity of a fireball is the velocity at its terminal height, and this velocity is represented as V_t.

For the sake of clarity in the following discussion we would use the dimensional height values for the mentioned points. We use the subscripts I, II, III, etc., to indicate the different ways of expressing the terminal height according to the simplifications made in (10) and (11). The resulting terminal height for fast meteors (10) will be called hereafter h_I, and for the simplified solution where some deceleration is considered (11) we will use subscript II, h_{II}:

$$h_I = h_0 y_t = h_0 \cdot ln2\alpha\beta \tag{12}$$

$$h_{II} = h_0 y_t = h_0 \cdot ln\frac{2\alpha\beta}{\left(1 - e^{\beta(v_t^2 - 1)}\right)} \tag{13}$$

We shall remark here that, as explained in Gritsevich (2007), for small β values ($\beta < 2$), we recommend the use of the asymptotic expression suggested by Kulakov and Stulov (1992) and Stulov et al. (1995), which provides very good results.

Given (12) and (13), we can point out that, for fast meteors, where deceleration is not accounted for, the terminal height (h_I) is a function of the dimensionless parameters α and β (Gritsevich and Popelenskaya 2008). As we have stated this simplification is not always true and may be only applied in some well-studied cases. Fireballs do decelerate before disintegrating or starting its dark flight when they are meteorite-droppers, therefore a second approximation is suggested for terminal heights (h_{II}). This new terminal height depends on α, β and the terminal velocity.

The terminal velocity is obtained from observations. Sometimes the final part of the trajectory could not be visible or even recorded, but the derivation of α and β only needs three observed (h, v) points, one of which should be the entry point (the entry velocity). Using these three points it is possible to obtain the remaining (h,v) trajectory points from the adjusted fireball trajectory (see Whipple and Jacchia 1957). Though the terminal point of the trajectory would not be exactly determined using this adjustment, a combination of this adjustment with other methodologies and/or hypotheses shall lead to a good estimation. Consequently, the dimensionless methodology allows us to calculate the terminal height depending only on two (α and β) parameters for h_I, or three (α, β and V_t) parameters for h_{II}.

Regarding the entry velocity, which is required to obtain α and β, and to scale velocity values, in principle it is possible to consider it as another unknown and derive it along with α and β values as discussed by Gritsevich (2009). We foresee a future study on this subject in order to improve the methodology.

As for the derivation of α and β parameters, it is done via a least-squares method applied to the observed height and velocity values using (7). As mentioned, it requires at least three points of the trajectory, including the entry velocity. A detailed explanation of this derivation can be found in Gritsevich (2007) and Gritsevich (2008a, b). Although α and β derivation only requires three observed (h,v) values, since parameters α and β mainly describe the meteor deceleration, and this is remarkably present in the last part of the luminous trajectory of the meteor flight, it is highly recommendable to include (h,v) values of the part where the main deceleration is present. This will generally decrease the error in the results which, otherwise, may differ from the real behavior of the meteor; it is particularly convenient for meteors that penetrate deeper into the atmosphere due to the great amount of deceleration that they suffer.

In Moreno-Ibáñez et al. (2015) the accuracy of (12) and (13) was tested against the observed values of 143 meteoroids recorded by the MORP during atmospheric flight. It is important to recall that one of these bodies was recovered as a meteorite, thus proving the validity of the methodology for meteorites as well. The standard deviation of the results decreased from h_I (standard deviation is 4.11 km) to h_{II} (standard deviation is 1.52 km). Nonetheless, the results obtained with h_{II} showed a lack of agreement between observed and calculated terminal heights at low heights, which, on average, are related to low β values. This was assumed to be related to the combination of (6) and (10), which used simplified functions of the general solutions and are thought for high β values.

In order to get a better performance of h_{II} compared to the analytical solution of the problem, Moreno-Ibáñez et al. (2015) made use of the mathematical analysis carried out in Gritsevich et al. (2016). This analysis sought for the possibility of including an approximation function which slightly modified this equation. Both, the analytical solution (7) and the simplified calculated height for decelerated meteors (11) have no singularities and are monotonous on the interval $0<v<1$. Besides, the dependency on α is the same (through $ln\alpha$) for both equations. Thus, the use of an approximation function is possible and it shall only affect parameter β. It shall be remarked that this approximation function is thought to improve accuracy in those cases where $\beta > 3$, otherwise the previous simplified solutions and the asymptotic solution are, in principle, more reliable.

By means of these approximation functions we try to adjust the mathematically derived results just in the range of meteor velocities values that we are usually interested in. This is $v \in [0.3, 1)$. These functions are meant to work for fixed α and β (although α is not strictly required in this analysis) values but the error analysis carried out by Gritsevich et al. (2016) prove that they work more efficiently for determined β values, depending on the approximated function used. The approximation function suggested by Gritsevich et al. (2016) is introduced through the β parameter where parameter β is substituted by β-A in (13); A represents the approximation function. We seek that expression (11) approaches, within the range of values mentioned, the analytical solution (7):

$$ln\left(\frac{2\alpha\,(\beta - A)}{1 - e^{(\beta - a)\left(v^2 - 1\right)}}\right) \approx ln\alpha + \beta - ln\frac{\Delta}{2} \tag{14}$$

We will not describe the whole process here, but it is worth mention that function A is a function of both β and *v*, $A = A(\beta, v)$, and its final shape is quite complex. The direct use of $A(\beta,v)$ in (11) would stand up against the search of simplification we are looking for. Nonetheless, after analyzing different efficient simplifications for function $A(\beta,v)$ and their attached errors, two reliable possibilities came out (Gritsevich et al. 2016):

- $A_0 = 1.1$
- $A_1 = 1.0 + (1.0 - v) \cdot (2.5)/\beta$

We shall recall here, that we are approximating mathematically (11) to the exact solution (7). Hence, the new expression of the terminal heights that we will introduce are still simplified solutions of (7) and their results should be considered in terms of fast meteors where deceleration has been accounted for.

The error analysis performed when using A_0 and A_1 in (14) suggests that the optimal performance of these approximations (given 20–30 meteor trajectory points observed) occurs at $\beta \approx 2.89$ for A_0, and at $\beta \approx 2.1$ for A_1. The average statistical deviation for any derived parameters using A_0 is 5–10 % and 1–2 % for A_1 (Gritsevich et al. 2016).

Then, if we apply this function approximation to the derivation of the terminal height (13), our accuracy should increase. We first start using $A = A_0$, we decided to call this new terminal height h_{III}:

$$h_{III} = h_0 \cdot ln\left(\frac{2\alpha\,(\beta - 1.1)}{1 - e^{(\beta - 1.1)\left(v_t^2 - 1\right)}}\right) \tag{15}$$

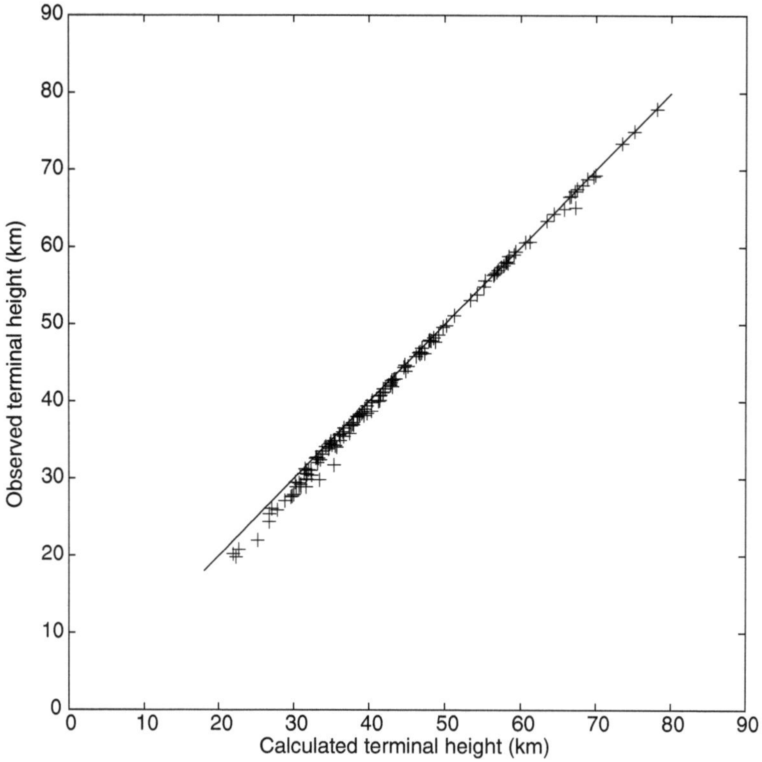

Fig. 2 Observerd terminal height values (MORP) vs. calculated values (h_{III}). The line $h_{observed} = h_{calculated}$ is also plotted (from Moreno-Ibáñez et al. 2015)

Its validity has been tested by means of the fireball data gathered by the MORP network (Moreno-Ibáñez et al. 2015). The graphical representation of these results is plotted in Fig. 2.

The resulting accuracy is quite good. The standard deviation is reduced down to a value of 0.75 km. It can be pointed out that most of the error come from the lowest height values, which are again mainly associated with small β values. Since the approximation A_0 is supposed to show a better performance for higher values of the mass-loss parameter, these differences were expected to appear. Note that h_{III} is the result of a simplification made on the analytic solution of the equations of motion. Thus, despite including the mathematical modification suggested by Gritsevich et al. (2016) we may still appreciate a residual error due to the original simplification assumed. All in all, the adjustment proved to be good and promising.

Alternatively, we present here the analysis of the terminal height (13) for the MORP database using $A = A_1$. Let's call the new expression for the terminal height h_{IV}:

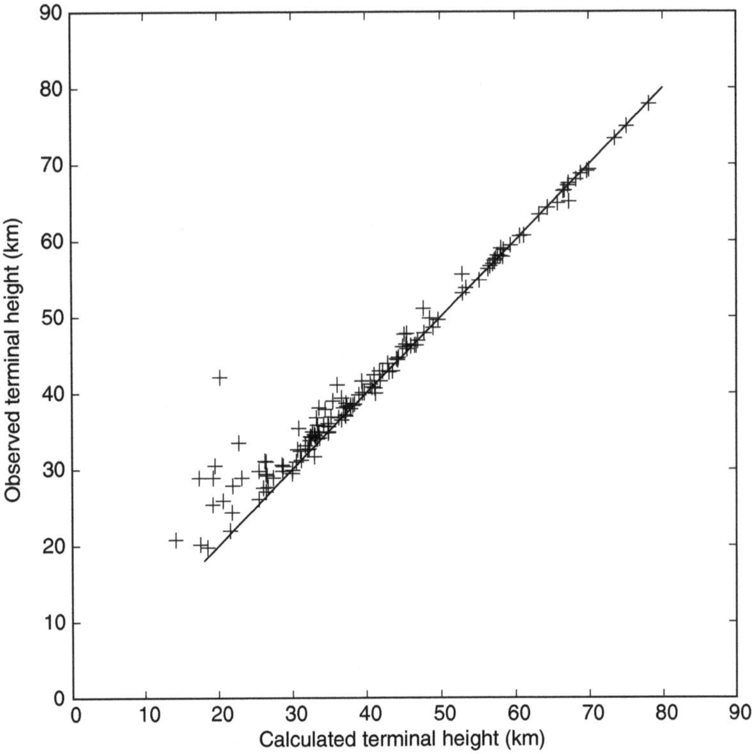

Fig. 3 Observerd terminal height values (MORP) vs. calculated values (h_{IV}). The line $h_{observed} = h_{calculated}$ is also plotted

$$h_{IV} = h_0 \cdot ln\left(\frac{2\alpha\left(\beta - (1.0 + (1.0 - v_t))\,\frac{2.5}{\beta}\right)}{1 - e^{\left(\beta - \left(1.0 + (1.0 - v_t)\frac{2.5}{\beta}\right)\right)(v_t^2 - 1)}}\right) \tag{16}$$

Results of this new analysis are plotted in Fig. 3.

This case should be analyzed in a more careful way. But, as we can observe in Fig. 3, the correlation between the observed and calculated terminal heights is broken again for low height values (low β values). Besides, now the differences between the calculated and the observed terminal heights are negative. The explanation can be found in β values. As we have discussed, these approximations were thought to work efficiently for $\beta > 3$, and we should only consider its accuracy for that range of results. Graphically, Figs. 4 and 5 show the relationship between the mass-loss parameter and the h_{III} and h_{IV} respectively. The sudden change in accuracy is quite clear at the right side of the dashed line indicating $\beta = 3$. Lower β values show different levels of terminal height accuracy.

Additionally, it is quite interesting to note that, for the MORP database, the use of h_{III} lead to better global results (including those fireballs with $\beta < 3$) than h_{IV};

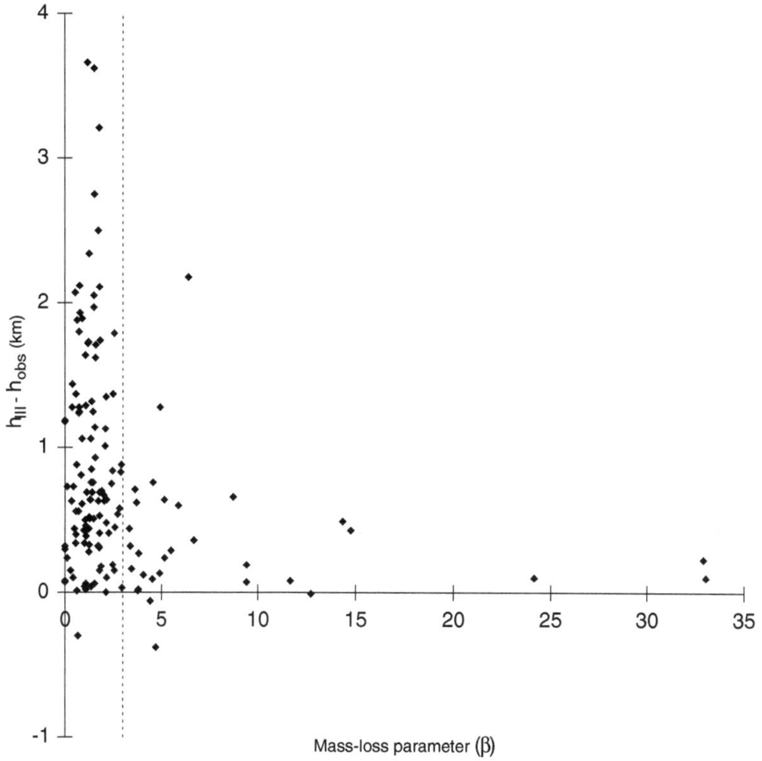

Fig. 4 Mass-loss parameter (β) against $h_{III}-h_{obs}$. The *dashed line* indicates $\beta = 3$

conversely, h_{IV} shows better adjustment if consider only meteors with $\beta > 3$. The global results for h_{IV} are biased by five cases at very low β values; hence, avoiding the contribution of these events to the global accuracy the global accuracy enhances dramatically.

To summarize, the approximation functions proposed in Gritsevich et al. (2016) are apparently capable to improve the general behavior of the dimensionless methodology when we consider the simplifications to the analytic solution described (10) and (11). In particular, the main objective is to solve the problems with the accuracy derived for high β values. We have used them here for the specific case of the terminal height and focusing on the problems which arise at low β values of h_I and h_{II}. In these last cases, further study should be addressed. However, we have proved that the use of A_0 enhances the global accuracy of (11). This is explained by the improved accuracy at moderated β values, which may include some of the meteors that are able to penetrate to lower heights with such moderate β values.

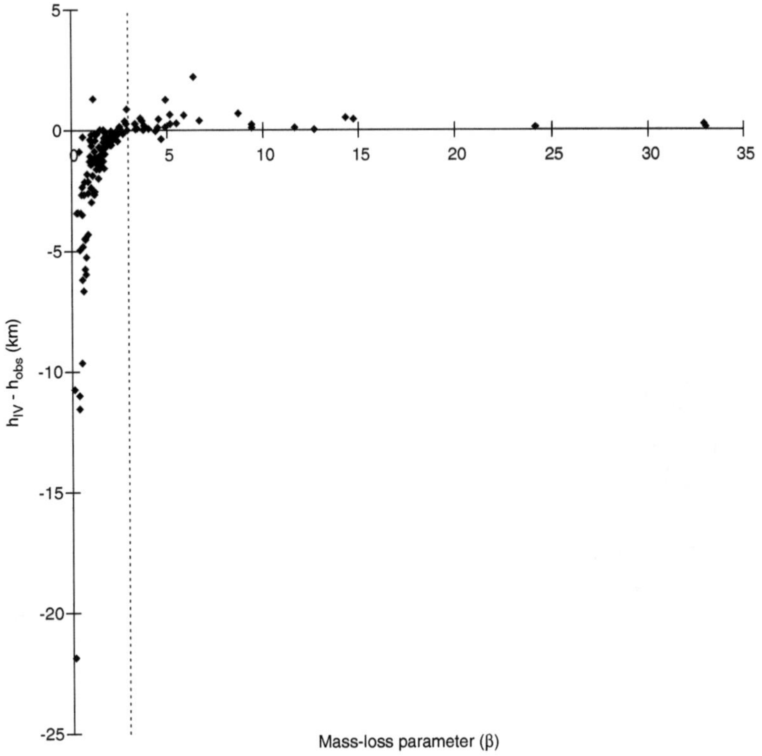

Fig. 5 Mass-loss parameter (β) against $h_{IV}-h_{obs}$. The *dashed line* indicates $\beta = 3$

4 Conclusions

Along this chapter we have presented the utility of the terminal height for meteor science and the mathematical adjustment provided by the values derived from the dimensionless methodology. The results shown in this chapter are summarized in following discussion:

1. The dimensionless terminal height expressions presented in this chapter provide useful tools to tackle previously analyzed problems. The methodology discussed in Ceplecha and McCrosky (1976) in order to distinguish between ordinary and carbonaceous chondrite fireballs recordings, could be alternatively approached by using dimensional analysis. Furthermore, the terminal heights introduced here largely resemble the PE criterion suggested by Ceplecha and McCrosky (1976). The mathematical definition of α depends on the ratio of preatmospheric cross-section to preatmospheric mass (a ratio easily convertible to bulk density, preatmospheric mass and shape coefficient, all of these parameters are used in the PE criterion), and on the trajectory slope γ related to Z_R of the PE criterion. Respectively, the mass-loss parameter is proportional to preatmospheric velocity

with a power of two and inversely proportional to the effective destruction enthalpy. Thus, the degree of penetration of fireballs into the Earth's atmosphere is correctly described with the definition of terminal heights discussed in this chapter.

2. The dimensionless methodology is able to describe in a simple way the physical event. For example, the ablation coefficient is easily derived from the mass-loss parameter, $\sigma = 2\beta/(1-\mu)V_e^2$. Note that generally $\mu = 0$ or 2/3 (see Bouquet et al. 2014), this constraints the value of μ when deriving the ablation parameter and, hence, the derivation is quite straightforward. Given the difficulties of deriving the exact values for some physical properties (i.e. bulk density, shape, etc.) from the observation, the reduction of unknowns achieved with this methodology could be used as a powerful tool to pursue a classification based on α and β parameters (see previous chapter by Gritsevich et al. 2017). In some cases it could be quite convenient to use h_{III} or h_{IV} and β to characterize different events instead of using a combination of α and β. Particularly, members of meteor showers (generally carbonaceous chondrites) can be classified using these two parameters, given the excellent behavior of h_{III} and h_{IV} for high β values.
3. The discrepancies between observed and calculated terminal heights found at low β values have to be studied in more detail. Typically, meteorite-droppers have low β values, which mean low ablation and thus, higher chances of survival. In other words, tough bodies (such as ordinary chondrites) may be affected by this error, and any further study should be aware of it. We already mentioned that these discrepancies could be due to simplifications arising from the analytical solution. Though the mathematical modification introduced by means of the approximation function A_0 is able to correct the global accuracy of the results, the local deviation at low β is still present. This is also of particular relevance for any planetary defense study. At low β values the suggested calculated terminal heights have lower values than observed values. This would mean that any prediction about the atmospheric penetration of fireballs based on h_{II} or h_{III} would indicate higher values that the observed ones. On the contrary, the values suggested by h_{IV} would be lower than the real recordings. Anyway, given that observations also involve various errors (atmospheric conditions, whole trajectory recording, resolution of the camera, etc.), this subject should be studied in more detail.
4. For significantly decelerated bolides and a few well-studied cases, such as the Innisfree meteorite, published terminal heights may differ depending on the data-reduction approach used (e.g. 21 km in Halliday et al. (1981); 19.8 km in Halliday et al. (1996)). This not only affects the accuracy between calculated and observed terminal heights for any particular fireball, but also the global accuracy of the methodology described here. Nonetheless, this could be taken as an opportunity. The dimensionless methodology could set constraints on terminal heights and fireball flight duration values, which may help to put adequate restrictions on the recorded values.
5. It is worth noticing that a good estimation of terminal heights opens new fields of studies. First, it is possible to forecast terminal heights when the last part of the fireball trajectory has not been recorded, which happens quite often. The more

number of recorded points, the better the accuracy (α and β do strongly depend on the deceleration, and this is better described with an increasing number of (h, v) trajectory observations), but it is still possible to obtain α and β with only three recorded points. Depending on each event this may have little influence on their derivation. This is quite advantageous if we consider a fast meteoroid because, no more parameters are required (see (6)). On the contrary, for decelerated bodies with high β values, the missing (h,v) at the end of the trajectory could be adjusted based on the rest of the trajectory data (Whipple and Jacchia 1957). Thus, as discussed in this chapter, V_t (the terminal velocity) could be obtained (provided some assumptions or extra data from other observational techniques) for most of the registered fireballs and h_{III} or h_{IV} could be derived.
   Secondly, it is also notable, that meteor height may be expressed as a function of time. Thus, the ability of predicting terminal heights may be directly linked with the forecast of a total duration of meteor phase. This leads to a new class of problems, such as, for example, insights into determination of luminous efficiency based on meteor duration and calculation of critical kinetic energy needed to produce luminosity.
6. For the MORP data studied here, the use of the new implemented h_{III} provides more accurate global results than previous h_I and h_{II} terminal heights. Conversely, the use of h_{IV} shows some unexpected discrepancy that can be explained due to five cases with low β values. However, in this case, the adjustment using the approximated function A_1 is more precise for values of $\beta > 3$ (see Fig. 5). This is in agreement with the results discussed in Gritsevich et al. (2016). Mathematically, these authors concluded that close to $\beta \approx 2.89$ for A_0, and at $\beta \approx 2.1$ for A_1 (provided $v \in [0.3, 1)$) the difference between (13) and the resulting expression using approximated functions is optimized. This statement has been tested with a large amount of real cases in this chapter, supporting the analytical study.
7. Direct comparison between h_{III} and h_{IV} could be used for other purposes. The results presented here correspond only to one FN. It is still difficult to conclude whether h_{III} or h_{IV} would provide better general results for other FN data. It could be interesting to find out whether h_{III} is able to absorb better the widespread in results for different β values. According to MORP results, h_{III} achieves a better global accuracy. This is in part due to the five cases that bias the global accuracy achieved with h_{IV}. Nonetheless, it seems that terminal heights of fireballs showing moderate β values are more accurately determined using h_{III}. Resolving whether better global results are obtained either with h_{III} or h_{IV} might be quite useful in two senses, to detect and avoid systematic errors in database recordings, and to derive fast accurate results for large sets of data.

Globally, the terminal heights studied in this chapter have proved the dimensionless methodology to adequately describe the atmospheric flight of fireballs using three variables (α, β and V_e). Thus, it could be very interesting to use it with other FN. We foresee its application to the Finnish Fireball Network and the Spanish Meteor Network.

Acknowledgments We thank Eleanor Sansom for her valuable comments which helped to improve the content of this chapter. Dr. Trigo-Rodríguez and MMI acknowledge support from the Spanish Ministry of Sciences, research project AYA2015-67175-P. MG and MMI were supported, in part, by the Academy of Finland project No 260027. MG was also supported by the Magnus Ehrnrooth Foundation travel grant, and by the Russian Foundation for Basic Research, project N°s 14-08-00204, 16-05-00004 and 16-07-01072. This study was done in the frame of a PhD. on Physics at the Autonomous University of Barcelona (UAB).

References

Borovička, J., Spurný, P., Brown, P., Wiegert, P., Kalenda, P., Clark, D., Shrbený, L.: The trajectory, structure and origin of the Chelyabinsk asteroidal impactor. Nature **503**, 235–237 (2013)

Bouquet, A., Baratoux, D., Vaubillon, J., Gritsevich, M.I., Mimoun, D., Mousis, O., Bouley, S.: Simulation of the capabilities of an orbiter for monitoring the entry of interplanetary matter into the terrestrial atmosphere. Planet. Space Sci. **103**, 238–249 (2014)

Brown, P., Assink, J.D., Astiz, A., Blaauw, R., Boslough, M.B., Borovička, J., Brachet, N., Brown, D., Campbell-Brown, M., Ceranna, L., Cooke, W., de Groot-Hedlin, C., Drob, D.P., Edwards, W., Evers, L.G., Garces, M., Gill, J., Hedlin, M., Kingery, A., Laske, G., Le Pichon, A., Mialle, P., Moser, D.E., Saffer, A., Silber, E., Smets, P., Spalding, R.E., Spurný, P., Tagliaferri, E., Uren, D., Weryk, R.J., Whitaker, R., Krzeminski, Z.: A 500-kiloton airburst over Chelyabinsk and an enhanced hazard from small impactors. Nature **503**, 238–241 (2013)

Ceplecha, Z.: Geometric, dynamic, orbital and photometric data on meteoroids from photographic fireball networks. Bull. Astron. Inst. Czech. **38**, 222–234 (1987a)

Ceplecha, Z.: Classification of meteor orbits. Smithsonian Contr. Astrophys. **11**, 35–60 (1967)

Ceplecha, Z.: Discrete levels of meteor beginning heights. Smithsonian Astrophys. Obs. Spec. Rep. **279**, 1–54 (1968)

Ceplecha, Z.: Earth's influx of different populations of sporadic meteoroids from photographic and television data. Bull. Astron. Inst. Czech. **39**, 221–236 (1988)

Ceplecha, Z., McCrosky, R.E.: Fireball end heights—a diagnostic for the structure of meteoric material. J. Geophys. Res. **81**, 6257–6275 (1976)

Ceplecha, Z., ReVelle, D.O.: Fragmentation model of meteoroid motion, mass loss, and radiation in the atmosphere. Meteorit. Planet. Sci. **40**, 35 (2005)

Ceplecha, Z., Borovička, J., Elford, W.G., Revelle, D.O., Hawkes, R.L., Porubčan, V., Šimek, M.: Meteor phenomena and bodies. Space Sci. Rev. **84**, 327–471 (1998)

Chyba, F.C., Paul, J.T., Zahnle, K.J.: The 1908 Tunguska explosion: atmospheric disruption of a stony asteroid. Nature **361**, 40–44 (1993)

Collins, G.S., Melosh, H.J., Marcus, R.A.: Earth impact effects program: a web-based computer program for calculating the regional environmental consequences of a meteoroid impact on Earth. Meteorit. Planet. Sci. **40**, 817–840 (2005)

Gritsevich M., Koschny D.: Constraining the luminous efficiency of meteors Icarus, **212**(2), 877–884, (2011). http://dx.doi.org/10.1016/j.icarus.2011.01.033

Gritsevich, M., Dmitriev, V., Vinnikov, V., Kuznetsova, D., Lupovka, V., Peltoniemi, J., Mönkölä, S., Brower, J., Pupyrev, Y.: Constraining the pre-atmospheric parameters of large meteoroids: Košice, a case study. In: Trigo-Rodríguez, J.M., Gritsevich, M., Palme, H. (eds.) Assessment and mitigation of asteroid impact hazards, pp. 153–183. Springer, New York (2017). doi:10.1007/978-3-319-46179-3_8

Gritsevich, M.I.: Approximation of the observed motion of bolides by the analytical solution of equations of meteor physics. Solar Syst. Res. **41**(6), 509–516 (2007)

Gritsevich, M.I., Popelenskaya, N.V.: Meteor and fireball trajectories for high values of the mass loss parameter. Doklady Phys **53**(2), 88–92 (2008). http://dx.doi.org/10.1134/S1028335808020092

Gritsevich, M.I.: Identification of fireball dynamics parameters. Mosc. Univ. Mech. Bull. **63**(1), 1–5 (2008a)

Gritsevich, M.I.: The Pribram, Lost City, Innisfree, and Neuschwanstein falls: an analysis of the atmospheric trajectories. Solar Sys. Res. **42**, 372–390 (2008b)

Gritsevich, M.I.: Determination of parameters of meteor bodies based on flight observational data. Adv. Space Res. **44**, 323–334 (2009)

Gritsevich, M.I., Stulov, V.P., Turchak, L.I.: Consequences of natural cosmic bodies with the Earth's atmosphere and surface. Cosm. Res. **50**(1), 56–64 (2012)

Gritsevich, M.I., Lukashenko, V.T., Turchak, L.I.: Approximating the solution of meteor physics equations through the use of elementary functions. Math. Models Comput. Simul. **8**(1), 1–6 (2016) doi:10.1134/S2070048216010026

Halliday, I., Blackwell, A.T., Griffin, A.A.: Photographic observation and orbit of the Innisfree meteorite. Meteoritics **12**(3), 248–249 (1977)

Halliday, I., Blackwell, A.T., Griffin, A.A.: The Innisfree meteorite and the Canadian camera network. J. R. Astron. Soc. Can. **12**(1), 15–39 (1978)

Halliday, I., Griffin, A.A., Blackwell, A.T.: The Innisfree meteorite fall—a photographic analysis of fragmentation, dynamics and luminosity. Meteoritics **16**, 153–170 (1981)

Halliday, I., Griffin, A.A., Blackwell, A.T.: Detailed data for 259 fireballs from the Canadian camera network and inferences concerning the influx of large meteoroids. Meteorit. Planet. Sci. **31**, 185–217 (1996)

Hoppe, J.: Die physikalischen Vorgänge beim Eindringen meteoritischer Körpe in die Erdatmosphäre. Astron. Nachr. **262**, 169–198 (1937)

Jacchia, L.G., Whipple, F.L.: The harvard photographic meteor programme. Vistas Astron. **2**, 982–994 (1956)

Kulakov, A.L., Stulov, V.P.: Determination of meteor body parameters from observational data. Astron. Vestn. **26**(5), 67–75 (Issledovaniia Solnechnoi Sistemy (Sol. Syst. Res. (Engl. Transl.) **26**(5), 478–484)) (1992).

Levin, B.I.: Fizicheskaia teoriia meteorov i meteorne veshchestvo v Solnechnoi sisteme (Physical theory of meteors and meteorite substance in the solar system). Akad. Nauk SSSR, Moscow (in Russian, 1956)

Levin, B.I.: Physikalische Theorie der Meteore und die meteoritische Substanz im Sonnensystem. Akademie-Verlag, Berlin (1961)

Lyytinen, E., Gritsevich, M.: Implications of the atmospheric density profile in the processing of fireball observations. Planet. Space Sci. **120**, 35–42 (2016)

Moreno-Ibáñez, M., Gritsevich, M., Trigo-Rodriguez, J.M.: New methodology to determine the terminal height of a fireball. Icarus **250**, 544–552 (2015)

McCrosky, R.E., Posen, A., Schwartz, G., Shao, C.Y.: Lost City Meteorite—Its Recovery and a Comparison with Other Fireballs. SAO Spec. Rep. **336**, 41 (1971)

Popova, O.P., Jenniskens, P., Emel'yanenko, V., Kastashova, S., Biryukov, E., Khaibrakhamanov, S., Shuvalov, V., Rybnov, Y., Dudorov, A., Grokhovsky, V.I.: Chelyabinsk airburst, damage assessment, meteorite recovery, and characterization. Science **342**, 1069–1073 (2013)

Rendtel, J., Arlt, R., Mc Beath, A.: Handbook for visual meteor observers. International Meteor Organization, Potsdam (1995)

Revelle, D.O.: A predictive macroscopic integral radiation efficiency model. J. Geophys. Res. **85**, 1803–1808 (1980)

Revelle, D.O.: NEO fireball diversity: energetics-based entry modeling and analysis techniques. Proceedings IAU Symposium, vol. 236, pp. 95–106 (2007)

Revelle, D.O., Rajan, R.S.: On the luminous efficiency of meteoritic fireballs. J. Geophys. Res. **84**, 6255–6262 (1979)

Stulov, V.P.: Interactions of space bodies with the atmospheres of planets. Appl. Mech. Rev. **50**, 671–688 (1997)

Stulov, V.P., Mirskii, V.N., Vilsyi, A.I.: Aerodinamika bolidov (Aerodynamics of Bolides). Nauka, Moscow (1995) (in Russian)

Tancredi, G., Ishitsuka, J., Schultz, P.H., Harris, R.S., Brown, P., Revelle, D.O., Antier, K., Le Pichon, A., Rosales, D., Vidal, E., Varela, M.E., Sánchez, L., Benavente, S., Bojorquez, J., Cabezas, D., Dalmau, A.: A meteorite crater on Earth formed on September 15, 2007: the Carancas hypervelocity impact. Meteorit. Planet. Sci. **44**, 1967–1984 (2009)

Trigo-Rodríguez, J.M., Llorca, J.: The strength of cometary meteoroids: clues to the structure and evolution of comets. Mon. Not. R. Astron. Soc. **372**, 655–660 (2006)

Trigo-Rodríguez, J.M., Llorca, J.: Erratum: The strength of cometary meteoroids: clues to the structure and evolution of comets. Mon. Not. R. Astron. Soc. **375**, 415 (2007)

Trigo-Rodríguez, J.M., Madiedo, J.M., Williams, I.P., Castro-Tirado, A.J., Llorca, J., Vítek, S., Jelínek, M.: Observations of a very bright fireball and its likely link with comet C/1919 Q2 Metcalf. Mon. Not. R. Astron. Soc. **394**, 569–576 (2009)

Trigo-Rodríguez, J.M., Lyytinen, E., Gritsevich, M., Moreno-Ibáñez, M., Bottke, W.F., Williams, I., Lupovka, V., Dmitriev, V., Kohout, T., Grokhovsky, V.: Orbit and dynamic origin of the recently recovered Annama's H5 chondrite. Mon. Not. R. Astron. Soc. **449**, 2119–2127 (2015)

Vasilyev, N.V.: The Tunguska Meteorite problem today. Planet. Space Sci. **46**, 129–150 (1998)

Weisberg, M.K., McCoy, T.J., Krot, A.N.: Systematics and evaluation of meteorite classification. In: Lauretta, D.S., McSween, H.Y. (eds.) Meteorites and the Early Solar System II, pp. 19–52. University of Arizona Press, Tucson (2006)

Wetherill, G.W., Revelle, D.O.: Which fireballs are meteorites—a study of the prairie network photographic meteor data. Icarus **48**, 308–328 (1981)

Whipple, F.L., Jacchia, L.: Reduction methods for photographic meteor trails. Smithsonian Contrib. Astrophys. **1**, 183–206 (1957)

Constraining the Pre-atmospheric Parameters of Large Meteoroids: Košice, a Case Study

Maria Gritsevich, Vasily Dmitriev, Vladimir Vinnikov, Daria Kuznetsova, Valery Lupovka, Jouni Peltoniemi, Sanna Mönkölä, Jeffrey Brower, and Yuri Pupyrev

Abstract Out of a total around 50,000 meteorites currently known to science, the atmospheric passage was recorded instrumentally in only 25 cases with the potential to derive their atmospheric trajectories and pre-impact heliocentric orbits. Similarly, while observations of meteors generate thousands of new entries per month to existing databases, it is extremely rare they lead to meteorite recovery (http://www.meteoriteorbits.info/). These 25 exceptional cases thus deserve a thorough re-examination by different techniques—not only to ensure that we are able to match the model with the observations, but also to enable the best possible interpretation scenario and facilitate the robust extraction of key characteristics of a meteoroid based on the available data. In this study, we evaluate the dynamic mass of the

M. Gritsevich (✉)
Department of Physics, University of Helsinki, P.O. Box 64, Helsinki, 00014 Finland

Finnish Geospatial Research Institute, Geodeetinrinne 2, P.O. Box 15, Masala, 02431 Finland

Institute of Physics and Technology, Ural Federal University, Mira Street 19, Ekaterinburg, 620002 Russia

Department of Computational Physics, Dorodnicyn Computing Centre, Russian Academy of Sciences, Vavilova ul. 40, Moscow, 119333 Russia

Extraterrestrial Laboratory, State University of Geodesy and Cartography (MIIGAiK), Gorokhovskiy per. 4, Moscow, 105064, Russia
email: maria.gritsevich@helsinki.fi; gritsevich@list.ru

V. Vinnikov
Department of Computational Physics, Dorodnicyn Computing Centre, Russian Academy of Sciences, Vavilova ul. 40, Moscow, 119333 Russia
e-mail: vinnikov@ccas.ru

V. Dmitriev • V. Lupovka
Extraterrestrial Laboratory, State University of Geodesy and Cartography (MIIGAiK), Gorokhovskiy per. 4, Moscow, 105064 Russia
e-mail: v_dmitriev@miigaik.ru; v.lupovka@miigaik.ru

D. Kuznetsova
Observatoire de Midi-Pyrénées, Laboratoire d'Aérologie, Université Paul Sabatier, UMR 5560 CNRS, 14 av. Edouard Belin, Toulouse, 31400 France
e-mail: daria.kuznetsova@aero.obs-mip.fr

J.M. Trigo-Rodríguez et al. (eds.), *Assessment and Mitigation of Asteroid Impact Hazards*, Astrophysics and Space Science Proceedings 46,
DOI 10.1007/978-3-319-46179-3_8

Košice meteoroid using analysis of drag and mass-loss rate available from the observations. We estimate the dynamic pre-atmospheric meteoroid mass at 1850 kg. The pre-fragmentation size proportions of the Košice meteoroid are estimated based on the statistical distribution of the recovered meteorite fragments. The heliocentric orbit of the Košice meteoroid, derived using numerical integration of the equations of motion, is found to be in close agreement to earlier published results.

1 Introduction

On its journey around the Sun, the Earth is subjected to continuous bombardment by other bodies ranging in size from tiny dust particles to much larger asteroids and comets. Many attempts have been made to map the meteoroid environment in near-Earth space (e.g., Laurance and Brownlee 1986; Poppe et al. 2011; Brown et al. 2012), in order to estimate the extraterrestrial flux on Earth (Love and Brownlee 1993; Bland et al. 1996; Zolensky et al. 2006; Silber et al. 2009). Optical means have also been used to monitor the extraterrestrial flux in a more detailed way as the terrestrial encounters produce visible meteors or the larger fireballs (Oberst et al. 1998; Bland 2004; Trigo-Rodríguez et al. 2006, 2015; Weryk et al. 2008, 2013; Jenniskens et al. 2011; Blanch et al. 2017). There are still other observational methods being used as seen e.g. in works of Edwards et al. (2007), Fries and Fries (2010), Kero et al. (2012), Silber and Brown (2014), Silber et al. (2015), Vaubaillon et al. (2015), Räbinä et al. (2016) and being proposed for the future observations of meteors (Vinković et al. 2016).

J. Peltoniemi
Finnish Geospatial Research Institute, Geodeetinrinne 2, P.O. Box 15, Masala, 02431 Finland

Department of Physics, University of Helsinki, P.O. Box 64, Helsinki, 00014 Finland
e-mail: jouni.peltoniemi@nls.fi; jouni.peltoniemi@helsinki.fi

S. Mönkölä
Department of Mathematical Information Technology, University of Jyvaskyla, Jyvaskyla, 40014 Finland
e-mail: samonkol@jyu.fi

J. Brower
The Royal Astronomical Society of Canada, 4920 Dundas St W, Etobicoke, ON, Canada M9A 1B7
e-mail: jbrower@meteorchaser.net

Y. Pupyrev
Steklov Mathematical Institute, Russian Academy of Science, Gubkina str. 8, Moscow, 119991 Russia
e-mail: pupyrev@mi.ras.ru

Interpretation of this type of flux data yields important clues about the near-Earth object (NEO) population, helps identify and recover freshly fallen meteorites, and provides valuable insights for projects that are focused on space hazards (Shustov 2010; Perna et al. 2013; Harris et al. 2013; Emel'yanenko and Shustov 2013; Reinhardt et al. 2013). For all these applications it is crucial to evaluate the object's pre-atmospheric size and mass based on such observations in order to build a realistic model of atmospheric entry which takes into account the changes in mass along the trajectory, and to derive an impact scale that reflects the consequences of the collision with the planet and its corresponding level of threat (Collins et al. 2005; Gritsevich et al. 2012; Turchak and Gritsevich 2014; Moreno-Ibáñez et al. 2016).

In this study, we evaluate the pre-atmospheric parameters of the Košice meteorite, a recent fall, that has a well-derived trajectory and a large number of fragments that were quickly recovered after the event. The Košice fireball appeared over central-eastern Slovakia on February 28, 2010. The bolide reached an absolute magnitude of at least −18, enabling the radiometers of the European Fireball Network to record the fireball despite the cloudy and rainy weather. The details from the network's fireball images were insufficient to recreate its trajectory or predict a fall area. However, by using the data from the surveillance cameras operating in Hungary, the fall area was successfully calculated, which in turn led to the meteorites being quickly recovered (Borovička et al. 2013). The first reported fragment of the meteorite (27.2 g) was found on March 20, 2010 by Juraj Tóth. It was located northwest of the city of Košice in eastern Slovakia (Tóth et al. 2015).

In all, 218 fragments of the Košice meteorite, with a total mass of 11.285 kg, have been documented (Gritsevich et al. 2014a). Nearly 7 kg of the total are in the collections of Comenius University in Bratislava and at the Astronomical Institute of Slovak Academy of Sciences. About 1/3 of the recovered Košice fragments were thoroughly studied, including their magnetic susceptibility, bulk and grain density measurements as reported by Kohout et al. (2014). This analysis revealed that Košice meteorites are H5 ordinary chondrites that originated from a homogenous parent meteoroid.

Although the photometric pre-atmospheric mass estimate and the pre-impact orbit of the Košice meteoroid were published by Borovička et al. (2013), we find it to be a suitable candidate for our study as the results obtained by our approach can be directly compared to the independent result of calculations by other authors. Thus, in this study we describe complementary techniques, which are primarily focused on the trajectory-based analysis and the recovered meteorite fragments mass distribution. We determine the key dimensionless parameters responsible for the meteoroid's drag and mass loss rate along its visual path through the atmosphere by fitting the trajectory data obtained by Borovička et al. (2013), with the first integral of the drag and the mass-loss equations. These parameters allow to follow the changes in the dynamic mass of the main fragment as well as estimate the pre-atmospheric meteoroid mass. Additionally, we estimate the pre-fragmentation size proportions of the Košice meteoroid based on statistical analysis of the recovered meteorite fragments. Finally, we derive its heliocentric orbit using

Table 1 A summary of the employed parameters

Research milestones	Model parameters: Input	Output
Velocity determination	$t_i, L_i, n, i=\{1, \ldots, n\}$	$V_i, i=\{1, \ldots, n\}$
Trajectory fit	$t_i, h_i, V_i, n, i=\{1, \ldots, n\}$	α, β
Preatmospheric mass	$\alpha, \gamma, c_d, A_e, \rho_m$	M_e
Prefragmentation shape parameter	$m_i, N, i=\{1, \ldots, N\}$	B_0, d
Prefragmentation dimensions ratio	A_e, d	$a_x{:}a_y{:}a_z$
Terminal mass	$\beta, \mu, V_t, V_e, M_e$	M_t
Ablation coefficient	β, μ, V_e	σ
Orbit	$t_e, \overrightarrow{V_e}, \phi, h_e, \lambda$	$a, e, i, \Omega, \omega, M^0$

The input parameters include the data available from the observations (t_i, h_i, L_i, n, γ, ϕ, λ), data available based on measurements of the recovered meteorites (m_i, N, ρ_m), and three theoretically constrained parameters (μ, c_d, A_e) which are generally changing from case to case in a very limited value range. The meanings of these parameters are explained in Appendices 1 and 2 of this study

numerical integration of equations of motion. For simplicity, we summarize the input and output parameters for each subtask addressed in this study in Table 1.

2 The Inverse Problem of Matching the Trajectory

The change in mass and height for the main fragment of the Košice meteoroid can be approximated as functions of velocity using the following dependencies—cf. (Gritsevich 2008a):

$$m = \exp\left(\beta\left(1 - v^2\right) / (\mu - 1)\right) \tag{1}$$

$$y = \ln(2\alpha) + \beta - \ln\left(\overline{Ei}(\beta) - \overline{Ei}\left(\beta v^2\right)\right) \tag{2}$$

We provide detailed naming conventions and basic definitions for all variables and parameters used in this study in Appendices 1 and 2, and the description of the special function $\bar{Ei}(x)$ in Appendix 3. As a rule, we use capital letters for the dimensional variables (e.g. M for the mass in kg) and small letters for the dimensionless variables (e.g. m for the mass ranging between 0 and 1, referring to the mass M normalized by its pre-atmospheric value M_e).

Equations (1) and (2) rely on three dimensionless quantities, which are unique for the subgroup of meteor events with similar aerodynamic and friability properties. These parameters are introduced within a dimensional analysis concept (Stulov 1997). The ballistic coefficient, α, describes the drag properties of the meteoroid. It is proportional to the mass of the atmospheric column a meteoroid has to penetrate to reach the ground, to the pre-atmospheric meteoroid mass, M_e. The parameter β

corresponds to the mass loss rate and can also be linked to the ablation coefficient. The shape change coefficient, μ, may be estimated through light curve analysis (Gritsevich and Koschny 2011; Bouquet et al. 2014). It reflects the meteoroid's rotation rate and it takes into account the changes in the meteoroid's shape along its trajectory. In this section we are primarily focused on finding the (α, β) pair, that corresponds to the Košice meteoroid. This inverse problem can be solved by the least-squares method by matching Eq. (2) to the set of (y_i, v_i) values obtained at n trajectory points along the observed visual path of the fireball, as previously described by Gritsevich (2007, 2009); where this method was also applied to a number of observations. Moreno-Ibáñez et al. (2015, 2017) provide a grounded justification of the method by calculating terminal height values and comparing them to the observations. In addition, Lyytinen and Gritsevich (2016) describe how to use the method under condition of arbitrary atmosphere model.

Thus the mass loss parameter, β, is obtained numerically as a root for the following equation (i.e. using the Newton's method):

$$\sum_{i=1}^{n}\left[\left(\boldsymbol{\Delta}_i\left(\sum_{j=1}^{n}e^{-2y_j}\right)-\left(\sum_{j=1}^{n}\boldsymbol{\Delta}_j e^{-2y_j}\right)e^{-y_i}\right)\left(\boldsymbol{\Delta}_i-(\boldsymbol{\Delta}_i)'_{\beta}\right)\right]=0 \quad (3)$$

Then the value of the ballistic coefficient is uniquely obtained as:

$$\alpha = 0.5\left(\sum_{i=1}^{n}e^{-\beta-y_i}\boldsymbol{\Delta}_i\right)\Big/\sum_{i=1}^{n}e^{-2y_i} \quad (4)$$

Equation (3) has two different roots, with only one of them corresponding to the searched minimum. Thus, the necessary best-fit condition for the found pair (α, β) needs to be additionally checked as

$$\sum_{i=1}^{n}e^{-2y_i}\sum_{i=1}^{n}\left(\left((\boldsymbol{\Delta}_i)'_{\beta}-\boldsymbol{\Delta}_i\right)^2+\left(\boldsymbol{\Delta}_i-2\alpha e^{\beta-y_i}\right)\right)\left((\boldsymbol{\Delta}_i)''_{\beta}-2(\boldsymbol{\Delta}_i)'_{\beta}+\boldsymbol{\Delta}_i,\right)$$
$$>\left(\sum_{i=1}^{n}e^{-y_i}\left(\boldsymbol{\Delta}_i-(\boldsymbol{\Delta}_i)'_{\beta}\right)\right)^2 \quad (5)$$

The expressions for calculating $\Delta_i, (\Delta_i)'_{\beta}, (\Delta_i)''_{\beta}$ in Eqs. (3)–(5) are given in Appendix 2.

The actual height and length-along-trajectory values for the Košice fireball are provided by Borovička et al. (2013). The corresponding velocity values are calculated from these data as described by Whipple and Jacchia (1957) and Jacchia et al. (1967). The resulting values are summarized in Table 2. These data yield the following values for the Košice case: $\boldsymbol{\alpha} = 2.99$, $\boldsymbol{\beta} = 3.35$. For comparison, these results are shown along with the earlier derived corresponding values for the Lost City, Innisfree, Benešov and Neuschwanstein meteorite falls (Table 3).

Table 2 Details of the atmospheric trajectory for the Košice fireball

Time*, t_i (s)	Height*, h_i (km)	Length*, L_i (km)	Velocity, V_i (km/s)	Height, normalized by the scale height, y_i	Velocity, normalized by initial velocity, v_i
2.16	68.2764	0	14.97917	9.53581	0.999865
2.24	68.0288	0.2857	14.97882	9.501229	0.999841
2.32	66.5686	1.9701	14.97841	9.297291	0.999813
2.4	64.5812	4.263	14.97792	9.019721	0.999781
2.48	63.9505	4.9906	14.97734	8.931634	0.999742
2.56	63.7422	5.231	14.97667	8.902542	0.999698
2.64	62.6369	6.5063	14.97588	8.74817	0.999645
2.72	61.4596	7.8648	14.97495	8.583743	0.999583
2.8	60.4398	9.0416	14.973*86	8.441313	0.99951
2.88	59.4206	10.2178	14.97258	8.298966	0.999425
2.96	58.4696	11.3152	14.97108	8.166145	0.999324
3.04	57.3074	12.6566	14.96931	8.003827	0.999206
3.12	56.2359	13.8933	14.96724	7.854176	0.999068
3.16169	55.1798	15.1124	14.96602	7.706676	0.998986
3.2	54.9518	15.3756	14.9648	7.674832	0.998905
3.28	54.6753	15.6947	14.96194	7.636215	0.998714
3.36169	52.7791	17.8837	14.9585	7.371383	0.998485
3.36	52.2932	18.4446	14.95858	7.30352	0.99849
3.44	51.9038	18.8942	14.95463	7.249134	0.998227
3.52	51.2779	19.6169	14.95	7.161718	0.997917
3.56169	50.4611	20.5598	14.94727	7.04764	0.997735
3.6	49.6837	21.4574	14.94456	6.939064	0.997554
3.68	49.4245	21.7567	14.93816	6.902863	0.997127
3.76169	47.6025	23.8605	14.93048	6.648394	0.996614
3.96169	45.1692	26.6706	14.90538	6.308547	0.994939
4.16169	42.4232	29.8422	14.86787	5.925028	0.992435
4.4	39.9047	32.7514	14.79823	5.573282	0.987787
4.36169	39.0502	33.7386	14.81179	5.453939	0.988692
4.76169	35.2954	38.0768	14.60266	4.929525	0.974732
4.96169	32.9749	40.7583	14.41535	4.605433	0.962229

Table 2 (continued)

Time*, t_i (s)	Height*, h_i (km)	Length*, L_i (km)	Velocity, V_i (km/s)	Height, normalized by the scale height, y_i	Velocity, normalized by initial velocity, v_i
5.04	30.544	43.5677	14.31889	4.265922	0.955791
5.16169	30.4843	43.6367	14.13536	4.257584	0.94354
5.12	30.2927	43.8581	14.20335	4.230824	0.948078
5.2	29.7595	44.4745	14.06766	4.156355	0.939021
5.28	29.2252	45.092	13.90829	4.081732	0.928383
5.36169	27.817	46.7197	13.71683	3.885056	0.915603
5.36	27.3005	47.3168	13.72112	3.812919	0.915889
5.44	26.4961	48.2466	13.5013	3.700573	0.901216
5.52	25.6557	49.2182	13.24313	3.583198	0.883983
5.56169	25.4808	49.4204	13.09121	3.558771	0.873842
5.6	25.4054	49.5076	12.93992	3.54824	0.863744
5.68	23.78	51.3868	12.58382	3.321229	0.839974
5.76	22.6837	52.6543	12.1656	3.168115	0.812058
5.71304	22.6732	52.6664	12.41921	3.166648	0.828986
5.84	22.2856	53.1147	11.67442	3.112514	0.779271
5.92	21.7107	53.7794	11.09755	3.032221	0.740765
5.91304	21.3684	54.1752	11.1515	2.984413	0.744366
6	21.0391	54.556	10.42005	2.938422	0.695541
6.08	20.5059	55.1726	9.127726	2.888226	0.609279
6.11304	20.6797	54.9716	8.39702	2.863953	0.560504
6.16	19.9564	55.808	7.511363	2.787207	0.501386
6.24	19.4099	56.44	6.337642	2.71088	0.42304
6.31304	19.2934	56.5747	5.549147	2.694609	0.370407
6.32	19.1465	56.7446	5.485346	2.674092	0.366149
6.4	18.8756	57.0579	4.866452	2.636257	0.324837
6.48	18.6043	57.3716	4.450264	2.598366	0.297057
6.51304	18.0459	58.0175	4.393314	2.520377	0.293255
6.56	18.3224	57.6977	4.326966	2.558994	0.288826
6.64	18.3128	57.7088	4.244316	2.557654	0.283309
6.71304	17.4214	58.7398	4.192904	2.433156	0.279878

Time, height and length-along-trajectory values are provided according to (Borovička et al. 2013), marked with *. The corresponding velocity values as well as normalized height and velocity are derived from these data for further processing

Table 3 The derived dimensionless parameters and their comparison with the Lost City, Innisfree, Benešov and Neuschwanstein meteorite falls

Meteorite name	Country	Year	Mass found (kg)	Type	V_e (km/s)	γ°	α	β
Lost City	USA	1970	17.2[a]	H5	14.15	38	11.11[b]	1.16[b]
Innisfree	Canada	1977	4.58[c]	L5	14.55	68	8.25[b]	1.70[b]
Benešov	Czech Republic	1991	0.002[d]	LL3.5/H5	21.1	80	7.25[e]	1.75[e]
Neuschwanstein	Germany/Austria	2002	6.2[f]	EL6	20.95	49	3.92[b]	2.57[b]
Košice	Slovakia	2010	11.3[g]	H5	14.98	60	2.99	3.35

[a]McCrosky et al. (1971)
[b]Gritsevich (2008b)
[c]Halliday et al. (1981)
[d]Spurný et al. (2012)
[f]Spurný et al. (2003)
[e]Gritsevich (2008c)
[g]Gritsevich et al. (2014a)

3 Model Validation

To verify the described approach we introduce the obtained values for the ballistic coefficient α and the mass loss parameter β into Eq. (2) and plot the resulting curve (with the conversion back to dimensional height and velocity, as described in Appendix 2) along with experimental values given in Table 2.

In order to show the length-along-trajectory fit, we numerically solve the system of equations (Kuznetsova et al. 2014):

$$\begin{cases} \frac{dv}{dt} = -\frac{\alpha V_e \sin\gamma}{h_0} v^2 \exp\left(\beta\left(1 - v^2\right)\right) e^{-y}, \\ \frac{dv}{dt} = -\frac{V_e \sin\gamma}{h_0} v, \\ \frac{dL}{dt} = V_e v. \end{cases} \tag{6}$$

Here the first equation is obtained from the dimensionless equation of motion where the analytical solution for mass (Eq. (1)) and the parameters α and β are introduced. This approach allows us to avoid the use of the shape change coefficient μ, since its participation due to the combination of both equations cancels. The second equation is the general geometric relation along the visual path of the fireball (cf. Ceplecha et al. 1998). The third equation, the definition of the fireball's velocity, is taken as a time derivative from the length along the trajectory.

The slope γ between the Košice fireball trajectory and the horizon was 60° (Borovička et al. 2013). The initial conditions (v_0, y_0, L_0) used for solving Eqs. (6) are: $v_0 = V_1/V_e$, $y_0 = h_1/h_0$, $L_0 = 0$, where subscript "1" stands for the first values measured in the observations after the meteor's ignition point. The relation between height and velocity is presented in Fig. 1. The length and the velocity along trajectory as functions of time are shown in Figs. 2 and 3, respectively.

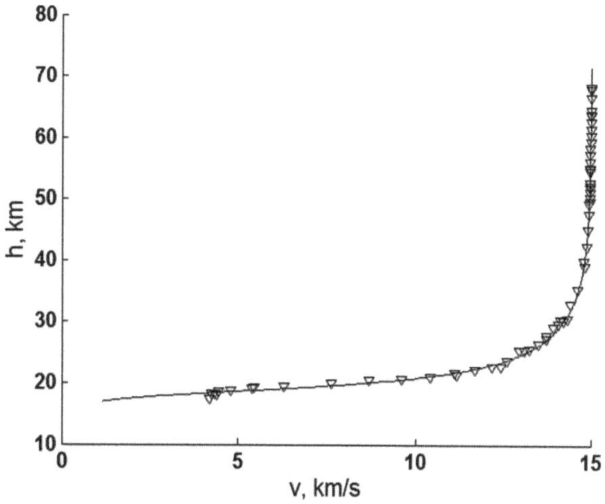

Fig. 1 Height vs velocity. *Solid line*: the analytical dependence according to the dimensional form of Eq. (2), *markers*: experimental data for the main body

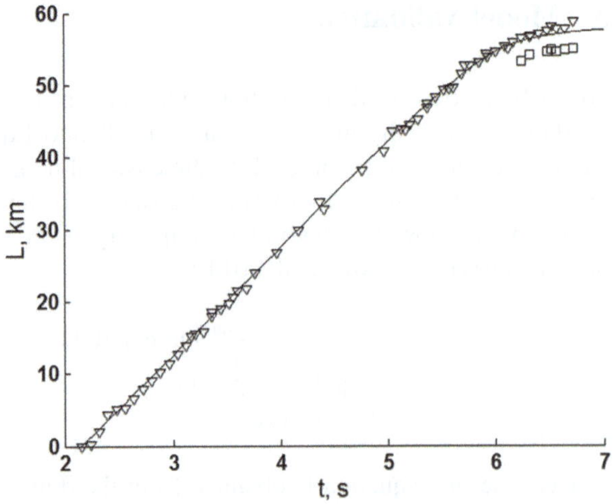

Fig. 2 Length along trajectory vs time. *Solid line*: the numerical solution, *triangles*: experimental data for the main body, *squares*: experimental data for the body fragment

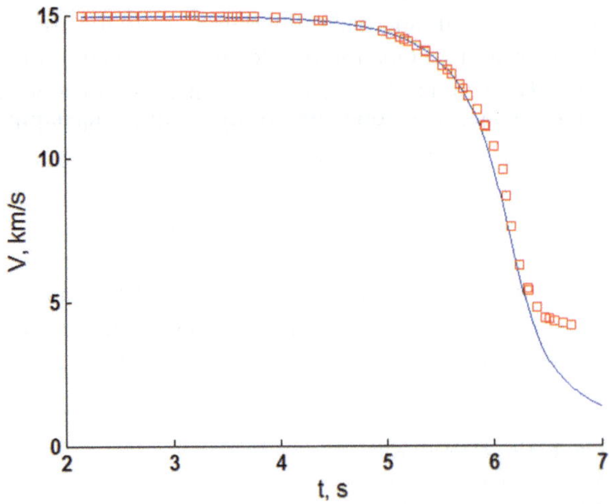

Fig. 3 Velocity vs time. *Solid line*: the numerical solution, *markers*: experimental data for the main body

The calculated functions $h(v), y(t), \text{and}\, L(t)$ fits closely to the corresponding values suggested by Borovička et al. (2013). All the figures show good agreement between the calculated and the observational data.

4 Pre-atmospheric Mass Estimation

The value of the ballistic coefficient α found according to Eq. (4), can be used to estimate the dynamic mass of the main fragment of the Košice meteoroid. Along with the definition for α (Appendix 2), one can introduce the shape factor, *A*, a dimensionless parameter which characterizes the "flatness" of an object. In meteor studies, the shape factor A is usually introduced as a ratio of the head cross section area to the volume of the object to the power of 2/3 (cf. Ceplecha et al. 1998). In our approach only the initial shape factor of meteoroid needs to be assumed at this stage of the study and further, it can freely change along the atmospheric trajectory. This substitution avoids the dependency on the unknown pre-atmospheric head cross-sectional area of the meteoroid and expresses its pre-atmospheric mass through pre-atmospheric shape factor and bulk density values which only vary within a limited range. In other words, combining the definitions for the ballistic coefficient α and the pre-atmospheric shape factor A_e the following expression is obtained:

$$M_e = \left(\frac{1}{2} c_d \frac{\rho_0 h_0}{\alpha \sin \gamma} \frac{A_e}{\rho_m^{2/3}} \right)^3 \tag{7}$$

In our calculation of pre-atmospheric mass the atmospheric density is taken as $\rho_0 = 0.00129\ \mathrm{g/cm^3}$ at sea level, and the atmospheric scale height as $h_0 = 7160$ m (the same value is also used to scale the observed heights to dimensionless quantities, Appendix 2). The slope γ between the horizon and the atmospheric trajectory is taken from the actual observations as referred to in the previous section. The product of the pre-atmospheric shape factor A_e and the drag coefficient c_d is set to 1.8, similarly as in the recent calculations for the Annama and Park Forest meteoroids (Trigo-Rodríguez et al. 2015, Meier et al. 2016).

5 Pre-fragmentation Meteoroid Shape

Despite the significant proliferation of applied analytical and experimental studies of scaling laws over the past decades (Andrews 1997; Aström et al. 2004; Aström 2006), little attention has been given to the reconstruction of the initial meteoroid's shape. Statistical analysis shows a similarity of laws that govern the size and mass distribution of fragments of the parent body as well as the fracturing energy and material strength values (Carpinteri and Pugno 2005). Multiple experiments (Iordache et al. 2005) demonstrate that the process of shattering is fractal in nature. Simply speaking, it means that any observed limited subset of small splinter masses does not give any information about their relative location in the overall distribution. This subset can be placed anywhere on the probability density curve, resulting only in scale change of the total mass. The same feature is exhibited by more classical fractals, such as clouds or mountain rocks. It is theoretically impossible to deduce the absolute size of fractal objects while observing their parts without comparison to surroundings. In practice, e.g., the CEB (European) concrete design code and the German standards code DIN, have adopted various versions of the scaling law (Bažant 2002).

The main idea of shape estimation is derived from the experiments that show that brittle fracturing produces multiple fragments of size lesser than or equal to the smallest dimension of the body. The number of fragments depends on fragment masses as a power law with exponential cutoff. The scaling exponent essentially indicates the initial form (dimensions ratio) of the fragmented body. This means that the fracture of rod-like, plate-like and sphere-like shapes produces corresponding mass distributions with different scaling exponents. However, the shape estimation technique relies on empirical coefficients obtained by fitting the not so numerous experimental results. Therefore, currently there are no approaches to consider error estimates of this analysis. While it is very tempting to use scaling analysis on meteorite fracturing, discussions still continue on the applicability of scaling analysis on the recovered fragments from a single meteorite dropping event. One of the main limitations of scaling analysis is that the range of fragment masses has to cover three to four orders of magnitude for an accurate analysis (Oddershede et al. 1998). A study which meets this criterion is the Jiddat al Harasis 073 meteorite (Gnos et al. 2009) which had suitable fragment masses ranging from 10 g to 10 kg. In contrast, some meteoritic mass distributions are much narrower in range, i.e., the appropriate fragment masses span just over one order, and in such cases under sampling can distort the true parameters of the scaling law. However in our case, the Košice meteorite has a suitable subrange that spans from 3 to 300 g (Gritsevich et al. 2014a), which should be sufficient for our analysis method.

There are only a few studies devoted to the initial shape estimation of a meteoroid, so any insight into the solution of this problem may be useful. Here we introduce an innovative approach, exploring the possibility of determining the dimensions of a pre-fragmented meteoroid based on an assumed value of its pre-fragmentation shape factor. A similar inversed meteoroid shape estimation technique has recently been implemented by Vinnikov et al. (2014, 2015, 2016a). The method follows the concepts proposed by Oddershede et al. (1993, 1998) and it is based on the following assumptions:

1. The meteoroid material is assumed to be brittle. The common definition of brittleness involves the practical absence of plastic deformation prior to fracturing. We also assume that the crack propagation speed is comparable to the speed of sound through a brittle media and is at least one order higher than the acoustic speed of the surrounding environment. However, we do not completely address the issue of supersonic fracture (Buehler et al. 2003),
2. The masses of the fragments recovered within one meteorite fall can be fitted via the power law (which is a special case of a Weibull distribution) with relatively small least squares error,
3. A complementary cumulative distribution function (CCDF) can be constrained by the scaling exponent B_0, which accounts for the largest recovered fragment mass via an exponential cutoff:

$$F_c(m) = Cm^{-B_0} \exp\left(-m/m_U\right), \tag{8}$$

4. For the value of B_0^* obtained by fitting Eq. (8) to the masses of recovered fragments as described below, we can estimate dimensionless shape parameter, *d*, (its definition is based on size proportions a_x, a_y, a_z in Appendix 2) from the

empirical equation

$$0.13d^2 - 0.21d + \left(1.1 - B_0^*\right) = 0 \tag{9}$$

In Eq. (8) C is a normalization constant, $m_U > m_L$ is the upper cutoff fragment mass, m_U corresponds to the threshold where the exponential decay starts to dominate over the power law, and $m_L > 0$ is an arbitrary lower mass limit acting as an additional constraint for under sampled tiny unrecoverable particles resulting from the fragmentation. These mass constraints are also among the sought parameters.

In other words, the studied meteorite sample consists of N fragments ranging in mass from m_0 to m_{N-1}. We assume that the fragment masses m_i are arranged in ascending order and the values m_L and m_U are intermediate in the initial range. If two or more fragments possess equal masses, then we add a small value corresponding to the fragment mass measurement error (e.g. 0.001 g) to one of them, to disambiguate the masses. The masses for the Košice fragments are taken from the study by Gritsevich et al. (2014a).

Since the density distribution $n(m)$ of fragment masses $\{m_i\}_{i=1}^N$ is discrete, we convert it to the piecewise complementary cumulative distribution $N^*(m)$ as suggested by Oddershede et al. (1998):

$$N^*(m) = \frac{1}{m}\int_m^\infty n(x)dx,\ m > m_L,\ n(m) = \sum\nolimits_{i=1}^N \delta\left(m - m_i\right), \tag{10}$$

where δ is the Dirac delta function.

In order to match $F_C(m)$ to $N^*(m)$ we conduct a normalization procedure, equaling their values at the point m_L:

$$F_C\left(m_L\right) = C\left(m_L\right)^{-B_0}\exp\left(-m_L/m_U\right) = N^*\left(m_L\right) = \frac{1}{m_L}\left(N - n_L\right),$$

$$n_L\ :\ m_{n_L-1} \le m_L \le m_{n_L}. \tag{11}$$

The normalization constant, C, can be expressed as follows:

$$C = \left(m_L\right)^{B_0-1}\exp\left(m_L/m_U\right)\left(N - n_L\right) \tag{12}$$

For simplicity, we assume that the lower bound, m_L, coincides with the mass of one of the fragments m_j, i.e. $n_L = j,\ \ m_L = m_j$.

Thus, the normalization constant C and the CCDF $F_C(m)$ become:

$$C = \left(m_j\right)^{B_0-1}\exp\left(m_j/m_U\right)\left(N - j\right) \tag{13}$$

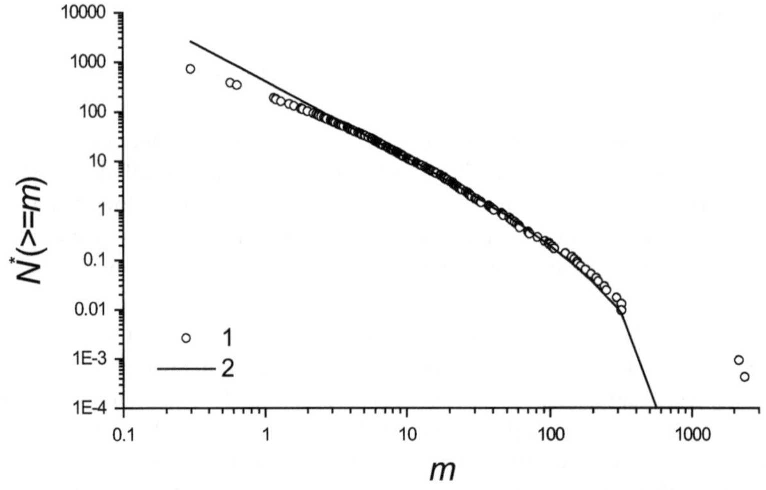

Fig. 4 Complementary cumulative distribution $N^*(m)$ and its approximation by $F_C(m)$ vs fragment mass m (in logarithmic scale). *Solid line*: $F_C(m)$, *markers*: values of $N^*(m)$ corresponding to experimental data points $m_i,\ i = 1, \ldots, N$

$$F_C(m) = \frac{N-j}{m_j}\left(\frac{m}{m_j}\right)^{-B_0} \exp\left(-\frac{m-m_j}{m_U}\right) \tag{14}$$

(see Fig. 4).

We note here that the scaling exponent B_0 is dimensionless and has no dependence on the dimensions of $N^*(m)$ and $F_C(m)$.

Next, the least-squares method is applied to minimize the expression

$$S\left(B_0, j, m_U\right) = \sum\nolimits_{i=j}^{N}\left[F_C\left(m_i\right) - \frac{N-j}{m_j}\right]^2 \tag{15}$$

We obtain the analytical expressions for the partial derivatives $\frac{\partial S}{\partial B_0}$, $\frac{\partial S}{\partial m_U}$, $\frac{\partial^2 S}{\partial B_0^2}$, $\frac{\partial^2 S}{\partial B_0 \partial m_U}$, $\frac{\partial^2 S}{\partial m_U^2}$ set the initial approximation for the optimal vector argument (B_0, m_U) and then apply Newton's method to derive (B_0, m_U) for the valid range of m_L.

The approach of Oddershede et al. (1993, 1998) for body shape estimation can be applied to multimodal distributions of the obtained masses. However, the multimodality of masses does not necessarily imply the distinguishable multimodality of the scaling law, since modes with higher mean masses contribute less significantly to the $F_C(m)$. The second mode for the Košice meteorite has a mean mass equal to 140 g, which corresponds to the $F_C(m)$ value about 0.1. Neglecting this mode does not introduce a large least squares error, so we consider the scaling law as unimodal.

Once B_0 is obtained, one can estimate the dimensionless shape parameter, d, from the empirical equation (9) and obtain the size proportions a_x, a_y, a_z based upon the calculated shape parameter value and the assumed shape factor of the pre-fragmented meteoroid (a solution to this problem is described in the Appendix 4 for a brick-like shaped objects).

6 Heliocentric Orbit

We obtained the heliocentric orbit for the Košice meteoroid based on the coordinates and velocity estimates given by Borovička et al. (2013). However, in contrast to the standard method, we used a slightly different orbit determination approach as implemented by Dmitriev et al. (2014, 2015). While the standard orbit computation method (Ceplecha 1987) is traditionally based on a set of corrections to the observed velocity vector (see e.g. Andreev 1990; Langbroek 2004; Zoladek 2011), we instead use strict transformations of the coordinates and velocity vectors as recommended by the IAU International Earth Rotation and Reference Systems Service (IERS Conventions 2010), as well as backward numerical integration (Everhart 1974) to the equations of motion. An approach similar to ours, which utilized the "mercury6" software described by Chambers (1999), was applied in several studies during the Chelyabinsk meteoroid's orbit reconstruction, e.g., (Zuluaga and Ferrin 2013) and (Zuluaga et al. 2013). Also the approach described below has been applied to the data collected by the Finnish Fireball Network, including recently recovered Annama meteorite (Gritsevich et al. 2014b; Dmitriev et al. 2015; Trigo-Rodríguez et al. 2015).

Specifically, our approach used the following transformations:

1. Transformation of velocity vector from topocentric to geocentric coordinate system

$$\begin{pmatrix} V_x \\ V_y \\ V_z \end{pmatrix} = \mathbf{M}^T \begin{pmatrix} V_n \\ V_e \\ V_u \end{pmatrix}, \quad \mathbf{M} = \mathbf{Q}_1 \mathbf{R}_2 \left(\phi - 90^\circ\right) \mathbf{R}_3 \left(\lambda\right),$$

   where $\left(V_n\ V_e\ V_u\right)^{\mathrm{T}}$ and $\left(V_x\ V_y\ V_z\right)^{\mathrm{T}}$ topocentric and geocentric velocity vectors;

   $\mathbf{R}_2, \mathbf{R}_3$, and $\mathbf{Q}_1$ are the appropriate rotation matrices and mirror matrix, respectively;

   ϕ and λ—geodetic latitude and longitude of the beginning point of meteoroid trajectory.
2. Diurnal aberration

$$\begin{pmatrix} \Delta V_x \\ \Delta V_y \\ \Delta V_z \end{pmatrix} = -a_e \omega_\oplus \begin{pmatrix} \cos\phi \sin\lambda \\ \cos\phi \cos\lambda \\ 0 \end{pmatrix}$$

$$\begin{pmatrix} V_{x_{geo}} \\ V_{y_{geo}} \\ V_{z_{geo}} \end{pmatrix} = \begin{pmatrix} V_x \\ V_y \\ V_z \end{pmatrix} + \begin{pmatrix} \Delta V_x \\ \Delta V_y \\ \Delta V_z \end{pmatrix},$$

where $\omega_\oplus$—Earth rotation velocity, a_e—equatorial Earth radius.

3. Transformation of the beginning point coordinates and velocity vectors from the Earth-fixed geocentric coordinate system, ITRF2000, to the Geocentric Celestial Reference System (GCRS) realization ICRF2 (J2000) accordingly to Wallace (1998), IERS Conventions (2010), IAU SOFA (2013), and SOFA Astrometry Tools (2014):

$$\begin{pmatrix} X_{in} \\ Y_{in} \\ Z_{in} \end{pmatrix} = \mathbf{R}^{\mathrm{T}} \begin{pmatrix} X_{geo} \\ Y_{geo} \\ Z_{geo} \end{pmatrix}, \quad \begin{pmatrix} V_{x_{in}} \\ V_{y_{in}} \\ V_{z_{in}} \end{pmatrix} = \mathbf{R}^{\mathrm{T}} \begin{pmatrix} V_{x_{geo}} \\ V_{y_{geo}} \\ V_{z_{geo}} \end{pmatrix}, \qquad \mathbf{R} = \mathbf{PN\Pi S},$$

where **P**—precession matrix, **N**—nutation matrix, **Π**—polar motion matrix, and **S**—apparent Greenwich Sidereal Time matrix.

Contributions of polar motion and high order nutation are negligible in comparison to the observational errors, so these effects can be neglected.

4. JPL's ephemeris, DE421, is used for transformation of meteoroid position and velocity vectors from geocentric to heliocentric coordinate system as:

$$\vec{r}_{J2000} = \vec{R}_\oplus + \vec{R}_{in},$$

$$\dot{\vec{r}}_{J2000} = \vec{V}_\oplus + \vec{V}_{in},$$

here $\vec{r}_{J2000}$, $\dot{\vec{r}}_{J2000}$—required initial conditions for numerical integration,

$\vec{R}_\oplus$, $\vec{V}_\oplus$—Earth position and velocity vectors from JPL's DE421 ephemeris (Folkner et al. 2009),

$\vec{R}_{in}$, $\vec{V}_{in}$—meteoroid's position and velocity vectors in the celestial geocentric coordinate system ICRF2 (J2000).

The backward integration of the equations of perturbed meteoroid motion:

$$\ddot{\vec{r}} = \frac{GM_{Sun}}{r^3}\vec{r} + \ddot{\vec{r}}_{Earth}\left(C_{nm}, S_{nm}, \vec{r}, t\right) + \ddot{\vec{r}}_{Moon}\left(\vec{r}, t\right) + \sum \ddot{\vec{r}}_{planets}\left(\vec{r}, t\right) + \ddot{\vec{r}}_{atm}\left(\vec{r}, t\right)$$

was performed by the implicit single-sequence numerical method (Everhart 1974). The equations of perturbed meteoroid motion include a central body (Sun) attraction; perturbations from Earth's gravity field, the Moon, and other planets; and from the atmospheric drag. To obtain the undistorted heliocentric orbit, backward integration was performed for the part of the meteoroid's trajectory lying inside the Hill sphere (i.e. about 4 days backward in our case), see also a paper on this subject by Dmitriev et al. (2015) for further details.

The resulting orbital elements of the Košice meteoroid, obtained with developed software, are summarized in Table 4 and the corresponding discussion is given by Dmitriev et al. (2015). For a comparison, the orbital elements given in (Borovička et al. 2013, Table 7) are also presented. The estimated influences of the Earth's and Moon's gravity fields *(only the second zonal harmonic coefficient, which characterizes the Earth's flattening was taken into account)* on the meteoroid's orbit are presented in Table 5.

7 Results and Discussion

The initial main fragment mass M_e from Eq. (7), is estimated to be about 1500 kg. Taking into account the results obtained by Gritsevich et al. (2014a), the estimated pre-atmospheric meteoroid mass value was obtained by multiplying this estimate by a correction coefficient of 11/9 (since the most probable scenario suggests that the Košice meteoroid, prior to further extensive fragmentation in the lower atmosphere, was initially represented by two discrete pieces with cumulative residual masses of approximately 2 kg and 9 kg respectively). Thus, the estimated pre-atmospheric meteoroid mass value equals 1850 kg. This pre-atmospheric mass value corresponds well to the mass estimate that was based on photometry (Borovička et al. 2013).

The difference between the initial and recovered Košice meteorite fragment masses and their subsequent statistical analysis suggest that two to three larger fragments, in the range of 0.5–1 kg each, should exist but either they were not recovered or they were not reported by illegal meteorite hunters (Gritsevich et al. 2014a). On the other hand, it is quite possible that a considerable portion of total mass was converted into fine dust during fragmentation and was dispersed into the atmosphere. Once these dispersed particles fall on the ground, they are too small to be distinguished from surrounding ground content and therefore, are rarely recoverable.

Finally, to approach the topic of the initial meteoroid shape estimation, we applied the method described in Section "Pre-fragmentation meteoroid shape" to the Košice fragment masses dataset. The applied statistical analysis produced the following values for the scaling exponent, the minimum mass limit and the power-law cutoff mass: $B_0 = 1.53$, $m_L = 5.64$ g, $m_U = 155.17$ g. The approximate value of the dimensionless shape parameter was calculated as $d = 2.8$ and the calculated size proportions as $a_x : a_y : a_z = 2 : 1.69 : 1$. We emphasize that our described meteoroid shape estimation technique is limited to homogeneous

Table 4 The derived Košice meteorite orbit and its comparison to the data obtained by Borovička et al. (2013)

	a (AU)	*e*	*i*°	*Ω*°	*ω*°	*M*°
This study, 2010/02/24 UTC 22:24:47.0	2.72 ± 0.21	0.649 ± 0.022	2.06 ± 0.49	340.146 ± 0.013	204.07 ± 0.14	355.112 ± 0.007
Borovička et al. (2013)	2.71 ± 0.24	0.647 ± 0.032	2.0 ± 0.8	340.072 ± 0.004	204.2 ± 1.2	–

Table 5 Calculated influence of Earth and Moon gravity fields on meteoroid orbit

Perturbations	Δa (AU)	Δe	Δi°	$\Delta\Omega^\circ$	$\Delta\omega^\circ$	ΔM°
Earth gravity	2.5e-5	0.00021	0.00114	−0.000441	−0.01109	0.00675
Moon gravity	2.7e-5	−0.00016	0.00112	−0.000009	−0.00794	−0.00215

brittle solids that are not prestrained. It is known that prestrained materials (e.g. tempered glass), while being brittle, do not follow a power law mass distribution during fragmentation. We assume a scaling analysis is suitable for shape estimation, at least beyond the current state-of-the-art, in reconstructing meteoroid shapes since, for example, the mass distribution in the asteroid belt follows a power law (Hughes and Harris 1994).

On the other hand, the applicability of the scaling analysis to meteorite samples still need to be demonstrated in future studies. For example, results given by Oddershede et al. (1993) are questioned by Meibom and Balslev (1996). The latter authors performed fragmentation experiments with thick plates of dry clay. Meibom and Balslev (1996) observed different exponents for fragments larger and smaller than the plate thickness, made implications denying scale and material invariance and expressed the opinion that "the measured mass distribution tells little about the mechanisms of the fragmentation process". However, this reasoning can be countered by the following facts. The speed of sound in clay is about 1–2.8 km/s, in contrast, the respective values in gypsum are 2.3 km/s for transverse S-waves and 5 km/s for longitudinal P-waves. In comparison, the speed of sound in chondrites is about 3.5 km/s (Rivkin and Bottke 1996). This leads to the question does the clay really exhibits brittle behavior. Indeed, the universal scaling was not observed for thick large plates of dry clay, but the masses of these plates (540–920 g) were an order of magnitude greater than those of small gypsum plates (58–77 g). We suspect that the experiments with thick clay plates failed to provide energy sufficient to produce brittle fracture conditions. The brittle fracture is a dissipative process, which disperses locally supplied energy over the whole bulk. From this point of view, when all the initially channeled energy dissipates, then in absence of internal strains the cracks cease to propagate farther. The results of experiments with small thin clay plates (237–393 g) by Meibom and Balslev (1996) confirm exactly the ideas and the results of Oddershede et al. (1993). Linna et al. (2005) present another theory for additional fracturing mechanism for thin plates. However, recent studies by Renshaw and Schulson (2001) and by Aström (2009) cast a doubt on the existence of a separate scaling law fragmentation models for thin plates and volumetric bodies. Anyway, for meteoroid applications the buildup of energy flux from atmosphere drag (via ram pressure) is usually sufficient for fragmentation at all scales (with the exception of the heaviest iron meteorites).

Recent studies provide increasing evidence to support the theory of power law distributions for fragment masses. For example, Sotolongo-Costa et al. (2000, 2007) assume that high energy experiments with violent fragmentation processes produce power law distributions. On the contrary, low energy processes generate classical statistical functions like log-normal or Rosin-Ramler. The direct experiment on the Murchison CM2 chondrite with a high velocity impactor also yielded a simple power law distribution (Flynn et al. 2007). Computational experiments considered by Domokos et al. (2015) show that the calculated mass distribution is best fitted by a power law with an exponential cutoff. In addition, the transition from low energy distributions to high energy power law, is studied numerically in a paper by Hernández (2001) via a random stopping mathematical model as well as in a work of Spahn et al. (2014) via a random-walk model. Considering these findings, we assume that meteoroid's shape can be estimated from the value of the power law exponent.

8 Conclusions

One of the critical steps in the prediction of a potentially hazardous asteroid's impact threat to Earth is a clear understanding and modelling of the processes occurring during the object's entry into the atmosphere. This knowledge enables us to characterize, simulate, and to classify the various possible impact consequences. The reconstructed atmospheric trajectory, from well observed meteor events recorded by video or other instrumental means, is the key to deriving the pre-impact meteoroid's orbit in the solar system. The derived atmospheric trajectory is also required for dark flight simulations which enable us to reasonably predict the path of any surviving meteorite fragments all the way down to the ground.

In the present study we described several state-of-the-art techniques which can be used to reconstruct the pre-atmospheric parameters of large meteoroids, in our example, the data came from the recent fall of the Košice meteorite. In particular,

1. By using dimensionless expressions (see Appendix 2), which involve the pre-atmospheric meteoroid parameters, we have built a physically based parametrization to describe the changes in mass, height, and velocity of the object along its atmospheric path. The developed model is suitable to formulate well-posed inverse problem used to estimate a number of crucial unknown values including the ballistic coefficient, the mass loss parameter, the dynamic (i.e. obtained based on the observed deceleration rate) pre-atmospheric mass and surviving meteorite mass.
2. We have considered a pre-atmospheric meteoroid shape estimation technique that is based on statistical laws of distribution for recovered meteorite fragment masses. The idea of reconstructing the initial meteoroid shape is described in companion papers by Vinnikov et al. (2015, 2016a, b). It is based on experiments that demonstrates brittle fracturing produces multiple fragments of size less than or equal to the smallest dimension of the object. We demonstrate the application of a scaling analysis to meteorite cases which have a large number of recovered fragments is feasible and can form a solid basis for future theoretical studies on this subject.

3. We have implemented a precise technique to determine a meteoroid's orbit based on numerical integration of differential equations of motion. The technique takes into account the perturbations due to Earth's gravitational field (both the spherical part and the non-central part of the geopotential), perturbations from the atmospheric drag, perturbations from the Moon, and from other planets of the solar system. Based on our investigations, the attraction of the Moon and the effect of Earth flattening are seen as the main factors perturbing a meteoroid's orbit, second only to that of the Earth as a point mass. These perturbations are generally expressed in the orbital elements of the argument of the periapsis and by the mean anomaly of the meteoroid (see also Dmitriev et al. 2015).

Our methods are incorporated in a software package called "Meteor Toolkit". The software can be used to integrate the equations of motion backwards in time, preceding the impact, as well as forward in time, to the moment of impact with the Earth's surface. This software has a graphical user-friendly interface and uses freely distributed SPICE routines and kernels (Acton 1996) for coordinate transformation and computing ephemerides. In addition, it has a module for visualization of the computation results. Screenshots of this software are presented in Dmitriev et al. (2015).

Generally, the described research methods facilitate a better understanding of the processes associated with an object's atmospheric entry and create a focus on the connections between the properties and structure of the entering bodies (asteroids/meteoroids) and the recovered meteorites. Similar research methods and approaches were recently employed and led to the recovery of the Annama meteorite after the fireball's passage through the atmosphere was observed by three stations of the Finnish Fireball Network on 19 April 2014 (Gritsevich et al. 2014b; Trigo-Rodríguez et al. 2015; Dmitriev et al. 2015; Kohout et al. 2015; Lyytinen and Gritsevich 2016).

In summary, our approach significantly increases the knowledge of small solar system bodies, provides better insights of the potential impact hazards and consequences arising from the NEO population, and expedites the fast meteorite recovery from reliably identified fall sites.

Acknowledgments This study was supported, in part, by the Academy of Finland project No 260027, by the ERC Advanced Grant No 320773, by the Russian Foundation for Basic Research (project Nos 14-08-00204, 16-05-00004 and 16-07-01072), by the Magnus Ehrnrooth Foundation travel grant, and by the Act 211 of the Government of the Russian Federation (agreement No 02.A03.21.0006). The part of the trajectory reconstruction and the orbit computations were done by Maria Gritsevich, Vasily Dmitriev, Daria Kuznetsova, and Valery Lupovka at MIIGAiK under the support of the Russian Science Foundation, project No 14-22-00197 "Studies of Fundamental Geodetic Parameters and Topography of Planets and Satellites". The authors acknowledge being a part of the network supported by the COST Action TD1403 "Big Data Era in Sky and Earth Observation".

Appendix 1: List of Symbols

α	Ballistic coefficient
β	Mass loss parameter
γ	Slope between horizon and the trajectory
λ	Geodetic longitude of the beginning fireball point
μ	Shape change coefficient
ρ_0	Gas density at sea level
ρ_a	Gas density
ρ_m	Meteoroid bulk density
ϕ	Geodetic latitude of the beginning fireball point
a_x, a_y, a_z	Dimensionless sizes of prefragmented meteoroid
A	Shape factor of meteoroid
A_e	Pre-entry shape factor of meteoroid
B_0	Power-law scaling exponent
c_d	Drag coefficient
c_h	Heat-transfer coefficient
d	Shape parameter of meteoroid
$F_C(m)$	Power-law complementary cumulative distribution function, which is an approximation of meteorite fragment distribution
h	Height
h_0	Scale height
H^*	Effective destruction enthalpy
L	Length along atmospheric trajectory
m_i	Fragment masses
m_L	Minimum fragment mass limit
m_U	Power-law cutoff mass
M	Meteoroid mass
M_e	Pre-entry meteoroid mass
n	Number of considered points along the trajectory
N	Number of fragments
N^*	Piecewise complementary cumulative distribution of meteorite fragments
S	Head cross section area
S_e	Pre-entry middle section area of meteoroid
t	Time
V	Velocity
V_e	Pre-entry velocity

Orbital Elements

ω	Argument of periapsis
Ω	Longitude of the ascending node

a Semimajor axis
e Eccentricity
i Inclination
M Mean anomaly at epoch

Appendix 2: Dimensionless Quantities

$$y = h/h_0$$

$$v = V/V_e$$

$$m = M/M_e$$

$$A_e = \frac{S\rho_m^{2/3}}{M^{2/3}}$$

$$d = 1 + 2\left(a_x a_y + a_y a_z + a_z a_x\right)\left(a_x^2 + a_y^2 + a_z^2\right)^{-1}$$

$$\alpha = 0.5c_d \frac{\rho_0 h_0 S_e}{M_e \sin\gamma}$$

$$\beta = 0.5\,(1 - \mu)\,\frac{c_h V_e^2}{c_d H^*}$$

$$\mu = \log_m \frac{S}{S_e}$$

$$\Delta = \overline{\mathrm{Ei}}\,(\beta) - \overline{\mathrm{Ei}}\left(\beta v^2\right)$$

$$\Delta_i = \overline{\mathrm{Ei}}\,(\beta) - \overline{\mathrm{Ei}}\left(\beta v_i^2\right) = -2\ln v_i + \sum_{k=1}^{\infty} \frac{\beta^k}{k \cdot k!}\left(1 - v_i^{2k}\right)$$

$$(\Delta_i)_\beta{}' = \sum\nolimits_{k=1}^{\infty} \frac{\beta^{k-1}}{k!}\left(1 - v_i^{2k}\right)$$

$$(\Delta_i)_\beta{}'' = \sum\nolimits_{k=2}^{\infty} \frac{\beta^{k-2}}{(k-2)! \cdot k}\left(1 - v_i^{2k}\right)$$

Appendix 3: Special Mathematical Function $\bar{\mathrm{E}}\mathrm{i}(x)$

The exponential integral $\bar{\mathrm{E}}\mathrm{i}(x)$, which is defined for real nonzero values of x as:

$$\overline{\mathrm{Ei}}(x) = \int\limits_{-\infty}^{x} \frac{e^z dz}{z}.$$

The integral has to be understood in terms of the Cauchy principal value, due to the singularity in the integrand at zero.

Integrating the Taylor series for function e^{-z}/z, and extracting the logarithmic singularity, we can derive the following series representation for $\bar{\mathrm{E}}\mathrm{i}(x)$ for real values of x (see e.g. (Abramovitz and Stegun 1972)):

$$\overline{\mathrm{Ei}}(x) = c + \ln x + \sum\nolimits_{n=1}^{\infty} \frac{x^n}{n \cdot n!}, \quad x > 0,$$

where c is the Euler–Mascheroni constant (also called Euler's constant). It is defined as the limiting difference between the harmonic series and the natural logarithm:

$$c = \lim_{n \to \infty}\left(\sum\nolimits_{k=1}^{n} \frac{1}{k} - \ln n\right) \approx 0.5772.$$

Appendix 4: Relation Between the Shape Factor *A* and the Shape Parameter *d*

In our study we use dimensionless shape factor A and dimensionless shape parameter d, defined in Appendix 2. The first one is used in Eq. (7), the second one is derived from Eq. (9). Each of these parameters defines only a subset of objects with the appropriate shape properties. More insights onto the shape of the object may be obtained, if we combine both definitions into the nonlinear system of equations.

The shape factor A can be expressed as follows:

$$A = k\frac{a_x a_y}{\left(a_x a_y a_z\right)^{2/3}},$$

where the coefficient k shows how much shape of the considered object differs from the brick-like geometry, for which $k = 1$, e.g. a spherical object yields $k = 1.209$.

Since we are looking for the ratio $a_x : a_y : a_z$, and not for the absolute object size, it is convenient to assume that $a_z = 1$ is the smallest size dimension representing a unit length (i.e. the dimensions are normalized by the actual value a_z). Then for the remaining dimensions, a_x and a_y, we obtain:

$$\begin{cases} A = k\left(a_x a_y\right)^{1/3}, \\ d = 1 + 2\frac{a_x a_y + a_x + a_y}{a_x^2 + a_y^2 + 1}, \end{cases}$$

$$\begin{cases} a_x a_y = A^3 k^{-3}, \\ d = 1 + 2\frac{A^3 k^{-3} + a_x + a_y}{\left(a_x + a_y\right)^2 - 2A^3 k^{-3} + 1}, \end{cases}$$

$$\begin{cases} a_x a_y = A^3 k^{-3}, \\ \left(a_x + a_y\right)^2 (d-1) - 2\left(a_x + a_y\right) - 2A^3 k^{-3} d + d = 1, \end{cases}$$

Here we solve this problem for the brick-like shape of an object (as suggested as a general model for a meteoroid shape, e.g. by Halliday et al. 1989, 1996 based on the investigation of the Innisfree meteorite)

$$\begin{cases} (d-1)\,a_x^4 - 2a_x^3 + \left((d-1) - 2A^3\right)a_x^2 - 2A^3 a_x + (d-1)\,A^6 = 0, \\ a_y = A^3 a_x^{-1}. \end{cases}$$

If $A = 1.5$ and $d = 2.8$, then

$$0.9a_x^4 - a_x^3 + (0.9 - 3.375)\,a_x^2 - 3.375a_x + 10.2515625 = 0.$$

This equation has two real roots and two complex ones. The real sizes are:

$$a_x = 1.69,\ a_y = 2.0,\ \text{or}\ a_y = 2.0,\ a_y = 1.69.$$

Appendix 5: Results of analyses of cosmogenic radionuclides in the Košice meteorite

After this chapter was sent to the publisher, we discovered another important work with the relevant initial mass estimate by Povinec et al. (2015). The authors have estimated average radius of the meteoroid at 50 cm using both the ^{60}Co and ^{26}Al data, and they provide the pre-atmospheric meteoroid mass estimate of 1840 kg, bringing it extremely close to the presented in this chapter calculations.

References

Abramovitz, M., Stegun, I.A. (eds.): Handbook of Mathematical Functions with Formulas, Graphs, and Mathematical Tables. Dover Publications, New York (1972)

Acton, C.: Ancillary data services of NASA's navigation and ancillary information facility. Planet. Space Sci. **44**(1), 65–70 (1996)

Andreev, G.: The influence of the meteor position on the zenith attraction. In: Proceedings of the International Meteor Conference, Violau, Germany, 6–9 September 1990, pp. 25–27

Andrews, E.W.: Experimental Studies of Dynamic Fragmentation in Brittle Materials, 240 p. Brown University, Providence, RI (1997)

Aström, J.A.: Statistical models of brittle fragmentation. Adv. Phys. **55**, 247–278 (2006)

Aström, J.A.: Difference between fracture of thin brittle sheets and two-dimensional fracture. Phys. Rev. E Stat. Nonlin. Soft Matter Phys. **80**(4 Pt 2), 046113 (2009)

Aström, J.A., Linna, R.P., Timonen, J., Møller, P.F., Oddershede, L.: Exponential and power-law mass distributions in brittle fragmentation. Phys. Rev. E Stat. Nonlin. Soft Matter Phys. **70**(2 Pt 2), 026104 (2004)

Bažant, Z.P.: Scaling of Structural Strength. Hermes Penton Science, London (2002)

Bland, P.A.: Fireball cameras: The Desert Fireball Network. Astron. Geophys. **45**(5), 5.20–5.23 (2004)

Bland, P.A., Smith, T.B., Jull, A.J.T., Berry, F.J., Bevan, A.W.R., Cloudt, S., Pillinger, C.T.: The flux of meteorites to the Earth over the last 50,000 years. Mon. Not. Roy. Astron. Soc. **283**, 551 (1996)

Borovička, J., Tóth, J., Igaz, A., Spurný, P., Kalenda, P., Haloda, J., Svoreň, J., Kornoš, L., Silber, E., Brown, P., Husárik, M.: The Košice meteorite fall: Atmospheric trajectory, fragmentation, and orbit. Meteorit. Planet. Sci. **48**(10), 1757–1779 (2013)

Bouquet, A., Baratoux, D., Vaubaillon, J., Gritsevich, M.I., Mimoun, D., Mousis, O., Bouley, S.: Simulation of the capabilities of an orbiter for monitoring the entry of interplanetary matter into the terrestrial atmosphere. Planet. Space Sci. **103**, 238–249 (2014)

Brown, P.G., Weryk, R.J., Wong, D.K., Campbell-Brown, M.D.: The Canadian Meteor Orbit Radar II: a new facility for measurement of the dust environment in near-Earth space. American Astronomical Society, DPS meeting #44, #302.04 (2012)

Buehler, M.J., Abraham, F.F., Gao, H.: Hyperelasticity governs dynamic fracture at a critical length scale. Nature **426**, 141–146 (2003)

Carpinteri, A., Pugno, N.: Are scaling laws on strength of solids related to mechanics or to geometry? Nat. Mater. **4**, 421–423 (2005)

Ceplecha, Z.: Geometric, dynamic, orbital and photometric data on meteoroids from photographic fireball networks. Astron. Inst. Czech. Bull. **38**, 222–234 (1987)
Ceplecha, Z., Borovička, J., Elford, W.G., Revelle, D.O., Hawkes, R.L., Porubčan, V., Šimek, M.: Meteor phenomena and bodies. Space Sci. Rev. **84**, 327–471 (1998)
Chambers, J.E.: A hybrid symplectic integrator that permits close encounters between massive bodies. Mon. Not. Roy. Astron. Soc. **304**(4), 793–799 (1999)
Collins, G.S., Melosh, H.J., Marcus, R.A.: Earth Impact Effects Program: A Web-based computer program for calculating the regional environmental consequences of a meteoroid impact on Earth. Meteorit. Planet. Sci. **40**(6), 817–840 (2005)
Dmitriev, V., Lupovka, V., Gritsevich, M.: A new approach to meteor orbit determination. In: Rault, J.-L., Roggemans, P. (eds.) Proceedings of the International Meteor Conference, Giron, France, 18–21 September 2014. International Meteor Organization, ISBN 978-2-87355-028-8, pp. 157–159 (2014)
Dmitriev, V., Lupovka, V., Gritsevich, M.: Orbit determination based on meteor observations using numerical integration of equations of motion. Planet. Space Sci. **117**, 223–235 (2015)
Domokos, G., Kun, F., Sipos, A.Á., Szabó, T.: Universality of fragment shapes. Sci. Rep. **5**, 9147 (2015)
Edwards, W.N., Eaton, D.W., Brown, P.G.: Seismic observations of meteors: Coupling theory and observations. Rev. Geophys. **46**(4), RG4007 (2007)
Emel'yanenko, V.V., Shustov, B.M.: The Chelyabinsk event and the asteroid-comet hazard. Phys. Usp. **56**(8), 833–836 (2013)
Everhart, E.: Implicit single-sequence method for integrating orbits. Celest. Mech. **10**, 35–55 (1974)
Flynn, G.J., Durda, D.D., Kreft, J.W., Sitnitsky, I., Strait, M.: Catastrophic disruption experiments on the Murchison hydrous meteorite. In: Lunar and Planetary Science Conference, vol. 38. (2007)
Folkner, W., Williams, J., Boggs, D.: The Planetary and Lunar Ephemeris DE 421. IPN Progress Report, vol. 42–178, 34 p (2009)
Fries, M., Fries, J.: Doppler weather radar as a meteorite recovery tool. Meteorit. Planet. Sci. **45**, 1476–1487 (2010)
Gnos, E., Lorenzetti, S., Eugster, O., Jull, A.J.T., Hofmann, B.A., Al-Kathiri, A., Eggimann, M.: The Jiddat al Harasis 073 strewn field, Sultanate of Oman. Meteorit. Planet. Sci. **44**, 375–387 (2009)
Gritsevich, M.I.: Approximation of the observed motion of bolides by the analytical solution of the equations of meteor physics. Sol. Syst. Res. **41**(6), 509–514 (2007). http://dx.doi.org/10.1134/S003809460706007X
Gritsevich, M.I.: Estimating the terminal mass of large meteoroids. Dokl. Phys. **53**(11), 588–594 (2008a)
Gritsevich, M.I.: The Pribram, Lost City, Innisfree, and Neuschwanstein falls: an analysis of the atmospheric trajectories. Sol. Syst. Res. **42**(5), 372–390 (2008b)
Gritsevich, M.I.: Identification of fireball dynamic parameters. Moscow Univ. Mech. Bull. **63**(1), 1–5 (2008c). http://dx.doi.org/10.1007/s11971-008-1001-5
Gritsevich, M.I.: Determination of parameters of meteor bodies based on flight observational data. Adv. Space Res. **44**, 323–334 (2009)
Gritsevich, M., Koschny, D.: Constraining the luminous efficiency of meteors. Icarus **212**(2), 877–884 (2011)
Gritsevich, M.I., Stulov, V.P., Turchak, L.I.: Consequences for collisions of natural cosmic bodies with the earth atmosphere and surface. Cosmic Res. **50**(1), 56–64 (2012). http://dx.doi.org/10.1134/S0010952512010017
Gritsevich, M., Vinnikov, V., Kohout, T., Tóth, J., Peltoniemi, J., Turchak, L., Virtanen, J.: A comprehensive study of distribution laws for the fragments of Košice meteorite. Meteorit. Planet. Sci. **49**(3), 328–345 (2014a)
Gritsevich, M., Lyytinen, E., Moilanen, J., Kohout, T., Dmitriev, V., Lupovka, V., Midtskogen, V., Kruglikov, N., Ischenko, A., Yakovlev, G., Grokhovsky, V., Haloda, J., Halodova, P., Peltoniemi, J., Aikkila, A., Taavitsainen, A., Lauanne, J., Pekkola, M., Kokko, P., Lahtinen, P.,

Larionov, M.: First meteorite recovery based on observations by the Finnish Fireball Network. In: Rault, J.-L., Roggemans, P. (eds.) Proceedings of the International Meteor Conference, Giron, France, 18–21 September 2014. International Meteor Organization, ISBN 978-2-87355-028-8, pp. 162–169 (2014b)

Halliday, I., Griffin, A.A., Blackwell, A.T.: The Innisfree meteorite fall—a photo-graphic analysis of fragmentation, dynamics and luminosity. Meteoritics **16**(2), 153–170 (1981)

Halliday, I., Blackwell, A.T., Griffin, A.A.: Detailed records of many unrecovered meteorites in western Canada for which further searches are recommended. J. Roy. Astron. Soc. Can. **83**(2), 49–80 (1989)

Halliday, I., Griffin, A.A., Blackwell, A.T.: Detailed data for 259 fireballs from the Canada camera network and inferences concerning the influx of large meteoroids. Meteorit. Plan. Sci. **31**, 185–217 (1996)

Harris, A.W., Barucci, M.A., Cano, J.L., Fitzsimmons, A., Fulchignoni, M., Green, S.F., Hestroffer, D., Lappas, V., Lork, W., Michel, P., Morrison, D., Payson, D., Schäfer, F.: The European Union funded NEOShield project: A global approach to near-Earth object impact threat mitigation. Acta Astron. **90**(1), 80–84 (2013)

Hernández, G.: Discrete model for fragmentation with random stopping. Phys. Stat. Mech. Appl. **300**(1), 13–24 (2001)

Hughes, D.W., Harris, N.W.: The distribution of asteroid sizes and its significance. Planet. Space Sci. **42**(4), 291–295 (1994)

IAU SOFA Astrometry Tools, Release 10, 14 April, 2014, 81 p

IAU Division A: Fundamental Astronomy "Standards of Fundamental Astronomy Board", Release 10, 31 October, 2013

IERS Conventions: IERS Technical Note No. 36 (2010)

Jacchia, L.G., Verniani, F., Briggs, R.E.: An analysis of the atmospheric trajectories of 413 precisely reduced photographic meteors. Smithson. Contrib. Astrophys. **10**(1), 1–139 (1967)

Iordache, D.A., Chiroiu, V., Iordache, V.: Study of some theoretical descriptions of the dependence of the fracture parameters on the sample size. Rom. J. Phys. **50**(7–8), 847–858 (2005)

Jenniskens, P., Gural, P.S., Dynneson, L., Grigsby, B.J., Newman, K.E., Borden, M., Koop, M., Holman, D.: CAMS: Cameras for Allsky Meteor Surveillance to establish minor meteor showers. Icarus **216**(1), 40–61 (2011)

Kero, J., Szasz, C., Nakamura, T., Meisel, D.D., Ueda, M., Fujiwara, Y., Terasawa, T., Nishimura, K., Watanabe, J.: The 2009–2010 MU radar head echo observation programme for sporadic and shower meteors: radiant densities and diurnal rates. Mon. Not. Roy. Astron. Soc. **425**(1), 135–146 (2012)

Kohout, T., Havrila, K., Tóth, J., Husárik, M., Gritsevich, M., Britt, D., Borovička, J., Spurný, P., Igaz, A., Kornoš, L., Vereš, P., Koza, J., Zigo, P., Gajdoš, Š., Világi, J., Čapek, D., Krišandová, Z., Tomko, D., Šilha, J., Schunová, E., Bodnárová, M., Búzová, D., Krejčová, T.: Density, porosity and magnetic susceptibility of the Košice meteorites and homogeneity of its parent meteoroid. Planet. Space Sci. **93–94**, 96–100 (2014)

Kohout, T., Gritsevich, M., Lyytinen, E., Moilainen, J., Trigo-Rodríguez, J.M., Kruglikov, N., Ishchenko, A., Yakovlev, G., Grokhovsky, V., Haloda, J., Halodova, P., Meier, M.M.M., Laubenstein, M., Dimitrev, V., Lupovka, V.: Annama H5 meteorite fall: Orbit, trajectory, recovery, petrology, noble gases, and cosmogenic radionuclides. Meteorit. Planet. Sci. **50** (2015). MetSoc 2015 special issue, #5209

Kuznetsova, D., Gritsevich, M., Vinnikov, V.: The Kosice meteoroid investigation: From trajectory data to analytic model. In: Rault, J.-L., Roggemans, P. (eds.) Proceedings of the International Meteor Conference, Giron, France, 18–21 September 2014. International Meteor Organization, ISBN 978-2-87355-028-8, pp. 178–181 (2014)

Langbroek, M.: A spreadsheet that calculates meteor orbits. WGN J. Int. Meteor. Org. **32**(4), 109–110 (2004)

Laurance, M.R., Brownlee, D.E.: The flux of meteoroids and orbital space debris striking satellites in low earth orbit. Nature **323**, 136–138 (1986)

Linna, R.P., Åström, J.A., Timonen, J.: Dimensional effects in dynamic fragmentation of brittle materials. Phys. Rev. E **72**, 015601(R) (2005)

Love, S.G., Brownlee, D.E.: A direct measurement of the terrestrial mass accretion rate of cosmic dust. Science **262**, 550–553 (1993)

Lyytinen, E., Gritsevich, M.: Implications of the atmospheric density profile in the processing of fireball observations. Planet. Space Sci. **120**, 35–42 (2016)

Meibom, A., Balslev, I.: Composite power laws in shock fragmentation. Phys. Rev. Lett. **76**, 2492 (1996)

Meier, M.M.M., Welten, K.C., Riebe, M., Caffee, M.W., Gritsevich, M., Maden, C., Busemann, H.: Park Forest (L5) and the asteroidal source of shocked L chondrites. Meteorit. Planet. Sci. (2016)

McCrosky, R.E., Posen, A., Schwartz, G., Shao, C.-Y.: Lost City meteorite—its recovery and a comparison with other fireballs. J. Geophys. Res. **76**, 4090–4108 (1971)

Moreno-Ibáñez, M., Gritsevich, M., Trigo-Rodríguez, J.M.: New methodology to determine the terminal height of a fireball. Icarus **250**, 544–552 (2015)

Moreno-Ibáñez, M., Gritsevich, M., Trigo-Rodriguez, J.M., Lyytinen, E.: Current progress in the understanding of the physics of large bodies recorded by photographic and digital fireball networks. In: Roggemans, A., Roggemans P. (eds.) Proceedings of the International Meteor Conference, Egmond, The Netherlands, 2–5 June 2016, pp. 192–196

Moreno-Ibáñez, M., Gritsevich, M., Trigo-Rodríguez, J.M.: Measuring the terminal heights of bolides to understand the atmospheric flight of large asteroidal fragments. In: Trigo-Rodríguez, J.M., Gritsevich, M., Palme H. (eds.) Assessment and Mitigation of Asteroid Impact Hazards, pp. 129–151. Springer, New York (2017)

Oberst, J., Molau, S., Heinlein, D., Gritzner, C., Schindler, M., Spurny, P., Ceplecha, Z., Rendtel, J., Betlem, H.: The "European Fireball Network": Current status and future prospects. Meteorit. Planet. Sci. **33**(1), 49–56 (1998)

Oddershede, L., Dimon, P., Bohr, J.: Self-organized criticality in fragmenting. Phys. Rev. Lett. **71**(19), 3107–3110 (1993)

Oddershede, L., Meibom, A., Bohr, J.: Scaling analysis of meteorite shower mass distributions. Europhys. Lett. **43**(5), 598–604 (1998)

Perna, D., Barucci, M.A., Fulchignoni, M.: The near-Earth objects and their potential threat to our planet. Astron. Astrophys. Rev. **21**, 65 (2013)

Povinec, P.P., Masarik, J., Sýkora, I., Kováčik, A., Beňo, J., Meier, M.M.M., Wieler, R., Laubenstein, M., Porubčan, V.: Cosmogenic nuclides in the Košice meteorite: experimental investigations and Monte Carlo simulations. Meteorit. Planet. Sci. **50**, 880–892 (2015). doi:10.1111/maps.12380

Poppe, A., James, D., Horányi, M.: Measurements of the terrestrial dust influx variability by the Cosmic Dust Experiment. Planet. Space Sci. **59**(4), 319–326 (2011)

Räbinä, J., Mönkölä, S., Rossi, T., Markkanen, J., Gritsevich, M., Muinonen, K.: Controlled time integration for numerical simulation of meteor radar reflections. J. Quant. Spectrosc. Radiat. Transf. **178**, 295–305 (2016)

Reinhardt, J., Chen, X., Liu, W., Manchev, P., Paté-Cornell, M.: Project Fox: Assessing risks posed by asteroids. Amer. Geophys. Union, Fall Meeting 2013, abstract #NH23D-1547 (2013)

Renshaw, C.E., Schulson, E.M.: Universal behaviour in compressive failure of brittle materials. Nature **412**(6850), 897–900 (2001)

Rivkin, A.S., Bottke, W.F.: Hypovelocity impacts in the asteroid belt. Lunar Planet. Sci. **27**, 1077–1078 (1996)

Shustov, B.M.: On coordinated approach to the problem of asteroid-comet impact hazard. Cosmic Res. **48**(5), 378–391 (2010)

Silber, E.A., ReVelle, D.O., Brown, P.G., Edwards, W.N.: An estimate of the terrestrial influx of large meteoroids from infrasonic measurements. J. Geophys. Res. **114**, E08006 (2009). doi:10.1029/2009JE003334

Silber, E.A., Brown, P.G.: Optical observations of meteors generating infrasound – I: acoustic signal identification and phenomenology. J. Atmos. Sol. Terr. Phys. **119**, 116–128 (2014). doi:10.1016/j.jastp.2014.07.005
Silber, E.A., Brown, P.G., Krzeminski, Z.: Optical observations of meteors generating infrasound: weak shock theory and validation. J. Geophys. Res. Planet. **120**, 413–428 (2015). doi:10.1002/2014JE004680
Sotolongo-Costa, O., Rodriguez, A.H., Rodgers, G.J.: Tsallis entropy and the transition to scaling in fragmentation. Entropy **2**(4), 172–177 (2000)
Sotolongo-Costa, O., Gamez, R., Luzon. F., Posadas A., Weigandt Beckmann, P.: Non Extensivity in Meteor Showers. arXiv:0710.4963 (2007)
Spahn, F., Neto, E.V., Guimarães, A.H.F., Gorban, A.N., Brilliantov, N.V.: A statistical model of aggregate fragmentation. New J. Phys. **16**(1), 13031–13041 (2014)
Spurný, P., Oberst, J., Heinlein, D.: Photographic observations of Neuschwanstein, a second meteorite from the orbit of the Příbram chondrite. Nature **423**, 151–153 (2003)
Spurný, P., Haloda, J., Borovička, J.: Mystery of the Benesov bolide revealed after 20 years. In: Proceedings of the ACM 2012 in Niigata, Japan. LPI Contribution No. 1667, id.6143 (2012)
Stulov, V.P.: Interactions of space bodies with atmospheres of planets. Appl. Mech. Rev. **50**(11), 671–688 (1997). http://dx.doi.org/10.1115/1.3101678
Tóth, J., Svoreň, J., Borovička, J., Spurný, P., Igaz, A., Kornoš, L., Vereš, P., Husárik, M., Koza, J., Kučera, A., Zigo, P., Gajdoš, Š., Világi, J., Čapek, D., Krišandová, Z., Tomko, D., Šilha, J., Schunová, E., Bodnárová, M., Búzová, D., Krejčová, T.: The Košice meteorite fall: Recovery and strewn field. Meteorit. Planet. Sci. **50**(5), 853–863 (2015)
Trigo-Rodríguez, J.M., Llorca, J., Castro-Tirado, A.J., Ortiz, J.L., Docobo, J.A., Fabregat, J.: The Spanish fireball network. Astron. Geophys. **47**(2), 26–28 (2006)
Trigo-Rodríguez, J.M., Lyytinen, E., Gritsevich, M., Moreno-Ibáñez, M., Bottke, W.F., Williams, I., Lupovka, V., Dmitriev, V., Kohout, T., Grokhovsky, V.: Orbit and dynamic origin of the recently recovered Annama's H5 chondrite. Mon. Not. Roy. Astron. Soc. **449**(2), 2119–2127 (2015)
Turchak, L.I., Gritsevich, M.I.: Meteoroids interaction with the Earth atmosphere. J. Theor. Appl. Mech. **44**(4), 15–28 (2014)
Vaubaillon, J., Koten, P., Margonis, A., Tóth, J., Rudawska, R., Gritsevich, M., Zender, J., McAuliffe, J., Pautet, P.D., Jenniskens, P., Koschny, D., Colas, F., Bouley, S., Maquet, L., Leroy, A., Lecacheux, J., Borovicka, J., Watanabe, J., Oberst, J.: The 2011 Draconids: The first European airborne meteor observation campaign. Earth Moon Planet. **114**(3–4), 137–157 (2015)
Vinković, D., Gritsevich, M., Srećković, V., Pečnik, B., Szabó, G., Debattista, V., Škoda, P., Mahabal, A., Peltoniemi, J., Mönkölä, S., Mickaelian, A., Turunen, E., Kákona, J., Koskinen, J., Grokhovsky, V.: Big data era in meteor science. In: Roggemans, A., Roggemans, P. (eds.) Proceedings of the International Meteor Conference, pp. 319–329 (2016)
Vinnikov, V., Gritsevich, M., Kuznetsova, D., Turchak, L.: Empirical fragment distributions in meteorites, LPSC Abstract # 1439 (2014). http://www.hou.usra.edu/meetings/lpsc2014/pdf/1439.pdf
Vinnikov, V., Gritsevich, M., Turchak, L.: Shape estimation for Košice, Almahata Sitta and Bassikounou meteoroids. In: Proceedings of the International Astronomical Union (Cambridge Journals UK), vol. 10, pp. 394–396 (2015)
Vinnikov, V.V., Gritsevich, M.I., Kuznetsova, D.V., Turchak, L.I.: Estimation of the initial shape of meteoroids based on statistical distributions of fragment masses. Dokl. Phys. **61**(6), 305–308 (2016a). http://dx.doi.org/10.1134/S1028335816060021
Vinnikov, V., Gritsevich, M., Kuznetsova, D., Krivonosova, O., Zhilenko, D., Turchak, L.: Statistical approach to meteoroid shape estimation. In: Roggemans, A., Roggemans, P. (eds.) Proceedings of the International Meteor Conference, Egmond, The Netherlands, 2–5 June 2016b, pp. 330–332
Wallace, P.: SOFA: Standards of Fundamental Astronomy. Highlights Astron. **11A**, 191 (1998)

Weryk, R.J., Brown, P.G., Domokos, A., Edwards, W.N., Krzeminski, Z., Nudds, S.H., Welch, D.L.: The Southern Ontario All-sky Meteor Camera network. Earth Moon Planet. **102**, 241–246 (2008)

Weryk, R.J., Campbell-Brown, M.D., Wiegert, P.A., Brown, P.G., Krzeminski, Z., Musci, R.: The Canadian Automated Meteor Observatory (CAMO): System overview. Icarus **225**(1), 614–622 (2013)

Whipple, F.L., Jacchia, L.G.: Reduction methods for photographic meteor trails. SCOA **1**, 183–206 (1957)

Zoladek, P.: PyFN—multipurpose meteor software. In: Proceedings of the International Meteor Conference, Sibiu, Romania, 15–18 September, 2011, International Meteor Organization, pp. 53–55 (2011)

Zolensky, M., Bland, P., Brown, P., Halliday, I.: Flux of extraterrestrial materials. In: Meteorites and the Early Solar System II, pp. 869–888 (2006)

Zuluaga, J., Ferrin, I.: A preliminary reconstruction of the orbit of the Chelyabinsk meteoroid, arXiv:1302.5377 (2013)

Zuluaga, J., Ferrin, I., Geens, S.: The orbit of the Chelyabinsk event impactor as reconstructed from amateur and public footage. arXiv:1303.1796 (2013)

Detection of Nocturnal and Daylight Bolides from Ebre Observatory in the Framework of the SPMN Fireball Network

E. Blanch, Josep M. Trigo-Rodríguez, J.M. Madiedo, E. Lyytinen, M. Moreno-Ibáñez, M. Gritsevich, and D. Altadill

Abstract Bright daylight fireballs are rarely studied due to the lack of data, and the difficulty to calibrate the trajectory descriptions given by eyewitnesses. Only in few occasions casual recordings have been able to get orbital information of daylight

E. Blanch (✉) • D. Altadill
Observatori de l'Ebre (OE, CSIC — Universitat Ramon Llull), Horta Alta, 38, Roquetes, Tarragona 43520, Spain
e-mail: eblanch@obsebre.es

J.M. Trigo-Rodríguez
Institute of Space Sciences (IEEC-CSIC), Meteorites, Minor Bodies and Planetary Sciences Group, Campus UAB, c/Can Magrans s/n, 08193 Cerdanyola del Vallés (Barcelona), Catalonia, Spain
e-mail: trigo@ice.csic.es

J.M. Madiedo
Facultad de Física, Departamento de Física Atómica, Molecular y Nuclear, Universidad de Sevilla, Sevilla 41012, Spain

Facultad de Ciencias Experimentales, Universidad de Huelva, Huelva 21071, Spain

E. Lyytinen
Finnish Fireball Network, Kopernikuksentie 1, Helsinki 00130, Finland

M. Moreno-Ibáñez
Meteorites, Minor Institute of Space Sciences (IEEC-CSIC), Meteorites, Minor Bodies and Planetary Sciences Group, Campus UAB, c/Can Magrans s/n, 08193 Cerdanyola del Vallés (Barcelona), Catalonia, Spain

Finnish Geospatial Research Institute, P.O. Box 15, Masala 02431, Finland

M. Gritsevich
Department of Physics, University of Helsinki, Gustaf Hällströmin katu 2a, P.O. Box 64, FI-00014 Helsinki, Finland

Finnish Geospatial Research Institute, P.O. Box 15, Masala 02431, Finland

Institute of Physics and Technology, Ural Federal University, Mira Str. 19, Ekaterinburg 620002, Russia

Dorodnicyn Computing Centre, Russian Academy of Sciences, Department of Computational Physics, Vavilova str. 40, Moscow 119333, Russia

J.M. Trigo-Rodríguez et al. (eds.), *Assessment and Mitigation of Asteroid Impact Hazards*, Astrophysics and Space Science Proceedings 46,
DOI 10.1007/978-3-319-46179-3_9

bolides. Fortunately, modern video detectors can be also adapted for monitoring the sky during broad daylight and we present here some first results. In 2006 we started a continuous monitoring of large bolides over Catalonia during the night with the goal of recording meteorite-dropping bolides and recover new meteorites. Multiple station fireball monitoring in broad daylight was extended during early 2014. With such a goal we installed a new meteor video-detection station at the Ebre Observatory (URL-CSIC) to be complementary with the IEEC-CSIC stations operational at Montseny and Folgueroles, part of the Spanish Meteor and fireball Network (SPMN) initiative. The strategic location of this station will allow recording bolides during daytime over the Spanish north-east sky, and particularly over Catalonia. Just as an example of the station capabilities, here we describe two relevant bolides recorded using a new color video camera during the first year of continuous operation: a bright South Taurid bolide appeared on Nov. 6, 2015 and a daylight bolide SPMN100215 disrupted over Catalonia on Feb. 10, 2015.

1 Introduction

It is estimated by different techniques that every year about 50,000–100,000 tonnes of interplanetary matter penetrates into the Earth's atmosphere, mostly arriving in the size range in which they are known as meteoroids (Bland et al. 1996). These particles, with diameters from few microns up to 10 m, are usually rocks detached from asteroids, comets or, more rarely, from the Moon or Mars (Ceplecha et al. 1998). The dense atmosphere of the Earth is shielding us very efficiently and does not allow the majority to reach the Earth's surface (Lovell 1954; McKinley 1961; Gritsevich et al. 2012). As they enter the atmosphere at high speeds, in a geocentric velocity range between 11 and 73 km/s, they suddenly interact with the upper atmosphere and produce a meteor Bouquet et al. 2014). Their surfaces are progressively heated, and once they penetrate into the atmosphere undergo an abrupt collision with atoms and molecules of the dense atmosphere producing friction and subsequent heating. In this process, known as ablation, the meteoroid loses most of its mass, in the form of solid fragments or hot gas fluid matter (Ceplecha et al. 1998; Trigo-Rodríguez et al. 2004a). A small part of the kinetic energy is then transformed into heat and luminosity (Gritsevich and Koschny 2011; Bouquet et al. 2014), producing luminous phenomenon known as a meteor or, commonly, a shooting star. Those meteors brighter than Venus (magnitude –4 or lower) are called bolides or fireballs and announce the entry of larger particles, from a few centimeters to several meters. In that upper range, meteoroids with hundreds of kilograms in mass may partially survive its rough passage through the atmosphere and can reach the Earth's surface as meteorites, often very fragmented. As they come from other solar system bodies, their scientific study is extremely valuable because meteorites can give information about the processes of formation, collisional and dynamical evolution of asteroids, comets and planets (Burbine et al. 2002; Ceplecha et al. 1998). Moreover, it is of great importance, not only to recover the meteorite, but also to obtain the heliocentric orbit to better understand its origin and the dynamic pathways followed to reach our planet (for a review see e.g. Trigo-Rodríguez et al.

2015). The registration of these fireballs from several stations allows reconstruct the trajectories of these objects and deduce from them the places of falling meteorites. In addition, from the initial velocity and geocentric radiant, the heliocentric orbit can be determined and a dynamic association with its progenitor body can be achieved.

Consequently, establishing a good network of meteor and fireball detection is necessary for both, to recover the possible meteorite and to obtain as much information as possible of the bodies delivering meteorite-dropping bolides to Earth. In Spain, the SPanish Meteor and fireball Network (SPMN) has been continuously monitoring the sky for almost 20 years (Trigo-Rodríguez et al. 2006b). In this chapter we present a new daylight fireball monitoring program that has been undertaken in the framework of the SPMN to improve the detection of bolides occurred at the North-East Spain, especially bolides seen in broad daylight which are more difficult to observe, and that can produce meteorite falls (Trigo-Rodríguez et al. 2006a, b) without any chance to get orbital information. In order to exemplify the results that can be obtained, we describe two fireball events recorded during the first year of operation to remark the relevance of implementing these fireball monitoring initiatives all over the world.

2 The SPanish Meteor and Fireball Network (SPMN)

The SPMN was established in 1999 in order to study the orbits of meteoroids producing bright fireballs over the Iberian Peninsula and neighbouring areas and to accomplish the recovery of fresh meteorites for direct study in the laboratories. Since its establishment, the SPMN network made possible the study of several superbolides and promoted the recovery of two meteorites: the L6 ordinary chondrite Villalbeto de la Peña occurred in 2004 (Llorca et al. 2005, 2007; Trigo-Rodríguez et al. 2006a) and the eucrite Puerto Lapice in 2007 (Llorca et al. 2009; Trigo-Rodríguez et al. 2009). The researchers participating in the SPMN network have also identified and characterized an unnoticed meteorite fall occurred in 1931: Ardón L6 ordinary chondrite (Trigo-Rodríguez et al. 2014a) and are currently investigating another previously unnoticed meteorite fall occurred in 1916 (Trigo-Rodríguez et al. 2014b). During this time, the network has also contributed to popularize this field in Spain. To date, the SPMN homepage (www.spmn.uji.es) receives thousands of visits every month, and has become an updated reference of the fireball events that have occurred all over Spain, and bordering countries (see e.g. Trigo-Rodríguez et al. 2006b, 2008; Madiedo et al. 2014a, b, c, d). In the SPMN homepage a list of bright fireballs recorded over the Iberian Peninsula is continuously updated with additional information about their origin.

The network has been growing since the first stations set up during the Ph.D. thesis of Dr. Trigo-Rodríguez (Trigo-Rodríguez et al. 2002) using public funding received from different research projects, CSIC, IEEC and also private funds. During the last decade tens of first-quartile peer-reviewed papers have been published describing the meteor activity, obtaining meteoroid orbits, and computing meteor fluxes using this video systems described elsewhere (see e.g. the pioneer work by Madiedo and Trigo-Rodríguez 2008). Since then we have recorded hundreds of nighttime fireballs and several daytime fireballs (Madiedo et al. 2008). We are now reconstructing their atmospheric trajectories and orbits by using the stereoscopic images taken from other stations, just by using the accurate astrometric reduction methods described in Trigo-Rodríguez et al. (2004b) and other recent software developed by the team (Madiedo et al. 2011). In 2015 the network had 30 video and CCD stations monitoring the atmosphere for bright fireballs occurred over Portugal, Spain, north of Morocco and south of France. Thanks to the records of the network and with the use of new accurate techniques (Trigo-Rodríguez et al. 2004a) the SPMN team can obtain trajectory and orbital information of meteorite-dropping bolides that provide new clues regarding the dynamical processes that deliver meteorites to the Earth (Trigo-Rodríguez et al. 2006a). This research network is integrated by researchers of several universities and research centers around Spain and it is coordinated from the Institute of Space Sciences (CSIC-IEEC) (northern Spain stations), Universidad Complutense de Madrid (central stations) and Universidad de Huelva (southern stations).

3 The Ebre Observatory Meteor Detection Station

In order to improve the detection of meteor trails and to better help to locate the places of falling meteorites in the northeast of Iberian Peninsula, the Institute of Space Sciences (IEEC-CSIC) set up in 2014 the first meteor video detection station in the province of Tarragona, specifically at the Ebro Observatory (URL-CSIC), OE, located at the south of Catalonia, north-east Spain (40.8 N, 0.5 E).

The Ebre Observatory is a historic research institute founded by the Society of Jesus in 1904 to study the Sun-Earth relationships. The main activity of the OE is research, observation and outreach in the field of Earth Sciences and Space, Upper Atmosphere, Geomagnetism, Meteorology, and Seismology (Garcia and Roca 2007) and more recently Hydrology. The OE has previously contributed with the SPMN providing seismic records produced by the sonic booms generated by deep-penetrating bolides (Tapia and Trigo-Rodríguez 2012).

The Ebre Observatory video-detection station was installed on August 2014 and immediately became operational as a node of the SPMN network together with other stations installed by the IEEC-CSIC in northern Catalonia (Table 1). It consists of several video cameras, one observes the sky in south-west direction, recording the east of the Valencian Community and the center of the Iberian Peninsula, and the other observes the sky in north direction, recording Aragon and Catalonia. Both cameras use sophisticated software that allows detecting fireballs in the sky and recording the image that will be used astrometrically. Given the location of the OE, the orientation of the camera installed in the observatory pointing northward is strategic for SPMN researchers because it prevents the transit of the Sun and can provide coverage to all objects ablating over the sky of Aragon, and Catalonia, 24 h a day, 7 days a week.

Table 1 SPMN stations involved in the detections discussed in this chapter

Station #	Station (Province)	Longitude (E)	Latitude (N)	Alt. (m)	Imaging system
1	Montsec (Lleida)	00° 43′ 46″	42° 03′ 05″	1570	AS, WFV
2	Montseny (Girona)	02° 31′ 14″	41° 43′ 17″	300	WFV
3	Observatori de l'Ebre (Tarragona)	00° 29′ 44″	40° 49′ 16″	50	WFV

Acronyms for the different imaging systems are: AS (low-scan-rate CCD all-sky camera), WF (low-scan-rate CCD wide-field camera), and WFV (Wide field video cameras)

The camera data obtained here is a *Presentco* digital video camera with a half inch interline transfer CCD chip operating under a 2:1 interlaced scanning system. It provides images of 640×480 pixel^2. Our current development is a PAL video system, recording at 25 frames per second with a standard video lens allows getting a maximum horizontal field of view of $\sim 100°$ and providing a fireball limiting magnitude of about -6 in broad daylight. This video system, once orientated correctly with a software implemented to patrol the same atmospheric volume than nearby stations (see e.g. Trigo-Rodríguez et al. 2004a, b) allows obtaining tens of meteoroid orbits every night, increasing the coverage of the SPMN network and allowing the detection of elusive daytime fireballs.

Given that the meteor phenomenon is unexpected, the procedure involves monitoring continuously the night sky. OE video system monitors a common atmospheric volume with other camera systems available, from which stereoscopic images are obtained (see e.g. Figs. 1 and 2). An accurate astrometry of the images allows getting the meteor tracks measured in equatorial coordinates from which using the intersection of planes is possible to reconstruct the real atmospheric path of the fireball (Ceplecha 1987; Trigo-Rodríguez et al. 2004b).

From this collaboration, researchers of the SPMN may obtain a more detailed activity in the peninsular Northeast region and increase the level of their scientific contribution through the images captured by this station. Since its installation, the station has registered and analyzed 31 nighttime bright bolides and 2 daytime bright bolides which have only been registered from this station of the SPMN.

4 Results and Discussion

New color cameras tested in the framework of Spanish AYA2011-26522 research project were applied by our team to record bright fireballs and monitor the sky continuously. Just as an example of night and day capabilities we include the detection of a nocturnal and a daylight fireball.

The nocturnal event described here was a -12 magnitude Taurid fireball recorded on Nov. 6, 2015 at 5 h 12 m 45 s $\pm$ 1 s UTC from Observatori de l'Ebre (see Fig. 1) and Montsec Astronomical Observatory (OAdM, www.oadm.cat). The meteoroid suffered different flares associated with progressive fragmentation during

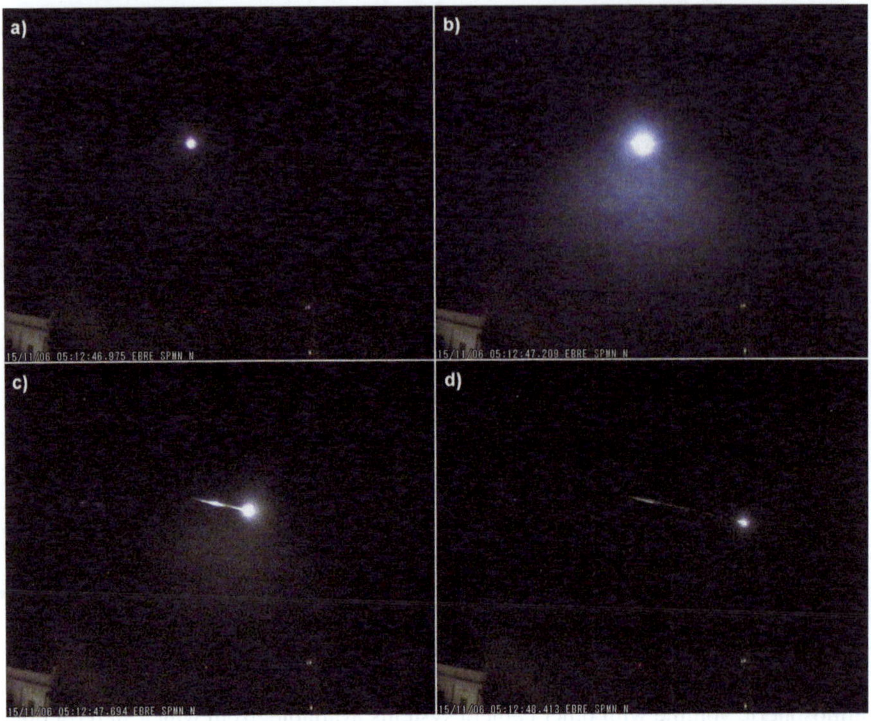

Fig. 1 Image sequences of the video recorded from North faced camera installed at the Ebre Observatory of a fireball (SPMN061115D) occurred on November 6th, 2015. (**a**) Frame #58 at 5:12:46.9 UT, (**b**) #64 at 47.2 s, (**c**) #76 at 47.7 s and (**d**) #94 at 48.4 s

atmospheric deepening, including one with a magnitude similar to the full Moon (Fig. 1b). After the bright flare shown in the second selected frame, a persistent train was left behind. Successive flares occurred before the total extinction of the luminous phase. The full bolide sequence in a .gif archive format can be seen at our continuously updated SPMN list of bright bolides: http://www.spmn.uji.es/ESP/SPMNlist.html

The SPMN061115 fireball was also imaged by the all-sky CCD camera operated at Observatori Astronòmic del Montsec (OAdM) and a part of the image is shown in Fig. 2. The astrometric application of this all-sky camera is described elsewhere, so it is not described here (Trigo-Rodríguez et al. 2004a, 2008). According with the trajectory and orbital data given in Tables 2 and 3, this event was clearly associated with the South Taurids stream from 2P/Encke comet.

A second example selected here is a spectacular daylight bolide named SPMN100215 and disrupted over Catalonia on Feb. 10, 2015 at 14 h 6 m 22 s $\pm$ 0.5 UTC whose selected frames are shown in Fig. 3. This event was widely seen by eyewitnesses when crossing Catalonia skies, and even allowed a casual picture of the persistent train from the Pyrenees. The fireball appeared almost at midday so

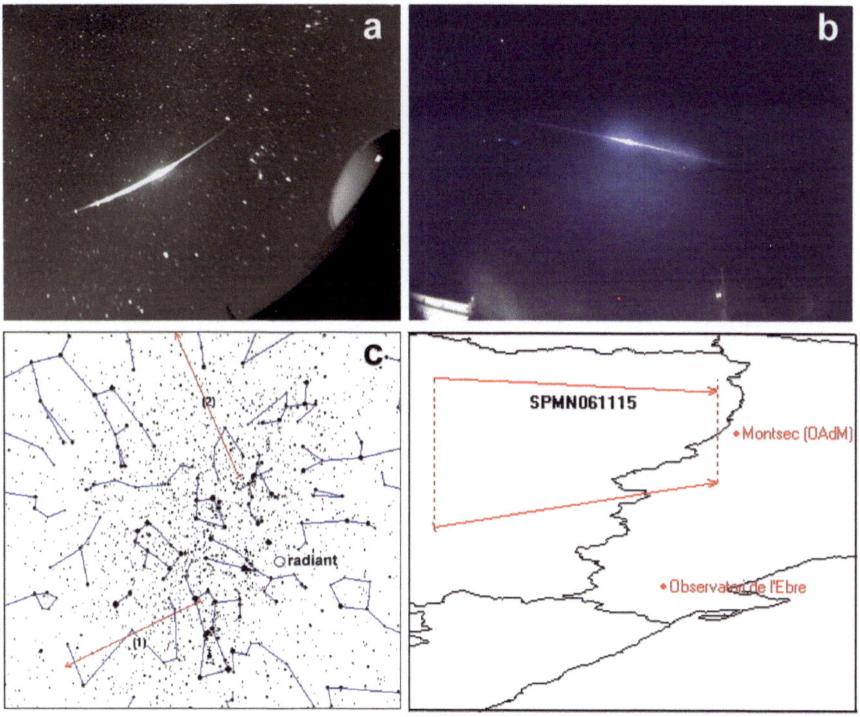

Fig. 2 Image of the SPMN061115 Taurid fireball recorded from (**a**) OAdM all-sky camera (IEEC-CSIC), (**b**) Observatori de l'Ebre, (**c**) radiant determination, and (**d**) trajectory reconstruction

Table 2 Trajectory, radiant and velocity data for the SPMN061116 Taurid bolide

	H_b	H_{max}	H_e	α_g (°)	δ_g (°)	V_g (km/s)	V_g	V_h
	88.4	76.2	65.4	55.2 ± 0.4	$+16.5 \pm 0.4$	31.5 ± 0.5	29.8 ± 0.5	38.8 ± 0.4
STA [1]	–	–	–	52.2	+14.0	–	27	–

H_b, H_{max} and H_e are the height for the beginning, maximum, and ending parts of the computed trajectory. Radiant is given for equinox (J2000.0). Last three columns are the velocity at infinity, geocentric and heliocentric. The average radiant and velocity for the South Taurids (STA) is taken from Cook (1973) and Jenniskens (2006)

Table 3 Orbital elements of SPMN061116 Taurid bolide

q (AU)	a (AU)	e	i (°)	ω (°)	Ω (°)
0.366 ± 0.008	3.1 ± 0.3	0.882 ± 0.013	4.9 ± 0.5	110.3 ± 0.9	43.9788 ± 0.0001

Equinox (J2000.00)

the Montseny station looking south had been turned off to avoid having the Sun in the field of view. For that reason the event was recorded only from Observatori de l'Ebre, but still in such condition the fireball detection was able to provide interesting results.

Fig. 3 Image sequences of the first daytime fireball (SPMN100215) recorded in broad daylight with the North faced camera on 10 February at 14:06 UT. It was observed the progressive fragmentation of the meteoroid as it entered the Earth's atmosphere. (**a**) Frame #28 at 14:06:22.7 UT, (**b**) #54 at 23.8 s, (**c**) #66 at 24.3 s, (**d**) #72 at 24.5 s, (**e**) #79 at 24.8 s, (**f**) #82 at 24.9 s, (**g**) #84 at 25.0 s, (**h**) #87 at 25.1 s and (**i**) #96 at 25.5 s

Table 4 Trajectory geometry and ablation coefficient (σ) as a function of the meteoroid velocity at infinity (V_∞)

V_∞ (km/s)	α	β	σ (s^2/km^2)	M_{beg} (kg)	M_{end} (kg)
20	11.8	3.17	0.048	270	0.077
25	38.5	2.97	0.028	8	0.003
30	113	2.77	0.018	0.3	0

Table 5 Range of orbital elements of SPMN100215, Equinox (J2000.00)

V (km/s)	q (AU)	a (AU)	e	i (°)	ω (°)	Ω (°)
20	0.593	1.33	0.554	2.19	263.11	141.32
25	0.540	2.13	0.746	1.97	266.63	141.32
30	0.497	7.37	0.933	1.80	268.36	141.32

We measured 18 individual frames of the daylight fireball, along the entry track. With these we derived a one station solution, using an approximated analytic formula for velocity and the standard atmosphere density values (Lyytinen and Gritsevich 2013). Unfortunately the very beginning of the bolide is not visible in the video to try to scale with this, so only a range of orbital values can be obtained (Tables 4 and 5). In that table three different entry velocity 20, 25 and 30 km/s were considered. The fireball came from an azimuth-direction of 210° (with uncertainties smaller than 2° for each velocity value alternative solutions), and with a slope of about 29.5° at the beginning of the luminous flight. Such geometry corresponds to a radiant in RA: 328°, Dec: −13°. The different velocity solutions differ only by a few tenths of a degree between them.

The solutions for each velocity allow deriving the ballistic coefficient α and the mass loss parameter β described in Gritsevich (2009). Both parameters are used to 'scale' a fireball's flight as recorded by the cameras. The ballistic coefficient, α, characterizes the drag intensity since it is proportional to the mass of the atmospheric column with the cross section along the trajectory divided by the pre-atmospheric mass (M_{beg}). The mass loss parameter, β, which if multiplied and divided by M_{beg}, is proportional to the pre-atmospheric kinetic energy of the meteoroid divided by the energy required for its destruction. Once these parameters are found based on observations, they can be used to mathematically describe the changes in height, mass, and velocity along the meteor's trajectory. Furthermore, these values are the key to predicting the probable consequences of the fireball (see e.g. Gritsevich et al. 2012; Moreno-Ibáñez et al. 2015) so it is indeed critical to define these parameters using the most rigorous approach available.

In this work we have found these parameters by applying the latest modification of the method implemented by Lyytinen and Gritsevich (2016) which allows considering different atmospheric models. Also the meteoroid ending mass is inferred. Certainly the computed values for the meteoroid initial mass are strongly dependent of the velocity because bigger velocity means higher in the atmosphere and less dense atmosphere. These inferred ablation coefficient values are given in Table 4, together with the computed beginning and ending masses (M_{beg} and M_{end}). In any case, a meteorite recovery is highly unlikely because the quite fast velocity means that the meteoroid ablated practically to end. In the 20 km/s alternative the terminal mass would have been something like 77 g, a few grams in the second alternate velocity and below one gram in the 30 km/s alternative.

In the 25 km/s alternative the height at the start of measurements, frame 25 is 68 km, which is quite low for this velocity beginning height but the fireball it is already quite bright at this stage and may have started a lot earlier. In fact, that is one of the problems of daylight fireball detection: the beginning and ending parts of the bolides are usually undetectable because the intrinsic brightness of the sky during the day. The last measurable frame 102 had the height of 33.9 km.

From the trajectory and orbital data obtained we can conclude that SPMN100215 fireball was of sporadic origin, but particularly coming from the so-called ‘helion’ source. We note that the ‘sporadic source regions’ within the radiant distribution comprise useful groupings of orbital types. The main radiant groupings for the sporadic meteor background associated with the decay of cometary meteoroids are concentrated about the ecliptic in the directions of the apex of the Earth’s way, the antihelion and the helion (Hawkins 1956). The only difference between the orbits detected in the antihelion and helion regions is whether the meteoroid impacted our planet pre- or post-perihelion in its orbit as described by Galligan and Baggaley (2005). In any case, we must to remark that in general daytime bolides and shower meteors will not necessarily come from those specific regions of the sky. For example, the Cheliabinsky fireball had the entry from about the direction of the Sun, but the velocity was a little bit low for typical (anti)helion. Other energetic event, the 2015 Sept. 7 superbolide (one of the biggest energy fireballs in the NEO-JPL list http://neo.jpl.nasa.gov/fireballs/) also came from roughly the Sun direction with the velocity of 23 km/s, lower than expected for an helion type meteoroid.

5 Conclusions

Digital video cameras are currently providing valuable information about the sources of large meteoroids reaching the Earth. We have tested a new color digital camera to understand its applicability to fireball detection day and night. We have reached the following conclusions:

- Color video cameras allow implementing a continuous fireball monitoring, achievable 24 h a day, 7 days a week. We have found that, over the limiting meteor magnitude, there is not other illumination limitation, but those originated by cameras pointing south that are interfered by Sun transit through the FOV.
- Multi-station data provides the most accurate trajectory, radiant and orbital determination. From the subtended atmospheric volumes by the cameras flux calculations can be made.

– Fireballs detected by well-calibrated video-cameras from only one station still can provide useful information to decipher the meteoroid source and understand their nature. A list of bright fireball events over Spain is being constantly updated in order to provide public information about the origin of large meteoroids producing these impressive and unexpected events.
– Fireball monitoring cameras should be installed in research centers and observatories, not only because of their scientific outcome, but also for raising the social interest on these research topics.

Acknowledgements We acknowledge funding support from the Consejo Superior de Investigaciones Científicas (CSIC, project: #201050I043), Institut d'Estudis Espacials de Catalunya (IEEC), and the Spanish Ministry of Science (research project AYA2015-67175-P, PI: Dr. Trigo-Rodríguez). MG was supported, in part, by the Academy of Finland project No 260027, by the Magnus Ehrnrooth Foundation travel grant, and by the Russian Foundation for Basic Research, project Nos 14-08-00204, 16-05-00004 and 16-07-01072.

The work of DA and EB is supported by Universitat Ramon Llull projects 2016-URL-Trac-001 supported by "Obra Social la Caixa" and 2016-URL-IR-001 supported by "Generalitat de Catalunya".

References

Blanch, E., Trigo-Rodríguez, J.M., Madiedo, J.M., Lyytinen, E., Moreno-Ibáñez, M., Gritsevich, M., Altadill, D.: Detection of nocturnal and daylight bolides from Ebre Observatory in the framework of the SPMN fireball network. In: Trigo-Rodríguez, J.M., Gritsevich, M., Palme H. (eds.) Assessment and Mitigation of Asteroid Impact Hazards, pp. 185–197. Springer, New York (2017)

Bland, P.A., Smith, T.B., Jull, A.J.T., Berry, F.J., Bevan, A.W.R., Pillinger, C.T.: The flux of meteorites to the Earth over the last 50 000 years. Mon. Not. R. Astron. Soc. **283**, 551–565 (1996). doi:10.1093/mnras/283.2.551

Bouquet, A., Baratoux, D., Vaubaillon, J., Gritsevich, M.I., Mimoun, D., Mousis, O., Bouley, S.: Simulation of the capabilities of an orbiter for monitoring the entry of interplanetary matter into the terrestrial atmosphere. Planet. Space Sci. **103**, 238–249 (2014). http://dx.doi.org/10.1016/j.pss.2014.09.001

Burbine, T.H., McCoy, T.J., Meibom, A., Gladman, B., Keil, K.: Meteoritic parent bodies: their number and identification. In: Bottke Jr., W.F., et al. (eds.) Asteroids III, pp. 653–667. University of Arizona, Tucson (2002)

Ceplecha, Z.: Geometric, dynamic, orbital and photometric data on meteoroids from photographic fireball networks. Bull. Astron. Inst. Czech. **38**, 222–234 (1987b) (ISSN 0004–6248)

Ceplecha, Z., Borovička, J., Elford, W.G., Revelle, D.O., Hawkes, R.L., Porubčan, V., Šimek, M.: Meteor phenomena and bodies. Space Sci. Rev. **84**, 327–471 (1998)

Cook, A.F.: A working list of meteor streams. In: Hemegway, C.L., Millman, P.M., Cook, A.F. (eds.) Evolutionary and Physical Properties of Meteoroids, Proc. IAU Colloq. 13, vol. SP-319, pp. 183–191. NASA, Washington (1973)

Galligan, D.P., Baggaley, W.J.: The radiant distribution of AMOR radar meteors. Mon. Not. R. Astron. Soc. **359**, 551–560 (2005)

Garcia, M., Roca, A.: Observatori de l'Ebre, un segle d'història (1904–2004). Roquetes (Tarragona), Observatori de l'Ebre. ISBN: 978-84-88253-05-7, 310 p, il. fot., 28 cm (2007)

Gritsevich, M.I.: Determination of parameters of meteor bodies based on flight observational data. Adv. Space Res. **44**(3), 323–334 (2009). http://dx.doi.org/10.1016/j.asr.2009.03.030

Gritsevich, M., Koschny, D.: Constraining the luminous efficiency of meteors. Icarus **212**(2), 877–884 (2011)

Gritsevich, M.I., Stulov, V.P., Turchak, L.I.: Consequences for collisions of natural cosmic bodies with the earth atmosphere and surface. Cosm. Res. **50**(1), 56–64 (2012). http://dx.doi.org/10.1134/S0010952512010017

Hawkins, G.S.: A radio echo survey of sporadic meteor radiants. Mon. Not. R. Astron. Soc. **116**, 92 (1956)

Jenniskens, P.: Meteor Showers and Their Parent Comets, p. 790. Cambridge University Press, Cambridge, UK (2006)

Llorca, J., Trigo-Rodríguez, J.M., Ortiz, J.L., Docobo, J.A., Garcia-Guinea, J., Castro-Tirado, A.J., Rubin, A.E., Eugster, O., Edwards, W., Laubenstein, M., Casanova, I.: The Villalbeto de la Peña meteorite fall: I. Fireball energy, meteorite recovery, strewn field and petrography. Meteorit. Planet. Sci. **40**, 795–804 (2005)

Llorca, J., Gich, M., Molins, E.: The Villalbeto de la Peña meteorite fall: III. Bulk chemistry, porosity, magnetic properties, 57Fe Mössbauer spectroscopy, and Raman spectroscopy. Meteorit. Planet. Sci. **42**, 177–182 (2007)

Llorca, J., Casanova, I., Trigo-Rodríguez, J.M., Madiedo, J.M., Roszjar, J., Bischoff, A., Ott, U., Franchi, I.A., Greenwood, R.C., Laubenstein, M.: The Puerto Lápice eucrite. Meteorit. Planet. Sci. **44**(2), 159–174 (2009)

Lovell, A.C.B.: Meteor Astronomy. Oxford University Press, Oxford (1954)

Lyytinen, E., Gritsevich, M.: Implications of the atmospheric density profile in the processing of fireball observations. Planet. Space Sci. **120**, 35–42 (2016). http://dx.doi.org/10.1016/j.pss.2015.10.012

Lyytinen, E., Gritsevich, M.: A flexible fireball entry track calculation program. In: Proceedings of the International Meteor Conference 2012, La Palma, Spain, ISBN 978-2-87355-024-4, vol. 2, pp. 155–167 (2013)

Madiedo, J.M., Trigo-Rodríguez, J.M.: Multi-station video orbits of minor meteor showers. Earth Moon Planet. **102**(1), 133–139 (2008)

Madiedo, J.M., Trigo-Rodríguez, J.M., Castro-Tirado, A.J.: On the development of new SPMN diurnal video systems for daylight fireball monitoring. In: European Planetary Science Congress 2008, Proceedings of the Conference held 21–25 September, 2008 in Münster, Germany, Online, p. 738 (2008)

Madiedo, J.M., Trigo-Rodríguez, J.M., Lyytinen, E.: Data reduction and control software for meteor observing stations based on CCD video systems. In: Cooke, W.J., Moser, D.E., Hardin, B.F., Janches, D. (eds.) Meteoroids: The Smallest Solar System Bodies, Proceedings of the Meteoroids 2010 Conference held in Breckenridge, Colorado, USA, May 24–28, 2010, pp. 330–337. NASA/CP—2011–216469 (2011)

Madiedo, J.M., Trigo-Rodríguez, J.M., Zamorano, J., Ana-Hernández, L., Izquierdo, J., Ortiz, J.L., Castro-Tirado, A.J., de Miguel, A.S., Ocaña, F., Pastor, S., de los Reyes, J.A., Galadí, D., de Guindos, E., Organero, F., Fonseca, F., Cabrera-Caño, J.: Trajectory, orbit, and spectroscopic analysis of a bright fireball observed over Spain on April 13, 2013. Astron. Astrophys. **569** (2014a). id.A104, 8 pp

Madiedo, J.M., Trigo-Rodríguez, J.M., Williams, I.P., Konovalova, N., Ortiz, J.L., Castro-Tirado, A.J., Pastor, S., de los Reyes, J.A., Cabrera-Caño, J.: Near-earth object 2012XJ112 as a source of bright bolides of achondritic nature. Mon. Not. R. Astron. Soc. **439**, 3704–3711 (2014b)

Madiedo, J.M., Ortiz, J.L., Trigo-Rodríguez, J.M., Zamorano, J., Konovalova, N., Castro-Tirado, A.J., Ocaña, F., Sánchez de Miguel, A., Izquierdo, J., Cabrera-Caño, J.: Analysis of two superbolides with a cometary origin observed over the Iberian Peninsula. Icarus **233**, 27–35 (2014c)

Madiedo, J.M., Trigo-Rodríguez, J.M., Zamorano, J., Izquierdo, J., Sánchez de Miguel, A., Ocaña, F., Ortiz, J.L., Espartero, F., Morillas, L.G., Cardeñosa, D., Moreno-Ibáñez, M., Urzáiz, M.: Orbits and emission spectra from the 2014 Camelopardalids. Mon. Not. R. Astron. Soc. **445**, 3309–3314 (2014d)

McKinley, D.W.R.: Meteor Science and Engineering. McGraw-Hill, New York (1961)

Moreno-Ibáñez, M., Gritsevich, M., Trigo-Rodriguez, J.M.: New methodology to determine the terminal height of a fireball. Icarus **250**, 544–552 (2015)

Tapia, M., Trigo-Rodríguez, J.M.: Using seismic data to detect and study bolides: the case study of May 11th, 2011 bolide. In: European Planetary Science Congress 2012, held 23–28 September, 2012 in Madrid, Spain. id. EPSC2012-443. http://meetingorganizer.copernicus.org/EPSC2012/EPSC2012-443-1.pdf (2012)

Trigo-Rodríguez, J.M.: Spectroscopic analysis of cometary and asteroidal fragments during their entry to the terrestrial atmosphere. Ph.D. thesis (in Spanish), 356 p, Universitat de Valencia, Servei de Publicacions, Spain. http://hdl.handle.net/10803/9481 (2002)

Trigo-Rodríguez, J.M., Castro-Tirado, A.J., Llorca, J., Fabregat, J., Martínez, V.J., Reglero, V., Jelínek, M., Kubánek, P., Mateo, T., Postigo, A.D.U.: The development of the Spanish fireball network using a new all-sky CCD system. Earth Moon Planet. **95**, 553–567 (2004a)

Trigo-Rodríguez, J.M., Llorca, J., Lyytinen, E., Ortiz, J.L., Sánchez Caso, A., Pineda, C., Torrell, S.: 2002 Leonid storm fluxes and related orbital elements. Icarus **171**, 219–228 (2004b)

Trigo-Rodríguez, J.M., Borovicka, J., Spurny, P., Ortiz, J.L., Docobo, J.A., Castro-Tirado, A.J., Llorca, J.: The Villalbeto de la Peña meteorite fall: II. Determination of atmospheric trajectory and orbit. Meteorit. Planet. Sci. **41**, 505–517 (2006a)

Trigo-Rodríguez, J.M., Llorca, J., Castro-Tirado, A.J., Ortiz, J.L., Docobo, J.A., Fabregat, J.: The Spanish fireball network. Astron. Geophys. **47**(6), 6.26–6.28 (2006b). doi:10.1111/j.1468-4004.2006.47626.x

Trigo-Rodríguez, J.M., Madiedo, J.M., Gural, P.S., Castro-Tirado, A.J., Llorca, J., Fabregat, J., Vítek, S., Pujols, P.: Determination of meteoroid orbits and spatial fluxes by using high-resolution all-sky CCD cameras. Earth Moon Planet. **102**, 231–240 (2008)

Trigo-Rodríguez, J.M., Borovicka, J., Llorca, J., Madiedo, J.M., Zamorano, J., Izquierdo, J.: Puerto Lápice eucrite fall: Strewn field, physical description, probable fireball trajectory, and orbit. Meteorit. Planet. Sci. **44**, 175–186 (2009). doi:10.1111/j.1945-5100.2009.tb00726.x

Trigo-Rodríguez, J.M., Llorca, J., Weyrauch, M., Bischoff, A., Moyano-Cambero, C.E., Keil, K., Laubenstein, M., Pack, A., Madiedo, J.M., Alonso-Azcárate, J., Riebe, M., Wieler, R., Ott, U., Tapia, M, Mestres, N.: The Ardón L6 ordinary chondrite: a long-hidden Spanish meteorite fall. Meteorit. Planet. Sci. **49,** 1475–1484 (2014a)

Trigo-Rodríguez, J.M., Lyytinen, E., Gritsevich, M., Moreno-Ibáñez, M., Bottke, W.F., Williams, I., Lupovka, V., Dmitriev, V., Kohout, T., Grokhovsky, V.: Orbit and dynamic origin of the recently recovered Annama's H5 chondrite. Mon. Not. Royal Astron. Soc. **449**, 2119–2127 (2015)

Natural Hazard Associated to Shock Waves of Meter-Sized Meteoroids

Mar Tapia and Josep M. Trigo-Rodríguez

Abstract The Earth's atmosphere is a good barrier for protecting the Earth from meter-sized interplanetary projectiles but it is not fullproof. Large meteoroids or small asteroids under certain entry geometries and slow geocentric velocities, can penetrate deep into the atmosphere, and become a source of hazard to humans. The energy deposited in the atmosphere causes these bodies to ablate and the hazard can take several forms: sonic booms, airbursts, hot plasma or even crater excavation. By understanding these sources of hazards, strategies can be produced to mitigate their effects. The historic records of such unusual events are highly biased because our ancestors were probably unable to identify and classify many of them. An example is the airburst that occurred on June 30, 1908 over the Tunguska region of Siberia where several hypotheses were proposed to explain it. The recent Chelyabinsk meteorite fall, renewed our interest in the problem. This work introduces the physics and describes the main mechanisms adopted by this. It is concluded that an early detection system, capable of identifying the main sources of contemporary hazard, is needed together with planning general scenarios and safety actions.

1 Introduction

Understanding the atmospheric entry of projectiles in near-Earth space (meteoroids or space debris) requires a description of the physical processes that operate. When an object penetrates the atmosphere at high ablation processes take place due to the collisions with atmospheric molecules and exchange of kinetic energy. First, the object encounter air resistance to its motion and experiences aerodynamic heating

M. Tapia (✉)
Laboratori d'Estudis Geofísics Eduard Fontserè, Institut d'Estudis Catalans (LEGEF-IEC), C/Carme 47, Barcelona 08001, Spain
e-mail: mar.tapia@gmail.com

J.M. Trigo-Rodríguez
Institute of Space Sciences (IEEC-CSIC), Meteorites, Minor Bodies and Planetary Sciences Group, Campus UAB, c/Can Magrans s/n, 08193 Cerdanyola del Vallés (Barcelona), Catalonia, Spain

J.M. Trigo-Rodríguez et al. (eds.), *Assessment and Mitigation of Asteroid Impact Hazards*, Astrophysics and Space Science Proceedings 46,
DOI 10.1007/978-3-319-46179-3_10

until the vaporization temperature of the projectile constituents is reached and ablation starts (Ceplecha et al. 1998). The meteoroid head becomes surrounded by a cloud of hot gas that can be seen or recorded at great distances. Bright bolides can be observed from a distance of about 700 km on Earth (Trigo-Rodríguez et al. 2009). These bolides are sometimes brighter than the Moon and can visible during the day (see Blanch et al. 2017; Jenniskens et al. 2012). Sometimes, in these bright cases, witnesses heard sonic booms, or even some vibrations tens of seconds or minutes after the bolide appearance. This is because the supersonic entry creates infrasound waves and sonic booms. These waves can be recorded in several ways and their study adds valuable information to the investigation of the origin and energy deposition by the specific event. Often the bolide occurs over remote regions and these waves are the only source of information available from the event, being a clear example the Indonesian bolide described in the Silber et al. (2011) that is discussed in the next section.

As well as being useful for the waves generated by bolides, the waves generated by bolides can also become a danger to humanity if they are sufficiently energetic. The damage associated with the fall of Chelyabinsk, was mainly due to shockwaves and airbursts, and there is no doubt that bolides should be consider as a source of natural hazard. However, the public envisage a direct impact with a meteorite or the excavation of a crater, being both quite rare phenomena as the main hazard. A bolide bursting over a city probably poses a greater threat to its inhabitants than a direct impact on the ground excavating a crater, The Earth's atmosphere is an efficiently shielding protecting us from asteroids up to few hundred meters in diameter as they disrupt high in the atmosphere (Bland and Artemieva 2003; Trigo-Rodríguez and Williams 2017). Details about this will be explained in a following section.

It is highly probable that the next interplanetary projectile will be again a few tens of meters in size asteroid in direct collision route with the Earth. Such a high-velocity encounter could provoke an airbust, which could be far more destructive than its direct impact with the ground (see chapter by Trigo-Rodríguez and Williams 2017). The hazard, the exposure and the vulnerability make the shock wave of a bolide a natural risk to be seriously taken into account. Their assessment and the possibility of establishing an early warning system could be the way to tackle the problem. A discussion of this can be found in a later section.

At the end of this chapter, we discuss the SPanish Meteor Network (SPMN) contributions in the study of meteoroids and meteorites wave generation that started in 2010.

2 Bolide Airbusts: The Physics Behind the Phenomenon

2.1 *Infrasound and Seismic Wave Generation*

As mentioned in the introduction, during their entry into the atmosphere, meteoroids experience aerodynamic drag that generates heat that induces the meteoroid to ablate. This ablation produces column of ionized gas that can be observed as meteors. If a meteor is brighter than Venus, it is usually called a fireball. Meteors are used to compute the meteoroid trajectory using reference stars and triangulation methods (Ceplecha 1987; Trigo-Rodríguez et al. 2008).

The penetration of the meteoroid into the atmosphere generates shock waves along that cannot be directly observed, but sometimes produce audible phenomena. These waves are longitudinal mechanical waves, more exactly infrasound waves. The velocity of the infrasound waves is about a few hundreds meters per second. The speed of sound in the air, depends critically on temperature and composition, so that its value will vary as propagation takes place through the different layers in the atmosphere. Other nonlinear effects such as, atmospheric turbulence, gravity waves and winds may affect the infrasonic wave propagation (Silber et al. 2015), but its value is in the general ballpark of 340 m/s. In contrast, interplanetary meteoroids arrive with velocities between 11.2 and 72.8 km/s (Ceplecha et al. 1998), a hypersonic entry and very high Mach number of between $\sim$35 and 270 (Boyd 1998) (Table 1).

Meteoroids penetrating into the atmosphere with such high speeds will create a sonic booms or airbust as a result of the pile up of waves, breaking the sound barrier. The airbust wave trains are usually almost parallel to the meteoroid path due to its high entry velocity, making the Mach cone almost a cylinder (Fig. 1a). However, this is not the only way in which meteoroids create energetic waves, if they explode along, these explosions (point sources) generate quasi-spherical waves propagating in all directions (Fig. 1b). These and the impact on the ground (Fig. 1c), generating seismic waves, Fig. 1a, b and c exemplify the genuine generation of possible hazardous waves. Figure 1d–g are the cases where infrasonic wave coupled with the ground and transformed into seismic waves.

Recently, Silber and Brown (2014) and Silber et al. (2015) described in detail the mechanisms behind the meteor generated infrasound, taking into account the involved non-linear processes. Furthermore, these studies also provided an overview

Table 1 Mach number regimes and examples

	Mach number	Examples
Subsonic	<0.7	High velocity trains. Doppler effect
Transonic	0.7–1.2	Aircrafts Wings (mixed airflow)
Supersonic	1.2–5.0	Concorde
Hypersonic	5.0–25.0	Boeing X-51
	>25 (high hypersonic)	Bolides. Atmosphere entries

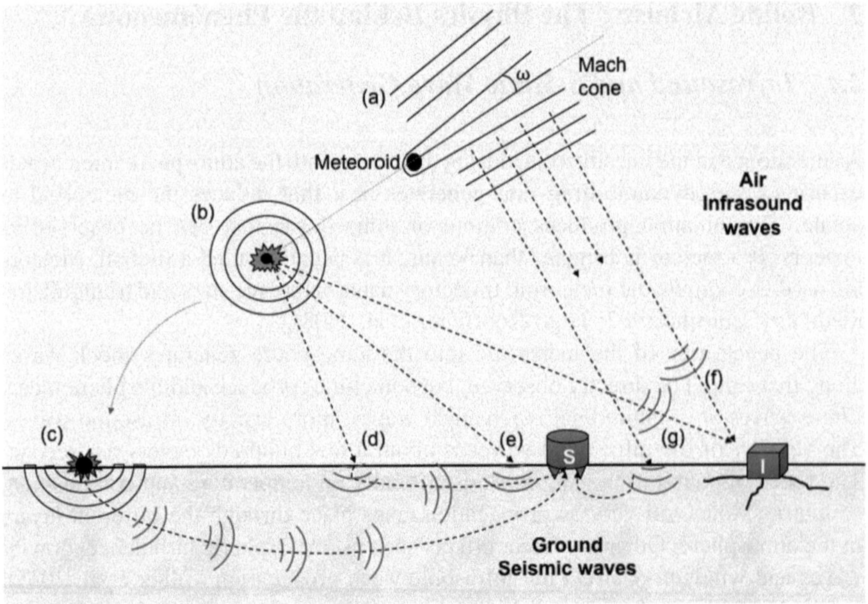

Fig. 1 Schematic diagram of meteor-generated waves. (*a*) Generation of air shock waves/airbursts during hypersonic entry. (*b*) Generation of air shock wave during meteoroid fragmentation. (*c*) Seismic wave generation during meteorite impacts. (*d*) Seismic precursor wave generation through matching of surface wave speeds (P, S, or Rayleigh). Travel via surface speeds allows these waves to arrive prior to (*e*) direct coupling of the atmospheric pressure wave with the surface at the site of the seismic station (*S*). The atmospheric pressure wave coming from the sonic boom or airburst (*f*) and the shock wave (*g*) from the ablation arriving to the infrasound station (*I*). Mach cone is almost a cylinder, ω is very small

of the cylindrical line source produced by meteors. For a comprehensive review of meteoroid generated shock waves and infrasound, see ReVelle (1976), Bronshten (1983), Edwards (2010), and Silber (2014).

2.2 *Listening to the Bolides: Bolide Imprints in Seismograms*

To record an infrasound wave, we need infrasound stations installed around the globe forming a network. The International Monitoring System (IMS) Infrasound network is a global monitoring network listening around the planet for any infrasound wave traveling through the atmosphere (Christie and Campus 2010). The network was basically established for nuclear verification after many countries signed the Comprehensive Nuclear-Test-Ban Treaty (CTBT) in 1996. Presently, about 60 infrasound stations in 35 countries around the world are either planned or operational (Fig. 2a). The IMS stations continuously detect micropressure changes (minimum detection around 1 mPa, maximum detection around 100 hPa and instrumental noise under 2 mPa) in the atmosphere due to infrasound waves.

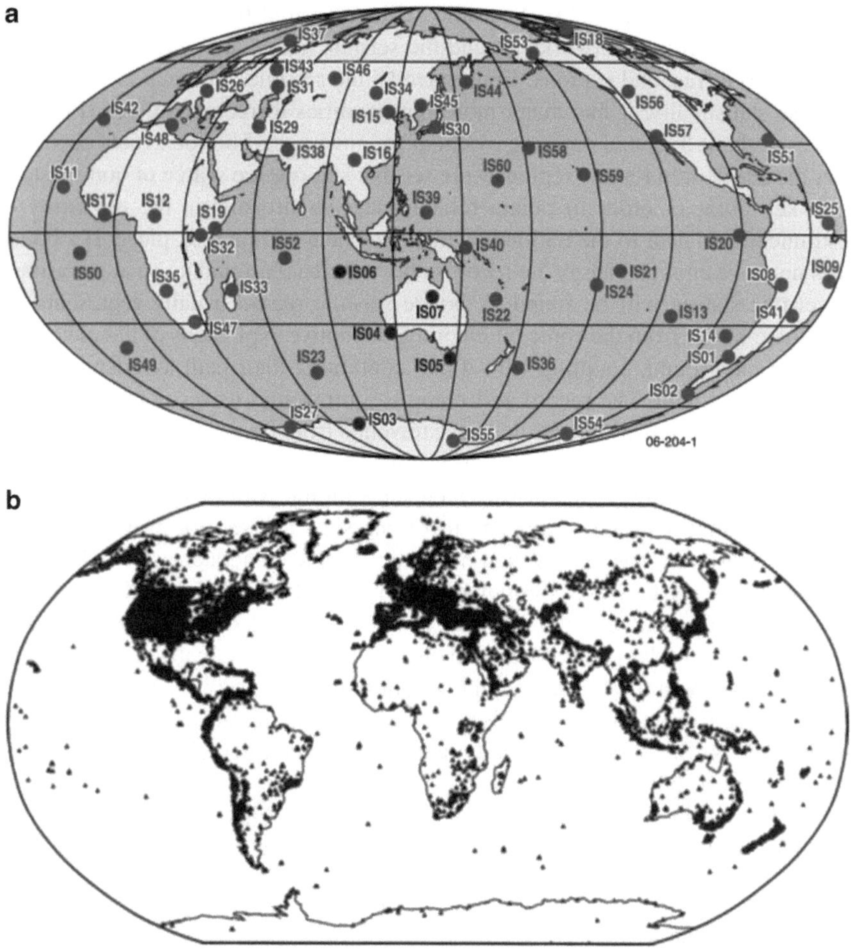

Fig. 2 (**a**) The International Monitoring System (IMS) infrasound network. ~60 stations to measure small changes in the air pressure. (**b**) International Registry (IR) of Seismograph Stations that correspond to more than 22,000 stations capable to measure small ground movements

However, if we take into account the seismic wave produced by the coupling of the infrasound wave with the ground (Fig. 1d, e) and the genuine seismic wave due to impacts on the ground (Fig. 1a), we can search for these specific data in seismograms. Infrasound waves can travel large distances without significant attenuation (because sound attenuates proportionally to the square of frequency, Bass et al. (1972). Consequently, infrasound station networks are not densely distributed around the world, while seismic wave propagating inside the Earth suffers from the higher attenuation (seismic attenuation is due to geometrical spreading and intrinsic attenuation that decays as an exponential function of frequency and ground properties) and so stations need to be packed more densely.

There are more than 22,000 seismic stations around the world, preferably in inland areas, forming networks around the seismic active areas. In contrast, there are only 60 infrasound stations. Consequently, the seismic approach is very useful because statistically, it has many more opportunities to record bolides (Fig. 2b). However, the infrasound ones are essential for coverage of bolide events occurring over the ocean or in remote regions were seismic stations are scarce or non-existent.

A key point in order to obtain reliable data on fireballs is the proximity of instrumentation near to the region in which the phenomenon took place. If a bolide does not have enough energy, no records above the background noise at infrasound or seismic station will be found. A bolide seismic record usually represents the infrasonic waves from the sonic boom or the disruptive explosions of the meteoroid reaching the ground, coupling with it and continuing their path as seismic waves through the ground. The record will comprise infrasound waves converted due to the coupling into seismic waves that have travelled from far away and seismic waves coming from the direct coupling of infrasound waves near the seismic station. The order of arrival of the different waves depends on the path of the source and the relative position between source and recording site. The seismic waves produced by the infrasonic wave interaction with the ground generate, as the theory of propagation of elastic waves through media predicts, P, S and Rayleigh waves with dependence of ground elastic properties (Ben-Menahem and Singh 1981; Edwards et al. 2007). Edwards et al. 2007 carried out a complete theoretical and experimental analysis describing the formation and coupling of the waves in these cases. Also, they calibrated infrasound and seismic conversion taking advantage of a known entry of the NASA's Stardust sample return capsule that generated shockwaves in the same way as bolides.

To illustrate these different types of seismic waves, Figs. 3, 4 and 9, show the examples of different bolide seismic records. Figure 3 corresponds to one of the

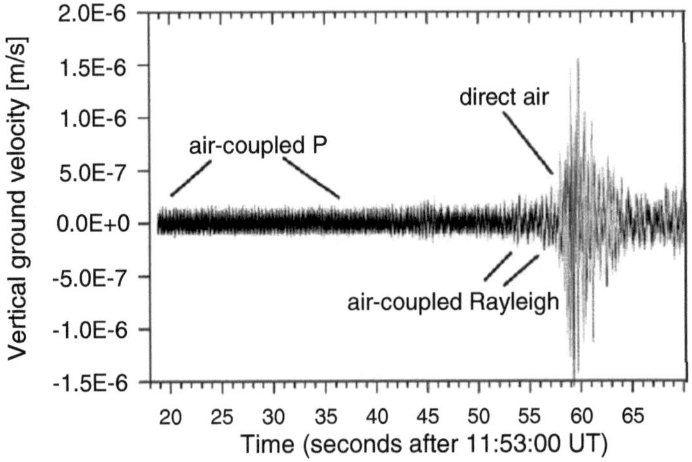

Fig. 3 Seismic record from the Morávka meteorite (modified from Brown et al. 2003)

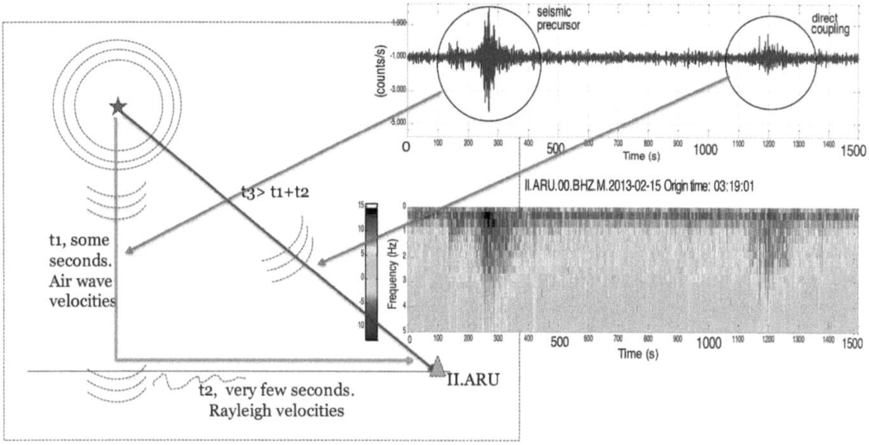

Fig. 4 Seismic record from together with its spectrogram and scheme of the different type of waves (Tapia and Trigo-Rodríguez 2013)

seismic records produced by the Morávka meteorite fall (Brown et al. 2003). This seismic record reflects that the bolide occurred at some distance (the height is around 30–40 km for the main fragmentation), with enough energy so that the first waves (P and Rayleigh) associated with the cylindrical blast wave coupled far away and travelled as seismic waves are noticeable. The second part, with higher amplitudes, is due to the direct arrival of airwaves. The Chelyabinsk seismic record from one of the closest seismic stations is shown in Fig. 4 where the first packet of waves coming from the seismic propagation can be distinguished, while the second one corresponds to the airwave travelling as in the previous case.

A second example, given in Fig. 9, shows a seismic record of another bolide recorded in the NE of the Iberian Peninsula where only the direct arrival of airwaves was noticed. This event shows a characteristic N shape that resembles a ballistic wave and the travel time matches well the distance to the more energetic meteoroid fragmentation according optical observations. No other types of waves are visible before this, probably because the bolide did not have enough energy to exceed the background noise of this seismic station.

These cases show typical bolide seismic records with different characteristics. Depending on the distance between the bolide path and the station, the bolide altitude, its velocity evolution and the number of fragmentations, the seismic record will look different. A seismic record of a bolide will be slightly different as seen from each one of the recording stations. If each signature in the record can be identified, a history of the bolide can be reconstructed even without other data.

3 The Chelyabinsk Meteorite: A Case Study

On 15 February 2013, an 18-m in diameter asteroid disrupted near the city of Chelyabinsk in Russia. This was the first time such an event occurred over a populated area, causing structural damage and personal injury (Brown et al. 2013; Borovička et al. 2013). The main damage was caused by the bolide shock wave, not to the impact of the object with the ground, thus indicating the destructive power of a shock wave is greater than was previously recognized. Until the Chelyabinsk event this type of phenomenon was not recognized as a potential source of natural hazard. Chelyabinsk was the largest meteoritic event over land since the poorly observed Tunguska impact in 1908. The Tunguska meteoroid was about 30–40 m in diameter (Boslough and Crawford 2008) and bolides of this size will impact the Earth once every 200–1000 years. Ben-Menahem (1975), using old seismograms, estimated that the energy release from the Tunguska event was about 15 Mt (6.3×10^{16} J). The impact at Chicxulub (Mexico), generally believed to be the cause of a mass extinction 65 million years ago was caused by an asteroid about 10 km in diameter, releasing 10^8 Mt (1.5×10^{23} J) of energy. Such an event occurs about once every 50–100 million years. The Chelyabinsk event, in contrast is likely to occur about once every 50–100 years.

Popova et al. (2013) has compiled all the available data for the Chelyabinsk event. Infrasound and seismic data were also recorded around the world. Le Pichon et al. (2013) analysed the infrasound records, while Tauzin et al. (2013), Heimann et al. (2013) and Antolik et al. (2014) the seismic records. The released energy by Chelyabinsk trajectory through atmosphere was so high (460 kT reported by Brown et al. 2013 and Le Pichon et al. 2013) that infrasound waves travelled twice round the globe, being the major event ever detected by IMS stations. Even stations located almost at the antipodes, IS24 at Tahiti and IS27 at Antarctica, recorded the event (Fig. 2a). In this case, the first arrival of the event to each station matches with a sound velocity between 280 and 300 m/s (IRIS 2013). The same applies for seismic records, the Rayleigh waves generated by the main blast travelled up to 4000 km with a velocity around 3.5 km/s, being recorded by several seismic stations around the world (Fig. 5).

Popova et al. (2013) compiled the distribution of damage in the region. They examined the reports from 50 villages to verify the extent of glass damage. The geometrical damage distribution matches with the expected meteorite generated waves and explains very well by itself when the sonic boom started and persisted, and even the involved physical mechanism. The pattern shows a strong cylindrical component, extending furthest perpendicular to the trajectory belonging to the hypersonic entry of the meteorite (Fig. 1a). And it can also appreciate a spherical component overlaying the cylindrical pattern belonging to the main ablation events (Fig. 1b). Popova et al. (2013) show in the supplementary material the performance of modelling that confirms that this shape corresponds to both superposed phenomena: the sonic boom and its subsequent non-negligible explosions.

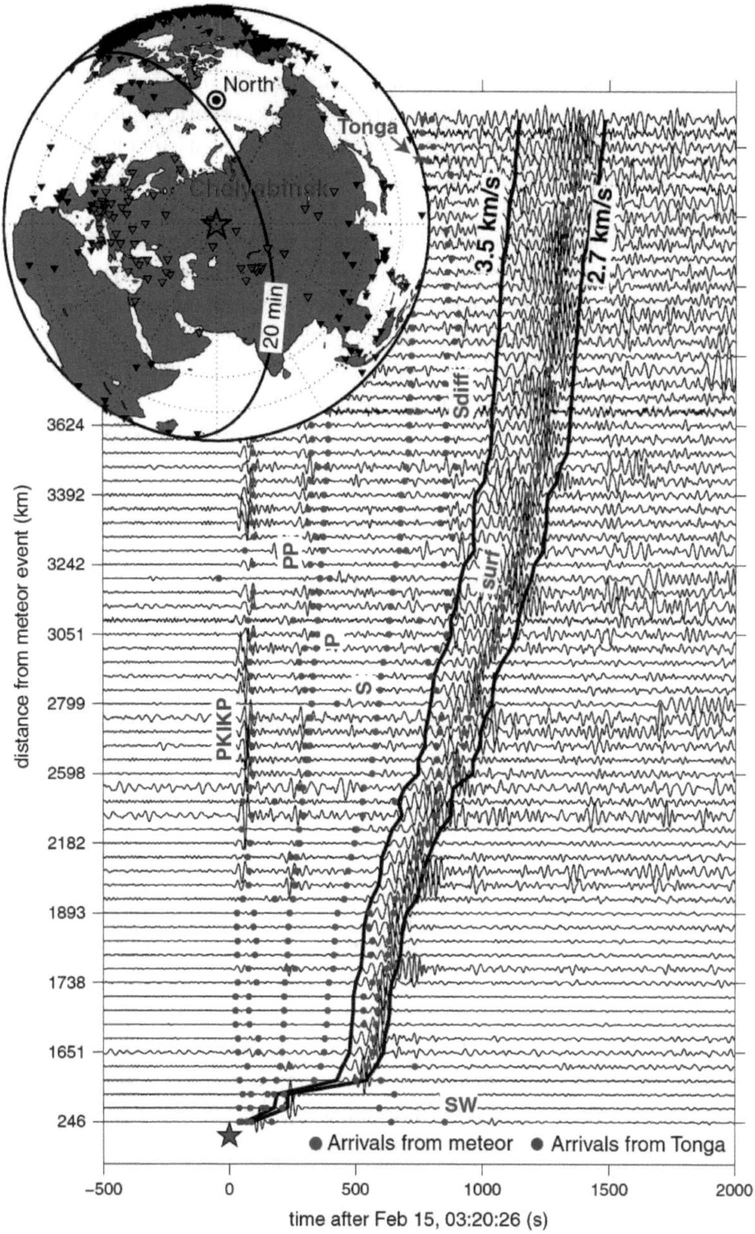

Fig. 5 Vertical component seismograms for stations located within 4000 km from the city of Chelyabinsk. The seismograms are band-pass filtered between 20 and 60 s. (Origin time = 03:20:36 UTC). Body waves (PKIKP, PP, and Sdiff) from the Tonga earthquake are indicated. The curves delineate inferred Rayleigh waves associated with the Chelyabinsk meteor. The map at the *top left* shows the location of the city of Chelyabinsk with all broadband seismological stations. The North Pole is indicated with a double *black circle* (from Tauzin et al. 2013)

In relation with the reported damages, the main damage due to the blast was associated with the lighter parts of buildings (broken windows, unstable roofs), as were not recorded as serious structural damage. The injuries were also mainly related to exposure to such damage, although some reported burned eyes due to the intense light (Yanagisawa 2015). The damage cost as much as $33 million dollars in repair (Lenta 2013). No direct damage to persons associated with the fall of meteorite was reported. Due to its relatively small diameter, the meteoroid disrupted in the atmosphere sufficiently high to not produce burns or fires as was reported in Tunguska. If Chelyabinsk meteorite was bigger, maybe some effects related to hot gases and winds would have been reported. Boslough and Crawford (2008) made 3D simulations related to Tunguska that show these effects and how a new model involving a smaller asteroid can perfectly explain Tunguska reported effects. The current evidence from Tunguska and Chelyabinsk meteorites indicate that more attention should be paid to these type of smaller size objects and their effects.

4 Sonic Booms from Bolides and Meteorites as Natural Hazard

It is clear that sonic booms from meter-sized bolides and meteorites must be taken into account, as these represent a natural hazard. The study and comprehension of them and their consequence could affect the course of action required to mitigate their effects.

It is difficult to protect against the occurrence of an unexpected meteoroid entry or meteorite fall due to its randomness. The reality is that little can be done to reduce the hazard from meter-sized bolides although some actions are planned for the larger potentially hazardous objects (Chapman 2015). The effects of such airburst and sonic booms can be considered as part of the rare and unexpected natural disasters as tornados or hurricanes that strike at random intervals. However, if is known that an impact is imminent, it is important to take airburst and sonic booms into account. The only way to reduce their effect is to receive an early warning combined with a knowledge of all possible scenarios so that appropriate action can be taken to mitigate damage. Also, informing and educating the general public is necessary so that any necessary rapid action can be taken to mitigate the associated damage. After witnessing the Chelyabinsk fireball, the eyewitnesses were not expecting sonic booms. Consequently, they did not move away from windows and fragile structures. This is a good example of how information could have significantly reduced the number of minor injuries.

The first step in any risk assessment is to identify and understand the sources of danger. In the situation being discussed, it is vital to know as much as possible about the meteoroid, their sizes, composition and trajectory characteristics (see Trigo-Rodríguez and Williams 2017).

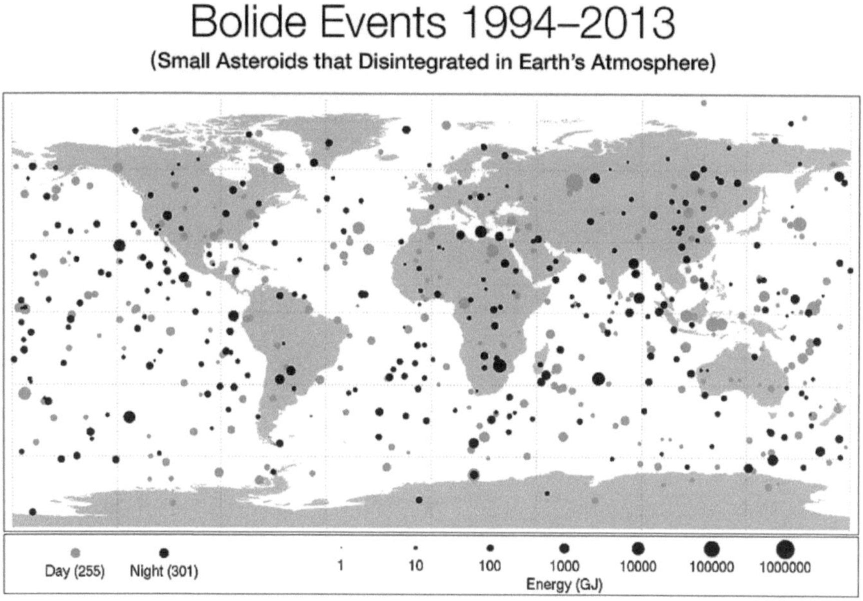

Fig. 6 This diagram maps the data gathered from 1994 to 2013 on small asteroids impacting Earth's atmosphere to create very bright meteors, technically called "bolides" and commonly referred to as "fireballs". Sizes of daytime impacts dots and nighttime impacts dots are proportional to the optical radiated energy of impacts measured in billions of Joules (GJ) of energy, and show the location of impacts from objects about 1 m to almost 20 m in size. Image Credit: JPL-NASA

Figure 6 is a map showing data from meter-sized meteoroids impacting Earth's atmosphere during the 1994–2013 period detected by visual techniques (telescopic surveys and satellite detections). It clearly exemplifies that these small impacts are frequent and that they are distributed randomly. In spite of this, we do not know with sufficient accuracy the impact frequency for objects around tens on meters in diameter that can cause damage on the ground such as the Chelyabinsk meteorite (Brown et al. 2013). Silber et al. (2009) measure the influx rate of 10-m sized meteoroids using infrasonic measurements and found that impacting rate for the 5–20 m object are underestimated by an order of magnitude in visual observations. The study of impact events on the Moon, can significantly add to the statistics of impacts by objects in this range (Madiedo et al. 2013; Ortiz et al. 2015).

Figure 7 compiles data from several studies to obtain a distribution of hazardous Near Earth objects that are candidate to collide with the Earth with a given probability (Boslough 2013). If a Tunguska event occurs every 1000 years, a Chelyabinsk one does every 50 years. We are facing a natural phenomenon with low-probability but high-consequences. Boslough (2013) concluded in his study that the risk associated with bolides and meteorites is mainly dominated by the sonic booms effects because the potential crater-forming objects are or will be soon discovered, whereas the vast majority of still unknown objects with diameter from

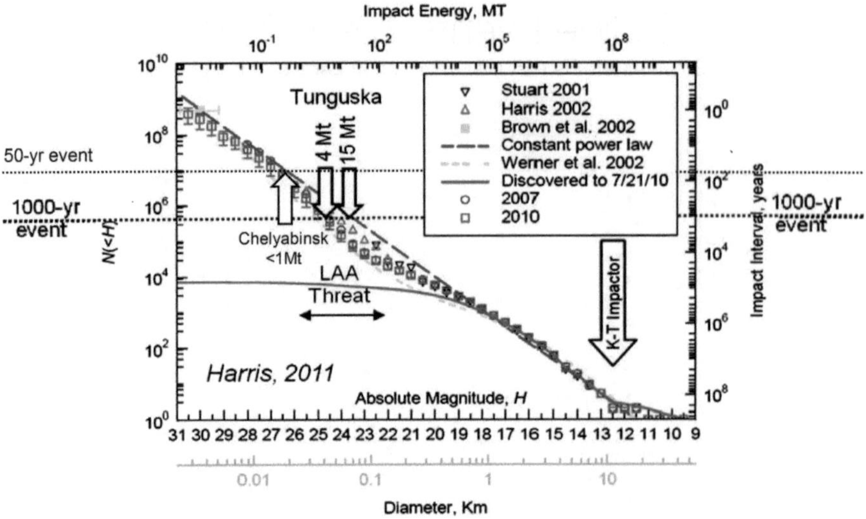

Fig. 7 Size distribution of Near-Earth objects (modified from Boslough 2013). Chelyabinsk and Tunguska events are marked on the plot. Chelyabinsk represents an event typical every 50 years

a few meters up to tens of meters are capable of producing a dangerous airbust will remain undiscovered. Also, nowadays it is known that the destructive power of an airburst is greater than was previously recognized, because the energy is directed and not isotropic (Boslough and Crawford 2008).

At present, Earth-based surveys of Near-Earth Objects (NEOs) are almost complete for object with a diameters larger than 100 m. Ground based observational searches pursue strategies biased to the discovery of the largest objects. The discovery of objects around 50 m could be achieved in reasonable time by combining space-based and ground-based measurements. Though there has been significant progress for objects less than 30 m completeness can never be achieved (NRC 2010). An effective early warning needs to detect objects with sizes around 20 m (Chelyabinsk size) with sufficient advanced time to evacuate affected and surrounding zones. This goal is still far from being achieved by ground-based surveys, and it is clear that a space-based facility is also necessary.

The second step is the evaluation of the risks and effects of the possibly hazardous object. This requires modeling of the fireball entry and good knowledge of the projectile composition and bulk properties. With this knowledge a realistic scenario can be produced, essential for decision-making purposes and taking actions to mitigate damage. Boslough (2013) proposed an airburst scale for hazard assessment and an early warning of asteroid impactors based on a 1–5 rating. Since this scale was first proposed at the 2011 Planetary Defense Conference in Romania, it has been called the Bucharest scale and is shown in Table 2. On this scale, the Chelyabinsk event is a Bucharest 3 (30 km detonation height and 0.5 Mt) and the recommended action for people in the locality is "take cover in basements".

Table 2 Proposed levels in the Bucharest scale for airburst effects assessment after Boslough (2013)

Warning levels		Proposed action
1	High-altitude airburst. No possible damage	No evacuation needed
2	High-altitude airburst. Minor damage	Avoid standing near windows
3	High-altitude airburst. Major damage	Take cover in basements
4	Low-altitude airburst. Heavy blast damage	Evacuate blast zone
5	Low-altitude airburst. Heavy thermal damage	Evacuate and take cover outside blast zone

Useful as they are these actions and recommendations are not the end of discussions. Chapman (2015) considers what should be done after all potentially hazardous asteroids are detected. A full understanding of the destructive effects of airbursts is some way off but improvements will be achieved in the coming years. Awareness of this problem is a very important step towards an improvement in its mitigation together with establishing new strategies to promote new research activities and experiments. Keeping the public aware of the situation is important, so that celebrating *Asteroid Day* every June 30th is an excellent initiative.

5 The SPMN Network Contribution: An Example of Bolide Seismic Analysis in the Iberian Peninsula

As consequence of the Villalbeto de la Peña meteorite fall (Llorca et al. 2005; Trigo-Rodríguez et al. 2006) the SPMN network started to study seismic data. It is currently compiling all available seismic data in order to create a database. A complete database could help to quantify better the hazard associated with meter-sized meteoroids. Another goal to achieve is obtaining an estimation of the detection threshold in the region of observation depending on the deployed seismic instrumentation and their characteristics in the same way as Ens et al. (2012) did for infrasound measurements. A good knowledge of the bolide-produced wave detection capabilities in the monitored area will improve the completeness of the seismic database. Also, SPMN wants to contribute to the knowledge of regional meteor signals that could reveal more characteristics of the source shock, having been less modified during their short propagation distances in comparison with the widely studied large bolides events (Silber and Brown 2014).

The study of all available seismic signals associated with bolides will provide key physical parameters complementing those extracted by astronomical methods. It is very important to calibrate the inferred physical parameters with seismic data (e.g. meteoroid energy, or mass knowing the velocity, directionality, …) with the values obtained from other techniques. For example, the energy estimates with empirical

Table 3 Trajectory data of the SPMN120511 bolide

Fireball point	Height (km)	Longitude (°)	Latitude (°)
Beginning	138 ± 3	-0.886 ± 0.001	42.157 ± 0.001
End	26 ± 2	-1.216 ± 0.001	41.955 ± 0.001

adjustment using ballistic waves is not well known and could underestimate the values due to different characteristic of wave formation, transformations and directivities.

One of the most important events for which SPMN obtained seismic data together as well as recovering the meteorite was Villalbeto de la Peña on Jan. 4, 2004 (Llorca et al. 2005; Trigo-Rodríguez et al. 2006). In this case the seismic data was found in only one seismic station. The seismic results were consistent with the results obtained by other astronomical techniques used (energy estimation and path).

Another important event in the SPMN database was a meteorite-dropping bolide occurring on May 12, 2011 called SPMN120511. The bolide trajectory obtained from four video stations is given in Table 3. Several searches were carried out a few weeks after the fall, but it was unfortunately no meteorite recovered.

This bolide was recorded from several seismic stations because it overflew an area near Pyreenes where a dense seismic network is deployed because of general seismic activity in the area. Six stations (CORG, CBRU, CLLI, CAVN, CEST and CTRE) belonging to the Catalan seismic network CA (ICGC 2000) recorded the event (Fig. 8). After correcting offset, trend and filtering the signal between 2 and 30 Hz, the signals were analysed in order to assure that all of them belong to the same event and was not a coincidence. The observed wave arrival times match with the main fragmentation experienced by this bolide. The energy released during the break-up seems differ at each station. The closest stations differ between 1 and 3 kT, and the farthest and less confidents due to more noisy signals (Table 4) around 0.1 kT. Seismic local effects for each station were not considered and could explain the differences, together with the correction due to atmosphere winds and temperature affecting propagation. The results point to a bolide with a released energy between some hundredths T to few kT and a meter-sized meteoroid. Studying the particle motion of the seismic records using hodograms (particle motion representations), it was found that one of the stations CAVN shows clearly different type of waves, P and Rayleigh wave (Fig. 9). The others stations do not show clear P waves, only Rayleigh waves (CORG, Fig. 10). The CAVN station with clear P waves allowed us to compute a back-azimut around 45° that matches the position of the main break-up, identified as source of the signal and a preliminary location.

The signals usually recorded show that stations located far away from the source have much lower frequency peaks that the nearest stations, maybe denoting a Doppler effect from the bolide moving away together with the attenuation effect of waves (Fig. 11).

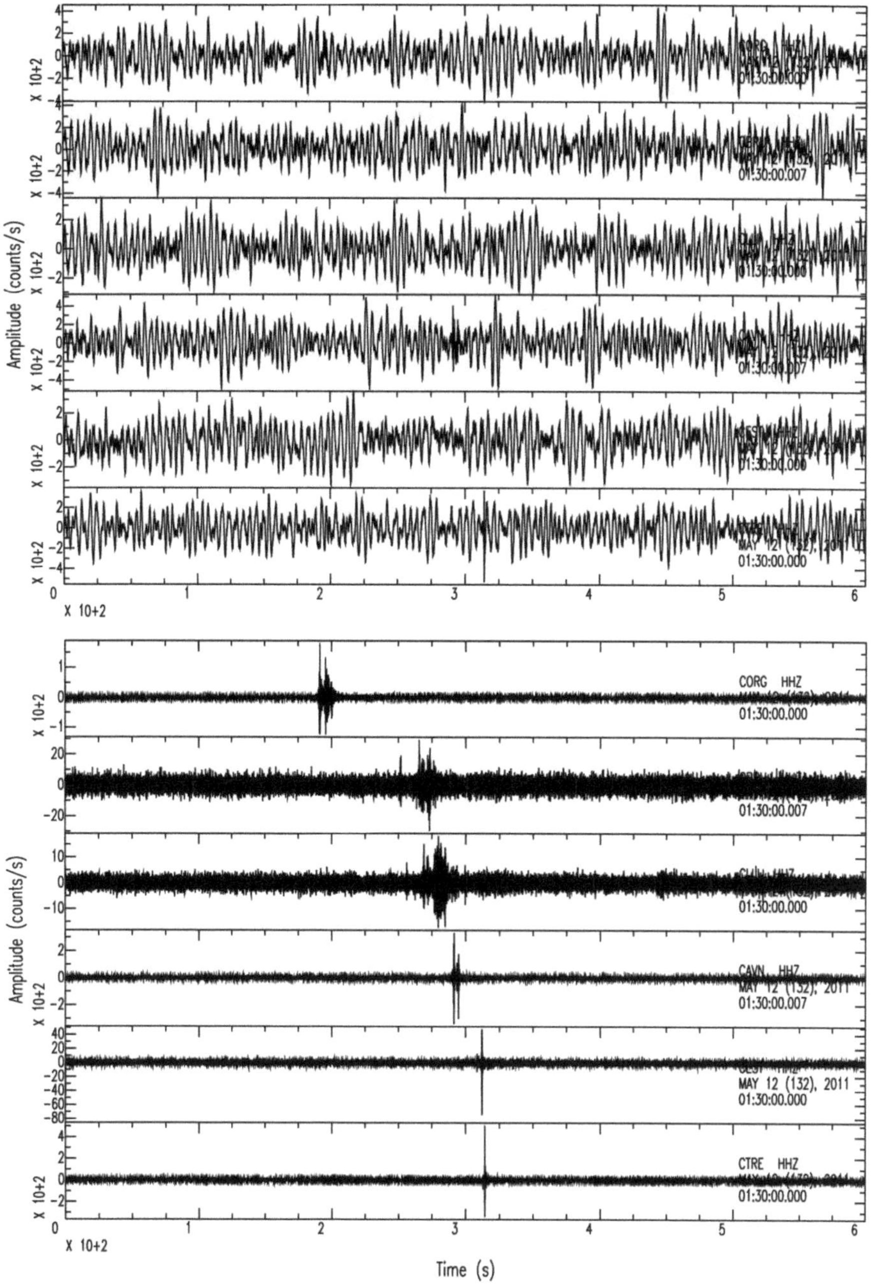

Fig. 8 CORG, CBRU, CLLI, CAVN, CEST and CTRE vertical raw seismic waveforms (*top*) and 2–30 Hz band-pass filtered seismic waveforms (*bottom*)

Table 4 Energy estimation, CTRE, CAVN and CORG are the most confidable values due to the proximity and signal quality

Station	Peak ground velocity (Nm/s)	W (kT)
CTRE	**475**	**2.7**
CAVN	**265**	**1.0**
CORG	**123**	**0.3**
CEST	59	0.1
CLLI	26	0.02
CBRU	42	0.1

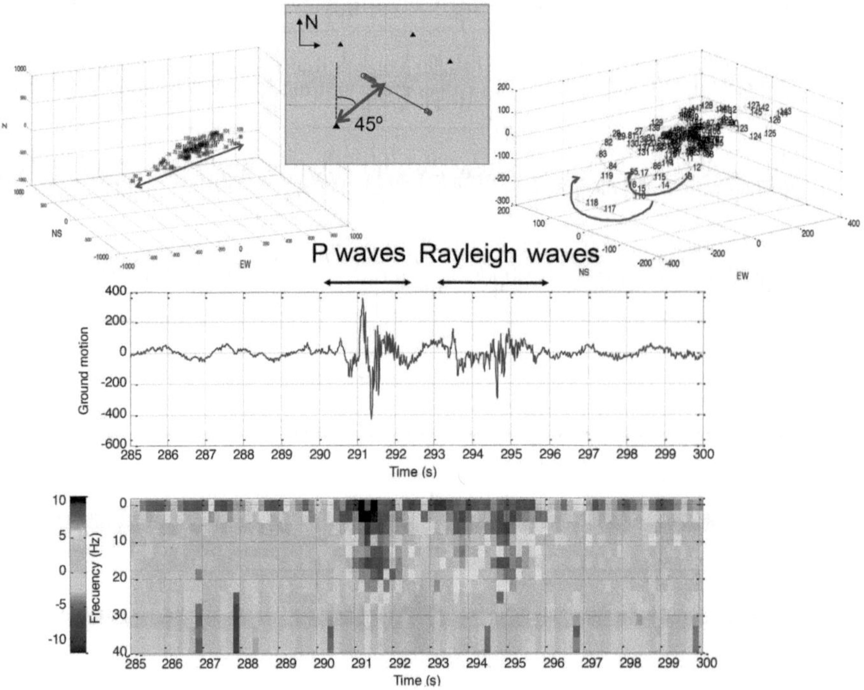

Fig. 9 CAVN seismic record from NE Iberian Peninsula bolide (SPMN120511 bolide, 2011, 12th May, origin time 01 h 30 m 41 $\pm$ 1 s T.U.C., SPMN database) together with its spectrogram (*bottom*) and hodograms (*top*). Adapted from Tapia and Trigo-Rodríguez (2012)

6 Conclusions

The hazard associated with the impact of interplanetary projectiles must be studied carefully. According to the current evidence about the magnitude of impact hazard by asteroids, the following conclusions were reached:

- The Earth is exposed to the influx of interplanetary projectiles that are the natural consequence of the disintegration taking place in the solar system bodies. Small asteroids and meter-sized meteoroids are a direct source of hazard to human beings.

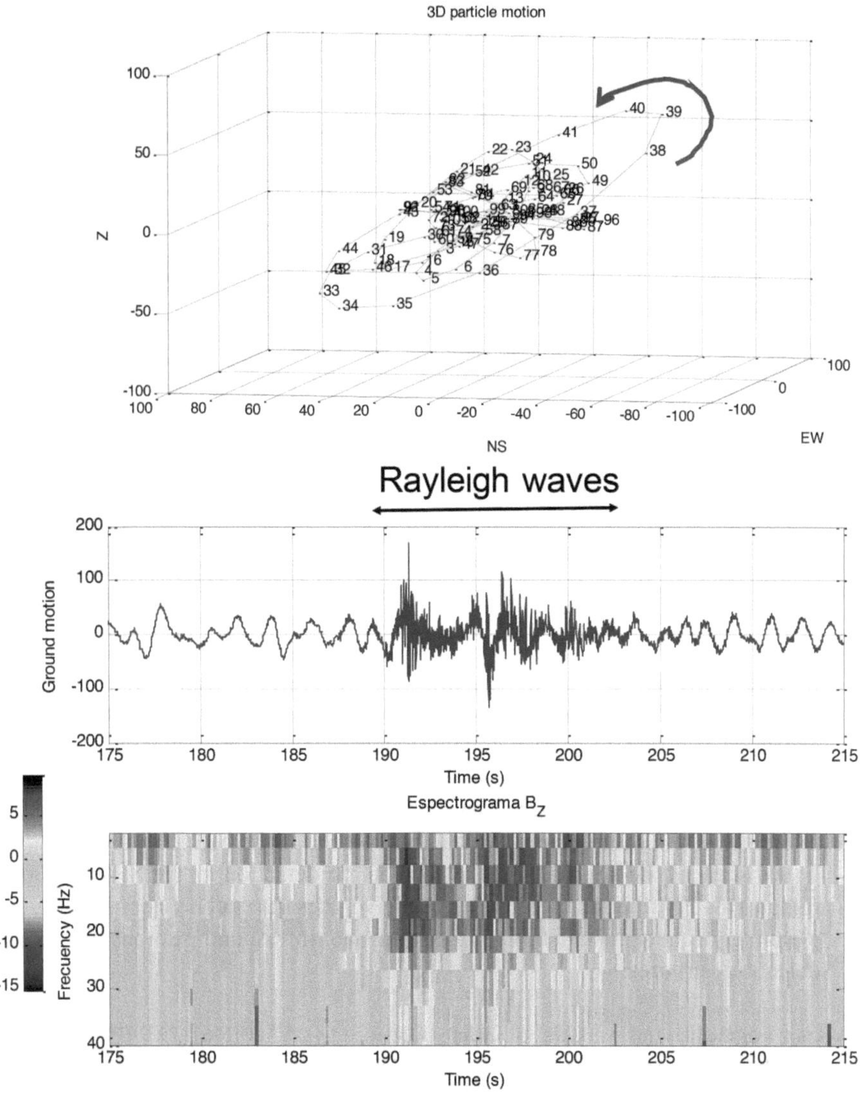

Fig. 10 CORG seismic record from NE Iberian Peninsula bolide (SPMN120511 bolide, 2011, 12th May, 01 h 30 m 41 ± 1 s T.U.C., SPMN database) together with its spectrogram(*bottom*) and hodograms(*top*). Adapted from Tapia and Trigo-Rodríguez (2012)

- Large disruptions produce airblasts and irradiative effects can be a source of direct hazard for humans as the Chelyabinsk event demonstrated.
- Bolide-induced shock waves cause indirect effects when reaching the ground that can be the most dangerous source of hazard.

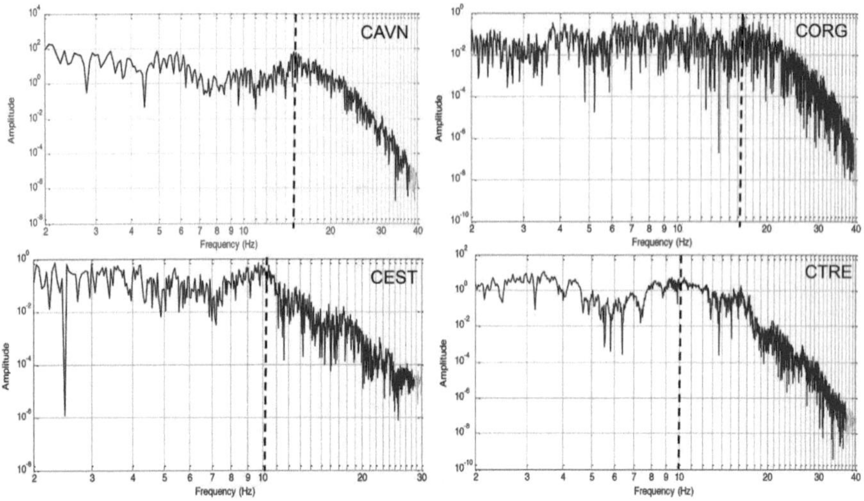

Fig. 11 CAVN, CORG, CEST and CTRE seismic Fourier spectra with its main frequency (*dashed line*) from NE Iberian Peninsula bolide (SPMN120511 bolide, 2011, 12th May, 01 h 30 m 41 ± 1 s T.U.C., SPMN database). Tapia and Trigo-Rodríguez (2012)

- Due to the importance of the generated waves from meter-sized bolides it is important that the study of seismic and infrasound data be continued and expanded in order to extract reliable parameters to complete our knowledge regarding their possibly dangerous effects.
- The SPMN network is studying available bolide seismic data to improve in the reliability of the complementary data that can add to the bolide studies.
- Issues such as the seismic threshold of bolide detection in the SPMN studied region (mainly Iberian peninsula) can give important conclusions for future planning of new installation of seismic and infrasound stations for bolide detection purposes.

Acknowledgments This work was financially supported by AYA2015-67175-P (PI. JMTR). We thank to the Institut Cartogràfic i Geològic de Catalunya for providing waveform data to this study. We thank E. Silber for her useful review and suggestions that helps to improve this manuscript. We also thank Prof. Iwan P. Williams for his careful English revision of the manuscript.

References

Antolik, M., Ichinose, G., Creasey, J., Clauter, D.: Seismic and infrasonic analysis of the major bolide event of 15 February 2013. Seismol. Res. Lett. **85**(2), 334–343 (2014)

Bass, H.E., Bauer, H.J., Evans, L.B.: Atmospheric absorption of sound: analytical expressions. J. Acoust. Soc. Am. **52**, 821–825 (1972)

Ben-Menahem, A.: Source parameters of the Siberia explosion of June 30, 1908, from analysis and synthesis of seismic signals at four stations. Phys. Earth Planet. In. **11**, 1–35 (1975)

Ben-Menahem, A., Singh, S.J.: Seismic Waves and Sources. Springer, New York (1981)

Bland, P.A., Artemieva, N.A.: Efficient disruption of small asteroids by Earth's atmosphere. Nature **424**, 288–291 (2003)

Borovička, J., Spurný, P., Brown, P., Wiegert, P., Kalenda, P., Clark, D., Shrbený, L.: The trajectory, structure and origin of the Chelyabinsk asteroidal impactor. Nature **503**(7475), 235–237 (2013)

Boslough, M.: Airburst warning and response. Acta Astronaut. **103**, 370–375 (2013)

Boslough, M., Crawford, D.A.: Low-altitude airbursts and the impact threat. Int. J. Impact. Eng. **35**, 1441–1448 (2008)

Boyd, I.: Computation of atmospheric entry flow about a Leonid meteoroid. Earth Moon Planet. **82**, 93–108 (1998)

Bronshten, V.A.: Physics of Meteoric Phenomena, D. Reidel Publ. Co., 1983

Brown, P.G., Kalenda, P., ReVelle, D.O., Borovicka, J.: The Morávka meteorite fall: 2. Interpretation of infrasonic and seismic data. Meteorit. Planet. Sci. **38**(7), 989–1003 (2003)

Brown, P.G., Assink, J.D., Astiz, L., Blaauw, R., Boslough, M.B., Borovička, J., Brachet, N., Brown, D., Campbell-Brown, M., Ceranna, L., Cooke, W., de Groot-Hedlin, C., Drob, D.P., Edwards, W., Evers, L.G., Garces, M., Gill, J., Hedlin, M., Kingery, A., Laske, G., Le Pichon, A., Mialle, P., Moser, D.E., Saffer, A., Silber, E., Smets, P., Spalding, R.E., Spurný, P., Tagliaferri, E., Uren, D., Weryk, R.J., Whitaker, R., Krzeminski, Z.: A 500-kiloton airburst over Chelyabinsk and an enhanced hazard from small impactors. Nature **503**(7475), 238–241 (2013)

Ceplecha, Z.: Geometric, dynamic, orbital and photometric data on meteoroids from photographic fireball networks. Bull. Astron. Inst. Czechosl. **38**, 222–234 (1987)

Ceplecha, Z., Borovicka, J., Elford, J.G., Revelle, D.O., Hawkes, R.L., Porubcan, V., Simek, M.: Meteor phenomena and bodies. Space Sci. Rev. **84**(3/4), 327–471 (1998)

Chapman, C.R.: Facing a hypervelocity asteroid impact disaster: to deflect or evacuate? Procedia Eng. **103**, 68 (2015)

Christie, D.R., Campus, P.: The IMS infrasound network: design and establishment of infrasound stations. In: Le Pichon, A., Blanc, E., Hauchecorne, A. (eds.) Infrasound Monitoring for Atmospheric Studies, pp. 29–75. Springer, New York (2010). doi:10.1007/978-1-4020-9508-5_2

ICGC: Institut Cartogràfic i Geològic de Catalunya: Catalan Seismic Network. International Federation of Digital Seismograph Networks. Other/Seismic Network. doi:10.7914/SN/CA (2000)

Edwards, W.N., Eaton, D.W., McCausland, P.J., ReVelle, D.O., Brown, P.G.: Calibrating infrasonic to seismic coupling using the Stardust simple return capsule shockwave: implications for seismic observations of meteors. J. Geophys. Res. **112**, B10306 (2007). doi:10.1029/2006JB004621

Edwards, W.N.: Meteor generated infrasound: Theory and Observation. Chapter 12 in Continuous infrasound monitoring for atmospheric studies. Ed LePichon, Brachet and Ceranna. J. Acoust. Soc. Am. **123**(5), 3827 (2010)

Ens, T.A., Brown, P.G., Edwards, W.N., Silber, E.A.: Infrasound production of bolides: a global statistical study. J. Atmos. Sol. Terr. Phys. **80**, 208 (2012)

Heimann, S., Gonzalez, A., Wang, R., Cesca, S., Dahm, T.: Seismic characterization of the Chelyabinsk Meteor's terminal explosion. Seismol. Res. Lett. **84**(6), 1021–1025 (2013)

IRIS: Special Event: Chelyabinsk, Russia Bolide (meteor). (2013). https://ds.iris.edu/ds/nodes/dmc/specialevents/2013/02/19/chelyabinsk-russia-bolide-meteor/ (consulted March 2016)

Jenniskens, P., Fries, M.D., Yin, Q.Z., Zolensky, M., Krot, A.N., Sandford, S.A., et al.: Radar-enabled recovery of the Sutter's Mill meteorite, a carbonaceous chondrite regolith breccia. Science. **338**(6114), 1583–1587 (2012)

Le Pichon, A., Ceranna, L., Pilger, C., Mialle, P., Brown, D., Herry, P., Brachet, N.: Russian fireball largest ever detected by CTBTO infrasound sensors. Geophys. Res. Lett. **40**(14), 3732–3737 (2013)

Lenta: Ущерб от челябинского метеорита превысит миллиард рублей [Damage from Chelyabinsk meteorite exceeds one billion rubles] (in Russian). Lenta.ru (2013)

Llorca, J., Trigo-Rodríguez, J.M., Ortiz, J.L., Docobo, J.A., García-Guinea, J., Castro-Tirado, A.J., Rubin, A.E., Eugster, O., Edwards, W., Laubenstein, M., Casanova, I.: The Villalbeto de la Peña meteorite fall: I. Fireball energy, meteorite recovery, strewn field and petrography. Meteorit. Planet. Sci. **40**, 795–804 (2005)

Madiedo, J.M., Ortiz, J.L., Morales, N., Cabrera-Caño, J.: A large lunar impact blast on 2013 September 11. Mon. Not. R. Astron. Soc. **439**(3), 2364–2369 (2013)

NRC: Committee to Review Near-Earth Object Surveys and Hazard Mitigation Strategies, National Research Council, Defending Planet Earth: Near-Earth Object Surveys and Hazard Mitigation Strategies, 152 pp (2010)

Ortiz, J.L., Madiedo, J.M., Morales, N., Santos-Sanz, P., Aceituno, F.J.: Lunar impact flashes from Geminids: analysis of luminous efficiencies and the flux of large meteoroids on Earth. Mon. Not. R. Astron. Soc. **454**(1), 344–352 (2015)

Popova, O.P., Jenniskens, P., Emel'yanenko, V., Kartashova, A., Biryukov, E., Jaibrajmanov, S., Shuvalov, V., Rybnov, Y., Dudorov, A., Grokhovsky, V.I., Badyukov, D.D., Yin, Q.-Z., Gural, P.S., Albers, J., Granvik, M., Evers, L.G., Kuiper, J., Kharlamov, V., Solovyov, A., Rusakov, Y.S., Korotkiy, S., Serdyuk, I., Korochantsev, A.V., Larionov, M.Y., Glazachev, D., Mayer, A.E., Gisler, G., Gladkovsky, S.V., Wimpenny, J., Sanborn, M.E., Yamakawa, A., Verosub, K.L., Rowland, D.J., Roeske, S., Botto, N.W., Friedrich, J.M., Zolensky, M.E., Le, L., Ross, D., Ziegler, K., Nakamura, T., Ahn, I., Lee, J.I., Zhou, Q., Li, X.-H., Li, Q.-L., Yu, L., Tang, G.-Q., Hiroi, T., Sears, D., Weinstein, I.A., Vokhmintsev, A.S., Ishchenko, A.V., Schmitt-Kopplin, P., Hertkorn, N., Nagao, K., Haba, M.K., Komatsu, M., Mikouchi, T., the Chelyabinsk Airburst Consortium: Chelyabinsk airburst, damage assessment, meteorite recovery, and characterization. Science **342**(6162), 1069–1073 (2013)

ReVelle, D.O.: On meteor-generated infrasound. J. Geophys. Res. **81**, 1217–1240 (1976)

Silber, E.A.: Observational and Theoretical Investigation of Cylindrical Line Source Blast Theory Using Meteors, University of Western Ontario - Electronic Thesis and Dissertation Repository. Paper 2112 (2014)

Silber, E.A., Brown, P.G.: Optical observations of meteors generating infrasound – I: Acoustic signal identification and phenomenology. J. Atmos. Solar-Terr. Phys. **119**, 116–128 (2014)

Silber, E.A., ReVelle, D.O., Brown, P.G., Edwards, W.N.: An estimate of the terrestrial influx of large meteoroids from infrasonic measurements. J. Geophys. Res. Planet. **114**(E8) (2009)

Silber, E.A., Le Pichon, A., Brown, P.G.: Infrasonic detection of a near-Earth object impact over Indonesia on 8 October 2009. Geophys. Res. Lett. **38**(12), L12201 (2011)

Silber, E.A., Brown, P.G., Krzeminski, Z.: Optical observations of meteors generating infrasound: weak shock theory and validation. J. Geophys. Res. Planet. **120**(3), 413–428 (2015)

Tapia, M., Trigo-Rodríguez, J.M.: Using seismic data to detect and study bolides: The case study of May 11th, 2011 bolide. European Planetary Science Congress, EPSC2012, Madrid (2012)

Tapia, M., Trigo-Rodríguez, J.M.: Seismic data detection of Chelyabinsk superbolide. In: European Planetary Science Congress, EPSC2013, London (2013)

Tauzin, B., Debayle, E., Quantin, Q., Coltice, N.: Seismoacoustic coupling induced by the breakup of the 15 February 2013 Chelyabinsk meteor. Geophys. Res. Lett. **40**, 1–5 (2013). American Geophysical Union

Trigo-Rodríguez, J.M., Llorca, J., Castro-Tirado, A., Ortiz, J.L., Docobo, J.A., Fabregat, J.: The Spanish fireball network. Astron. Geophys. **47**(6), 26–28 (2006)

Trigo-Rodríguez, J.M., Madiedo, J.M., Gural, P.S., Castro-Tirado, A.J., Llorca, J., Fabregat, J., Vítek, S., Pujols, P.: Determination of meteoroid orbits and spatial fluxes by using high-resolution all-sky CCD cameras. Earth Moon Planet. **102**, 231–240 (2008)

Trigo-Rodríguez, J.M., Madiedo, J.M., Williams, I.P., Castro-Tirado, A.J., Llorca, J., Vítek, S., Jelínek, M.: Observation of a very bright fireball and its likely link with comet C 1919 Q2 Metcalf. Mon. Not. R. Astron. Soc. **394**, 569–576 (2009)

Trigo-Rodríguez, J.M., Williams, I.P.: Dynamic sources of contemporary hazard from meteoroids and small asteroids. In: Trigo-Rodríguez, J.M., Gritsevich, M., Palme H. (eds.) Assessment and Mitigation of Asteroid Impact Hazards, pp. 11–32. Springer, New York (2017)

Yanagisawa, M.: Radiative characteristics of the Chelyabinsk superbolide. Planet. Space Sci. **118**, 79–89 (2015)

Chelyabinsk Meteorite as a Proxy for Studying the Properties of Potentially Hazardous Asteroids and Impact Deflection Strategies

Carles E. Moyano-Cambero, Josep M. Trigo-Rodríguez, Eva Pellicer, Marina Martínez-Jiménez, Jordi Llorca, Narcís Metres, and Jordi Sort

Abstract Most asteroids of the near-Earth population have experienced significant collisional processing since they formed, being disrupted and excavated and consequently producing smaller bodies that are delivered from the Main asteroid Belt to the near-Earth asteroid region, thanks to planetary resonances and non-gravitational forces. By studying meteorites arrived to Earth we can obtain clues on this processing through the study of shock metamorphism and brecciation of their rock constituents, among other features. The massive Chelyabinsk meteorite fall produced about one metric ton of meteorites that can be analyzed to decipher

C.E. Moyano-Cambero (✉) • J.M. Trigo-Rodríguez • M. Martínez-Jiménez
Institute of Space Sciences (IEEC-CSIC), Meteorites, Minor Bodies and Planetary Sciences Group, Campus UAB, c/Can Magrans s/n, 08193 Cerdanyola del Vallès (Barcelona), Catalonia, Spain
e-mail: moyano@ice.csic.es; trigo@ice.csic.es; mmartinez@ice.csic.es

E. Pellicer
Departament de Física, Universitat Autònoma de Barcelona, Cerdanyola del Vallès, Barcelona 08193, Spain
e-mail: eva.pellicer@uab.cat

J. Llorca
Institute of Energy Technologies, Universitat Politècnica de Catalunya, Diagonal 647, Barcelona 08028, Spain

Center for Research in NanoEngineering, Universitat Politècnica de Catalunya, Barcelona 08028, Spain
e-mail: jordi.llorca@upc.edu

N. Metres
Institut de Ciència de Materials de Barcelona, ICMAB-CSIC, Campus UAB, Cerdanyola del Vallès, Barcelona 08193, Spain
e-mail: narcis@icmab.es

J. Sort
Departament de Física, Universitat Autònoma de Barcelona, Cerdanyola del Vallès, Barcelona 08193, Spain

Institució Catalana de Recerca i Estudis Avançats (ICREA), Barcelona, Spain
e-mail: jordi.sort@uab.cat

J.M. Trigo-Rodríguez et al. (eds.), *Assessment and Mitigation of Asteroid Impact Hazards*, Astrophysics and Space Science Proceedings 46,
DOI 10.1007/978-3-319-46179-3_11

the physical processes affecting the surface of this Potentially Hazardous Asteroid. Here we describe physical properties of Chelyabinsk samples, and how impact processing has affected asteroid albedos, a first step in order to relate this data with the reflectance properties of near-Earth asteroids population. This information will be of major interest for future asteroid deflecting missions.

1 Introduction

Earth is subjected to a continuous flux of interplanetary material of approximately 40,000 tons per year, mostly reaching the top of the atmosphere as micron-sized dust particles up to cm-sized meteoroids (Brownlee 2001; Trigo-Rodríguez and Williams 2017). Meter-sized meteoroids to tens of meters-asteroids also contribute to this flux producing much rare (few per century) extremely bright bolide events, which denote the abrupt encounter and the subsequent ablation of most of the body mass when they penetrate into the atmosphere at supersonic velocity (Chapman and Morrison 1994; Tapia and Trigo-Rodríguez 2017; Trigo-Rodríguez and Williams 2017). Most of these events produce meteorites, which can provide a very good opportunity to comprehend the processes that deliver these extraterrestrial rocks to Earth, and the physico-chemical properties of the Near Earth Asteroids (NEAs).

Every asteroid has suffered a peculiar evolution and their bulk physical properties result from all kinds of processing (Britt et al. 2002). NEAs are usually asteroids that travelled from the Main asteroid Belt (MB) to the near-Earth region, mainly delivered via main-motion resonances and non-gravitational effects (Morbidelli and Nesvorný 1999). Asteroids crossing these resonances are typically small (less than 1 km in diameter), and are fragments produced in catastrophic impacts. Indeed, most of them have intense collisional histories due to the events occurring during their dynamic transport.

Among the NEA population, the Potentially Hazardous Asteroids (PHAs) are bodies larger than 200 m in diameter and whose Minimum Orbit Intersection Distance (MOID) with the Earth is 0.05 AU or less (Trigo-Rodríguez and Williams 2017). These encounters are not always close enough to become to hit the Earth and become catastrophic (Milani et al. 2002), but tidal effects can affect the weak structure of rubble piles, which are bodies formed from the re-aggregation of material after the collision and catastrophic disruption of large asteroids. These close approaches with terrestrial planets have been envisioned as a pathway to release meter-sized rocks from asteroidal surfaces that could produce meteorite-dropping bolides (Trigo-Rodríguez et al. 2007). Some surviving boulders detached from the surface of these asteroids can evidence a complex history of continuous impacts and breccification that is well exemplified in the heterogeneity of Almahata Sitta meteorites associated with the fall of the asteroid 2008 TC_3 (Bischoff et al. 2010; Goodrich et al. 2014; Horstmann and Bischoff 2014; Zolensky et al. 2010). The determination of the heliocentric orbit of meteorite-dropping bolides is an excellent way to identify the source of these rocks, usually coming from the main belt, but with the a probable origin in the near-Earth population (Dmitriev et al. 2015; Trigo-Rodríguez et al. 2015).

Most NEAs are S- (or Q-) class asteroids, typically associated with ordinary chondrites (hereafter OCs) (Binzel et al. 2006; Nakamura et al. 2011). These meteorites usually exhibit unequivocal features of shock metamorphism (Bischoff et al. 2006, 2013a, b; Bischoff and Stöffler 1992), defined as the mechanical deformation and transformation of rocks by shock wave compression without substantial movement of the rock constituents (e.g., Stöffler et al. 1991; Bischoff and Stöffler 1992). On the other hand, a significant number of chondritic meteorites have been found to be breccias (Bischoff et al. 2006). Actually, energetic impacts producing shock effects usually involve mass transport and substantial implantation of foreign lithologies in the surviving rock fragments that leads to progressive brecciation and ejection of materials present at the surfaces of asteroids (Bischoff et al. 2006). Also, collisions might compact loose surface materials to produce the frequent regolith breccias (e.g., Kieffer 1975; Bischoff et al. 1983). Therefore, the mineralogical study of these meteorites provides clues on the processes experienced by their parent bodies (e.g., Bischoff et al. 1993, 2006; Metzler et al. 2011).

Even NEAs being a few tens of meters in diameter are expected to have experienced a significant degree of shock and brecciation (Binzel et al. 2002). These processes make them to be weak, fractured bodies easily disrupted by the entrance through Earth's atmosphere (Bland and Artemieva 2003), and therefore meteorites reaching the Earth's surface are biased towards high-strength materials (Trigo-Rodríguez and Blum 2009). Despite of that, their entrance through Earth's atmosphere is still capable of sending a high-energy shock wave and a large amount of meteorites to the ground, being a significant source of hazard to humans (Brown et al. 2013; Moreno-Ibáñez et al. 2015).

A recent example of that is the Chelyabinsk superbolide occurred on Feb. 15, 2013, when an 18 m asteroid penetrated in the atmosphere at a velocity of ~19 km/s (Borovička et al. 2013). It was analyzed by different monitoring techniques, from which we know it delivered to the lower atmosphere an energy of ~500 kT of TNT (Brown et al. 2013). It exploded several times in the atmosphere and produced thousands of specimens with a currently Total Known Weight (TKW) of ~1000 kg of meteorites (Ruzicka et al. 2015). The largest piece of 540 kg was recovered by divers from the Chebarkul lake in fall 2013 (Kocherov et al. 2014). A body of such diameter encounters the Earth one per century (Brown et al. 2013; Trigo-Rodríguez and Williams 2017).

The Chelyabinsk meteorite shows a relatively high S4 (maximum is S6) degree of shock (Ruzicka et al. 2015), according to the weak mosaicism on olivine and the presence of crystalline plagioclase instead of maskelynite (Stöffler et al. 1991). It was originally classified as an LL5 OC (Ruzicka et al. 2015), but it was soon recognized to be a breccia with a mixture of different lithologies (Bischoff et al. 2013b). In contrast to Almahata Sitta, which contain foreign fragments of both chondritic and ureilitic origins (e.g., Bischoff et al. 2010; Goodrich et al. 2014; Horstmann and Bischoff 2014), and Kaidun (Zolensky and Ivanov 2003),

Chelyabinsk is a well-lithified polymict genomic breccia only containing lithologies related to LL ordinary chondrites (Bischoff et al. 2006, 2013b).

Most abundant Chelyabinsk specimens include the light-colored LL5-lithology described for classification purposes (Ruzicka et al. 2015), which has abundant shock veins, although light-colored fragments with very rare chondrule relicts and minor or no shock veins are also common (Righter et al. 2015). Some have been described as petrologic type 6, and therefore classified as LL6 (Bischoff et al. 2013b). Indeed, highly recrystallized LL6 lithologies with abundant shock veins also exist, in the form of much darker clasts than the ones mentioned above. In addition, Bischoff et al. (2013b) describe shock-darkened fragments in which fractures and interstitial spaces are filled up by opaques (e.g., mostly troilite). This feature may result from shock mobilization. In some of these shock-darkened areas all silicates are completely embedded by a FeS-metal network dominated by sulfides. Another specimens exhibit dark, fine-grained impact melt clasts with variable abundances of mineral and lithic clasts, which have also been described earlier (Ruzicka et al. 2015). The shock veins that can be found in both the light-colored and dark-colored lithologies, are less abundant or missing in the impact melt breccia fragments (Bischoff et al. 2013b).

In this study we compare reflectance spectra in the ultraviolet to near-infrared wavelengths of several samples from different lithologies of the Chelyabinsk meteorite. The goal is to understand how the different lithologies, processes and degree of shock metamorphism modify the spectra, and this is why we also compare those spectra with spectra from other ordinary chondrites and asteroids. Additionally we performed several nanoindentacion analyses in order to know and understand the physical properties of this meteorite and its progenitor asteroid. This kind of study might provide information of great interest for future missions such as the Asteroid Impact and Deflection Assessment (AIDA) mission concept (see Michel et al. 2015a, b), composed of two spacecrafts (AIM and DART) which will travel to the binary NEA (65803) Didymos, and impact its satellite with a solid projectile (DART), while collecting data about the collision and the asteroid (AIM), in order to test the concept of deflection of asteroids through kinetic impacts (Michel et al. 2015a, b). In fact, Chelyabinsk is particularly interesting to this end, since Didymos is thought to be of similar composition to this meteorite (Dunn et al. 2013).

2 Experimental Methods and Procedures

Two thin sections of Chelyabinsk (PL13049; PL13050), plus a thick section, were studied. High-resolution mosaics of the sections were created from separate ×50 images taken with a Zeiss Scope petrographic microscope. The mosaics allowed establishing target features to be characterized by petrographic microscopy, electron microscopy (SEM + EDS), nanoindentation, and micro-Raman techniques. Also, several ultraviolet to near-infrared spectra (~0.2–2.0 μm) were obtained from the

Fig. 1 Transmitted light optical mosaic of the PL 13049 thin section of the Chelyabinsk meteorite. The size of each square of the grid is 1 mm^2

different lithologies of the Chelyabinsk meteorite. Handy descriptions of these techniques are:

Optical microscopy: A Zeiss Scope Axio petrographic microscope was used working in reflected and transmitted light at ×50, ×100, ×200 and ×500. High-resolution mosaics of the thin sections were built and a grid superimposed to locate and naming the different features under study through the sample (Fig. 1).

Electron microscopy: A FEI Quanta 650 FEG (ICN2) was used, working in low-vacuum BSED mode. The EDS detector used to perform elemental analyses is an Inca 250 SSD XMax20 with Peltier cooling and an active area of 20 mm^2. Some selected areas were explored at different magnification, and SEM elemental mapping together with EDS spectra were obtained.

Micro-Raman study: Micro-Raman spectra were taken in backscattering geometry at room temperature using 5145 Å line of Argon-ion laser with a Jobin-Yvon T-64000 Raman spectrometer attached to an Olympus microscope and equipped with a liquid-nitrogen-cooled CCD detector (ICMAB-CSIC). The lateral spatial resolution was $\sim$1 μm and the laser power onto the sample was kept below 0.5 mW

Table 1 Ordinary chondrites from which the UV-Vis-NIR spectra were obtained

Meteorite	Chondrite type	Place	Recovery	Total known weight (kg)
Chelyabinsk	LL5-6	Chelyabinsk (Russia)	2013	~1000
Travis County (b)	H4	Travis County (USA)	1937	5.9
Bassikounou	H5	Hodhech Chargui (Mauritania)	2006	29.56
Tieschitz	H/L3.6	Olomoucký (Czech Republic)	1878	28
Gold Basin	L4	Arizona (USA)	1995	61
Kilabo	LL6	Jigawa (Nigeria)	2002	19

Although both Kilabo and Saint-Séverin are pure LL6 OCs, differing from Chelyabinsk with most of it being LL5, we see in Fig. 5 that they have comparable spectra

to avoid degradation due to sample overheating. The Raman spectrometer provided spectra in a working range between 100 and 1400 cm^{-1}.

UV-Vis-NIR spectrometry: A Shimadzu UV3600 UV-Vis-NIR spectrometer (UPC) was used to obtain the reflectance spectra of the meteorites analyzed here (Table 1) in the ultraviolet to near-infrared range (Moyano-Cambero et al. 2013; Trigo-Rodríguez et al. 2014a). The standard stage for the spectrometer is an Integrating Sphere (ISR) with a working range of ~0.2–2.0 μm. The sample beam interacts with the section to be studied at a phase angle of 8°. For calibration of the detector a standard baseline was created using a conventional $BaSO_4$ substrate. The area sampled corresponds to a slot of 2×1 cm^2. The scanned area corresponds to a 2×6 mm^2 area, which is below the size of the sample and therefore avoids any contribution from the epoxy or the glass in which the section is mounted. The spectra obtained this way show baseline noise between ~0.8 and 0.9 μm, and become too noisy after 2.1 μm to be used. Therefore, the region between 0.8 and 0.9 μm was deleted and only data up to 2.0 μm was used.

Nanoindentation: The mechanical properties of the Chelyabinsk thin section PL13049 were studied with an UMIS equipment from Fischer-Cripps Laboratories using a Berkovich pyramidal-shaped diamond tip (UAB). Large indentations with an applied force of 500 mN were carried out to assess the average mechanical properties of the three main lithologies of the meteorite. The thermal drift during nanoindentation was kept below 0.05 nm/s. Proper corrections for the contact area (calibrated with a fused quartz specimen), initial penetration depth and instrument compliance were applied (Fischer-Cripps 2004). The hardness (H) and reduced Young's modulus (E_r) values were determined from the load-displacement curves at the beginning of the unloading segment using the method of Oliver and Pharr (1992). The elastic recovery values, W_{el}/W_{tot} (where W_{el} and W_{tot} are the elastic and total indentation energies), were evaluated from the area between the unloading curve and the displacement axis (W_{el}) and between the loading curve and the displacement axis (W_{tot}), respectively.

3 Results

3.1 Micro-Raman Study

The thin sections studied here contain representative shock-darkening veins through their OC lithologies (see e.g., Fig. 1). After SEM observation of the different areas around the thin sections (Fig. 2) it was observed that, besides from the tiny melt troilite + metal veins that can be found all around the meteorite, the impact-melt lithology was a very good candidate for the study of shock melted minerals. Raman spectroscopy was applied on this lithology, on the thin sample PL 13049, in order to identify shock minerals. The spots analyzed are shown in Fig. 3, while the Raman spectra can be seen in Fig. 4. Also, the different minerals recognized by this technique are listed in Table 2. Spot C has been recognized as jadeite, previously reported in Chelyabinsk meteorite (Ozawa et al. 2014), with main frequencies at 379, 700 and 1039 cm^{-1}. In spot E we tentatively identified a mixture of oldhamite (or another Ca-rich sulfide), with main frequencies at 185, 215, and 285 cm^{-1}, and troilite, which has main frequencies at 160, 290, and 335 cm^{-1}. High-P merrillite ($Ca_9MgNa(PO_4)_7$) has been also identified nearby shock veins with its characteristic intense peaks at 956 and 972 cm^{-1} (Trigo-Rodríguez et al. 2014b). Spot F spectrum shows peaks close to those of an iron oxide such as hematite, but an alteration phase formed between the meteorite fall and the time of recovery (terrestrial weathering) cannot be ruled out. In several olivine spectra (like e.g., spot B in Table 2 and Fig. 4) there is a small shift in the olivine peaks.

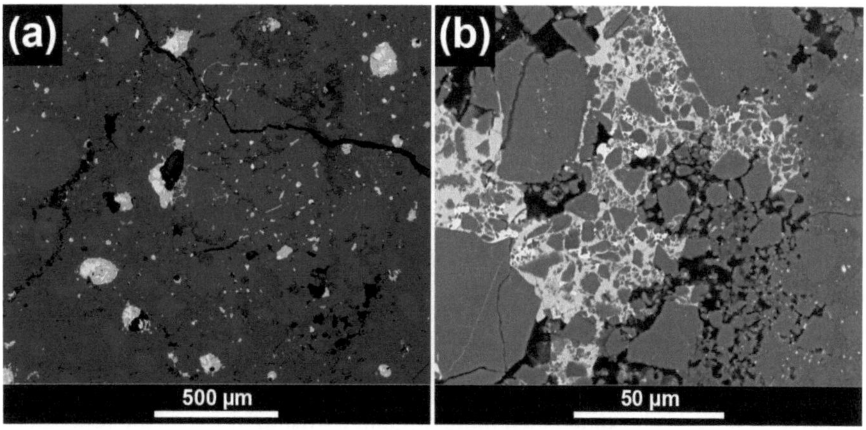

Fig. 2 Two SEM images in back-scattered electrons from the studied sections of the Chelyabinsk meteorite, showing: (**a**) texture of the interior of an impact melt breccia: Mafic silicate fragments (*medium grey*) are enclosed in plagioclase-normative interstitial material (*dark grey*). The light roundish objects consist of metal (*white*) and troilite (*light grey*), which are often intergrown. (**b**) In some shock-darkened areas silicates (*grey*) are completely embedded in (mainly) sulfides (*pale grey*) and metal (*white*)

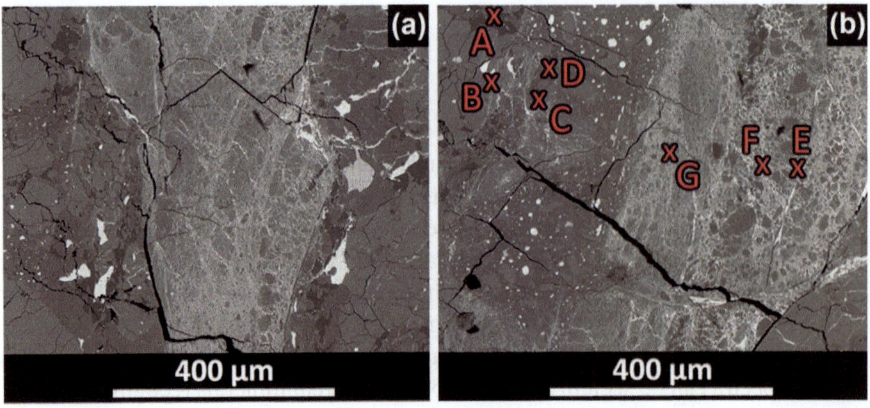

Fig. 3 Both (**a**) and (**b**) show SEM images in back-scattered electrons showing two selected areas of the analyzed thin section where the shock veins are evident. In (**b**) the selected Raman (*A–G*) points were marked. Their spectra and interpretation are shown at Fig. 4 and Table 2, respectively

3.2 UV-Vis-NIR Spectrometry

The distinction between the light and dark-colored lithologies of this meteorite was found to be consistent with previous studies (Britt and Pieters 1991, 1994; Keil et al. 1992). Thus, reflectance spectra of both lithologies were obtained from different regions of the thick and thin sections, in order to get clues on how shock metamorphism affects the reflectance of this meteorite, and hence the reflectance of related asteroids and possible NEAs. We show the several spectra obtained in Fig. 5, where they are compared with other OC. In ordinary chondrites the dominant rock-forming phase is olivine, with lower amounts of pyroxene, and actually Chelyabinsk is particularly olivine-rich and pyroxene depleted (Galimov et al. 2013). Therefore we also include the spectrum of a theoretical $Fo_{70} + En_{75}$ (in a proportion 2:1) with data from the RELAB database, which is close to the average composition of the Chelyabinsk meteorite according to the literature (Galimov 2013; Kohout et al. 2014; Righter et al. 2015). In most the OC and Chelyabinsk spectra the main feature in the working range presented here is the characteristic 1 μm olivine absorption band, together with the forsterite bands at ~0.5 μm. Afterwards 1.5 μm the pyroxene absorption band bends again the spectra. The similarity between the Chelyabinsk spectra from the light-colored lithology and the olivine + pyroxene spectrum is clearly seen. In the dark-colored lithology the olivine and pyroxene characteristic bands become much more difficult to distinguish, and the reflectance has been significantly attenuated. With respect to the other OCs compared here (see Table 1 for details), the spectra from Chelyabinsk are easily related to the spectra from other LL meteorites, as expected. For H and L chondrites the differences between the two exemplary specimens show a similar behavior to the variation between Chelyabinsk's dark- and light-colored lithologies. However, between the

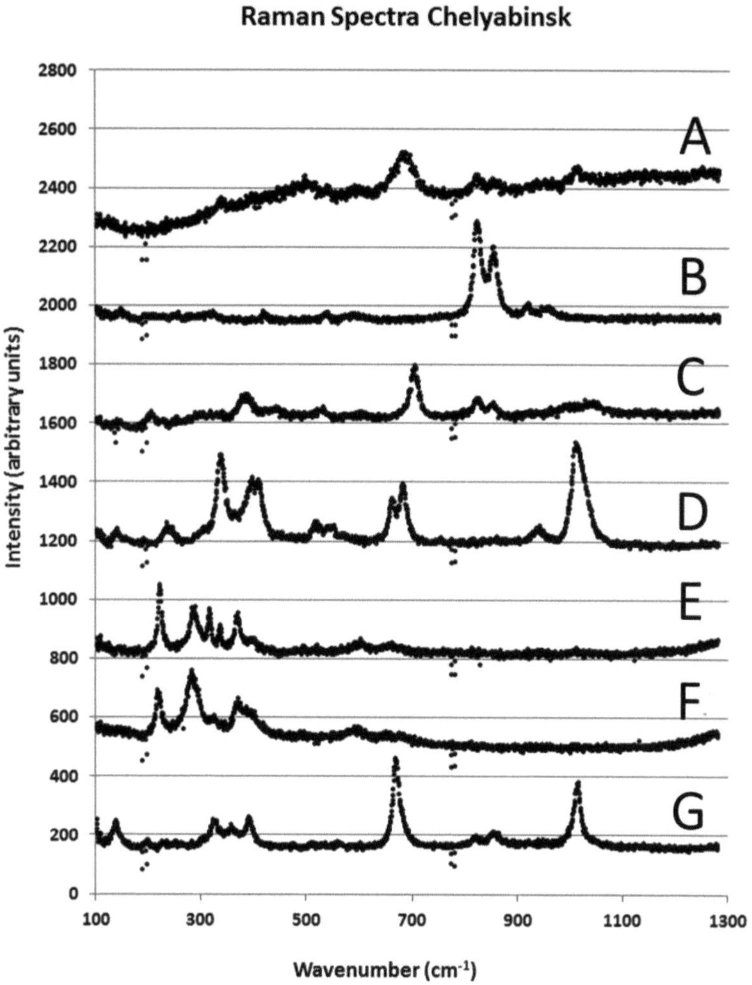

Fig. 4 Raman spectra obtained from Chelyabinsk sample PL 13049, measured at the points indicated in Fig. 3b. See also and Table 2 for the identification of the spectra

dark and light lithologies of the Kilabo LL6 meteorite the change implies only a decrease in reflectance, but not an attenuation of the olivine bands. Possibly it is the same effect happening in Chelyabinsk, although at a lower degree.

3.3 Nanoindentation

Finally a nanoindentation study was performed on the PL 13049 thin section, with the purpose of assessing the mean mechanical properties of the three main

Table 2 Main minerals found listed by alphabetical order, their theoretical frequencies in terrestrial unshocked rocks (except for merrillite and ringwoodite that are high-P phases), and our Full Width Half Maximum (FWHM) measurements in Chelyabinsk (see the spectra in Fig. 4)

Mineral	Chemical formula	Characteristic Raman frequencies (cm^{-1})	Measured frequencies in Chelyabinsk (cm^{-1})	Comments	Label on Fig. 3
Chromite[a]	(Fe, Mg)Cr_2O_4	493	497	High-P phase	A
		685	681	Peaks drifted +4 cm^{-1}	
Diopside[a]	Ca(Mg, Fe)Si_2O_6	139	139	High-P phase[a]	G
		325	324		
		357	357		
		560	561		
		667	667		
		855	854		
		1013	1014		
Low-Ca orthopyroxene[b]	(Mg, Fe)SiO_3	331	337	Peaks drifted +6 cm^{-1}	D
		654	660		
		673	679		
		1001	1008		
Hematite?[c, d]	α-Fe_2O_3	227	222	Peaks drifted	F
		377	370		
		411	400		
		497	498		
		612	595		

(continued)

Table 2 (continued)

Mineral	Chemical formula	Characteristic Raman frequencies (cm^{-1})	Measured frequencies in Chelyabinsk (cm^{-1})	Comments	Label on Fig. 3
Jadeite[c]	$NaAlSi_2O_6$	204	207	High-P phase ≥ 4 cm^{-1}	C
		370	389		
		429	444		
		522	528		
		695	699		
		982	989		
		1036	1038		
Merrillite[e, f, g]	$Ca_9MgNa(PO_4)_7$	956	956	High-P phase	–
		972	972		
Olivine[c]	$(Mg, Fe)_2SiO_4$	818	821	Peaks drifted +3–6 cm^{-1}	B
		846	852		
Ca- and Fe- sulfides?[c, h]	Diverse	165	108		E
		225	222		
		285	285		
		310	316		
		365	368		
Ringwoodite[a]	$(Mg, Fe^{2+})_2(SiO_4)$	794	794	High-P phase	–
		843	843		

[a]Handbook of Minerals Raman spectra hosted by University of Lyon (http://www.ens-lyon.fr/LST/Raman/)
[b]Chen and Xie (1992)
[c]Lafuente et al. (2015)
[d]Wachs and Routray (2012)
[e]Xie et al. (2002)
[f]Llorca and Trigo-Rodríguez (2004)
[g]Trigo-Rodríguez et al. (2014b)
[h]Avril et al. (2013)

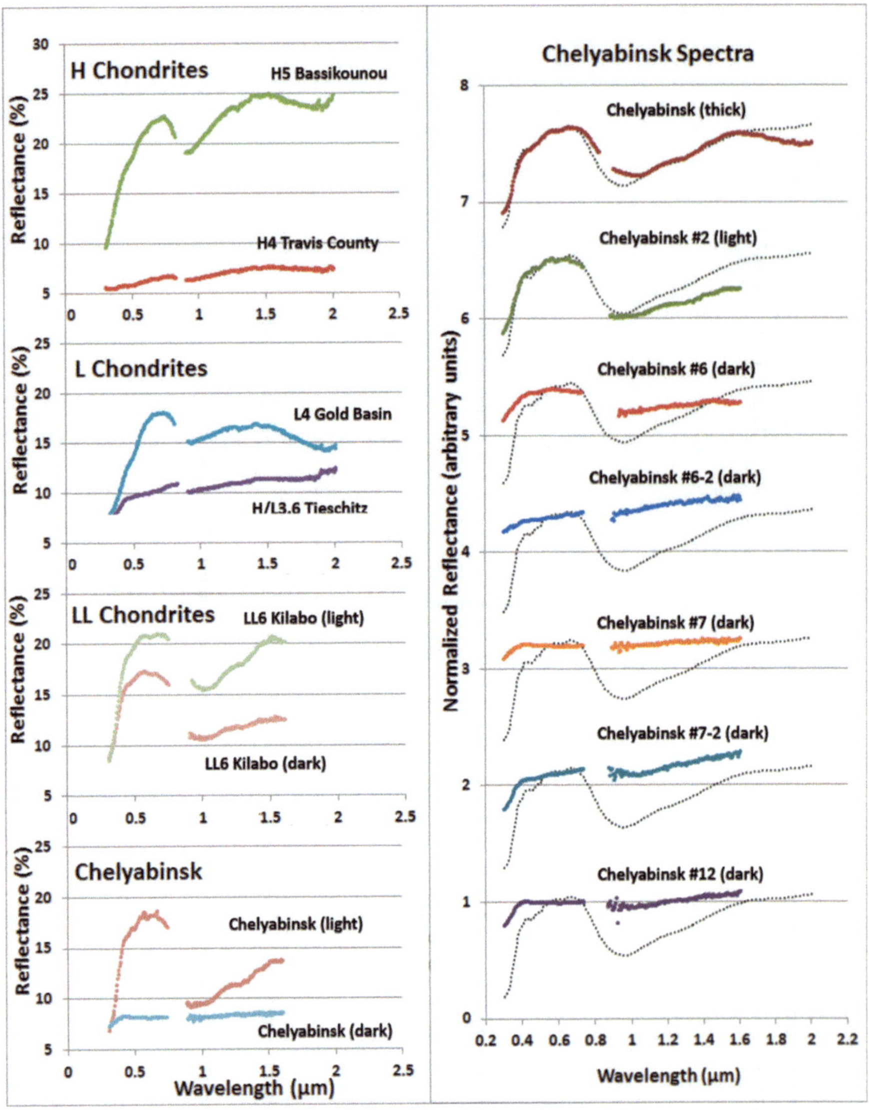

Fig. 5 Several reflectance spectra of Chelyabinsk and the other meteorites in Table 1. At *left*, spectra of H, L and LL chondrites, showing the relatively common changes in ordinary chondrites spectra, which imply lower reflectance and often, but not always, attenuation of the 1 and 2 µm bands. At *right*, we compare the Chelyabinsk spectra (shifted to avoid superposition) to a spectrum obtained from the 2 to 1 mixture of olivine (Fo_{70}) and pyroxene (En_{75}) spectra from the RELAB database

lithologies found on Chelyabinsk. Sixteen large indentations (maximum applied force of 500 mN) were performed on each lithology (Fig. 6) and the mean mechanical properties were calculated (Table 3). Although olivine, pyroxene and plagioclase conform ~90 % of the minerals in Chelyabinsk (Galimov et al. 2013), minor phases (such as troilite, kamacite, taenite and chromite) can have an important effect on the overall mechanical behavior. To take that into account, in both the light and the dark-colored lithologies at least one of the 16 indentations was done in troilite or metal, so its different mechanical properties have some effect on the mean properties of the lithology. Since the impact melt lithology shows an extensive network of thin metal and troilite melt veins all around, the mechanical properties of the large indentations are already affected by them, even without a direct indentation on these materials. The highest hardness (H) and Young's modulus (E_r) are found in the impact melt lithology, while the lowest values correspond to the light-colored. The elastic recovery (W_{el}/W_{tot}) is very similar for the three lithologies, considering the relatively low one sigma standard deviation. From the individual indentations performed on different minerals, some information was obtained about the effect that a certain amount of each mineral can have on the mechanical properties of a meteorite. In a mixture of silicates (olivine, pyroxene and plagioclase), olivine will contribute to higher H and E_r, while pyroxene, and particularly plagioclase, will decrease both. However, those minerals will also increase the W_{el}/W_{tot}. Chromite seems to be the stronger contributor to increase both H and E_r, while the presence of troilite will strongly decrease them. However, the minerals showing the lowest H and W_{el}/W_{tot} are the metals kamacite and taenite, specially the second one, although their E_r is intermediate to high. It is also worth to mention that in some of the indentations performed on the impact melt, small cracks appeared at the edges of the indentations. According to some studies (Fischer-Cripps 2004), the formation and length of these fractures can be correlated with the fracture toughness of the materials indented. Thus, in Chelyabinsk the impact melt lithology seems to be more propitious for fractioning.

4 Discussion

Laboratory studies concerning meteorite reflectance spectra have been particularly useful to know the general trends that might be characteristic of their progenitor asteroids. In fact, remote spectroscopic observations of Near Earth Objects (NEOs) are used to constrain their mineral compositions in order to identify possible meteorite analogs (Burbine 2002; Gaffey 1976; Rivkin et al. 2004; Trigo-Rodríguez et al. 2014a). Today we know that the MB is the direct dynamic source for NEAs (Binzel et al. 2002). The asteroids composing this group have a full range of sizes, and very distinctive origins.

The S-class is by far the most complex and populated group representing a significant fraction of MB asteroids that share similar reflectance spectra. The S asteroids have stony compositions revealed by their 1 and 2 μm bands, associated

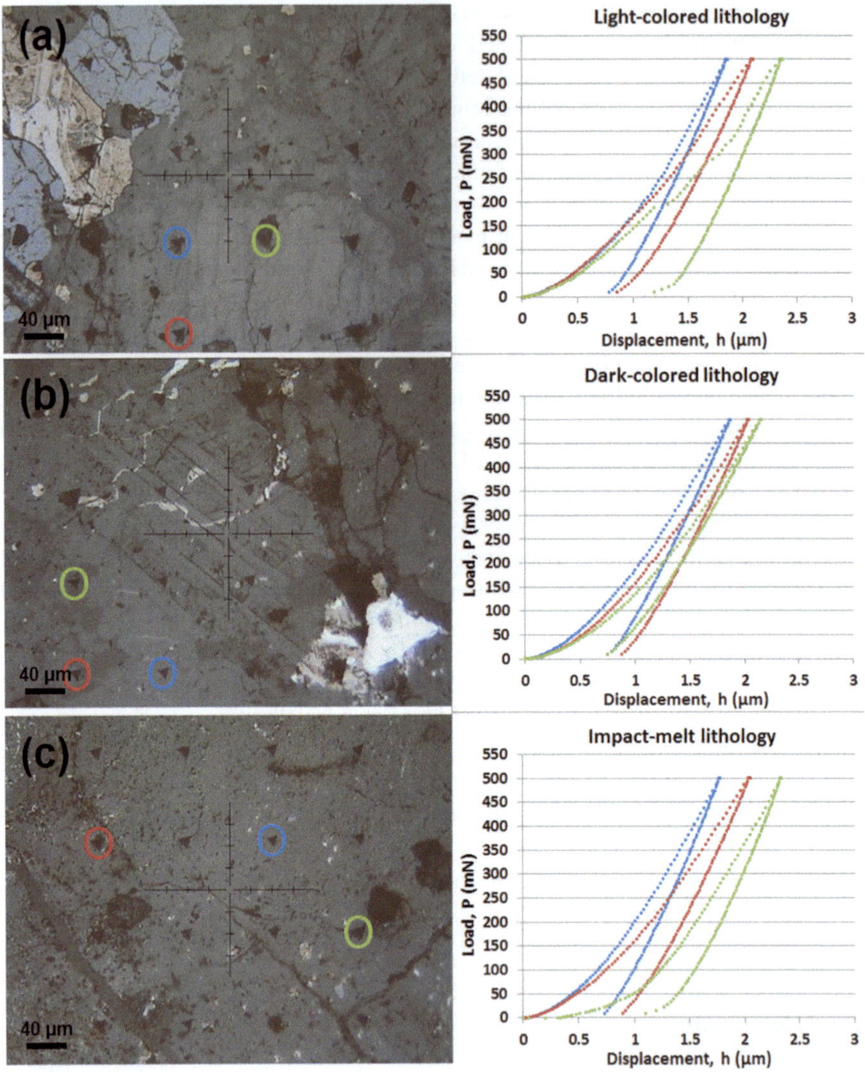

Fig. 6 Nanoindentations and curves. In the left-hand three optical images show arrays of 16 high-load indentations (up to 500 mN) on the three lithologies identified in the Chelyabinsk meteorite: (**a**) → light-colored; (**b**) → dark-colored; (**c**) → impact-melt. The *color circles* indicate the indentations from where the curves at the *right-hand side* of this figure were obtained. We selected indentations representative of the variability of mechanical properties on each lithology. The mechanical properties derived from this curves are shown in Table 3

Table 3 Nanoindentation results

Lithology	E_r (GPa)	H (GPa)	W_{el}/W_{tot}
Light-colored	65 ± 11	9.2 ± 2.4	0.65 ± 0.07
Dark-colored	65 ± 15	11 ± 4	0.66 ± 0.08
Impact melt	77 ± 7	11.8 ± 2.3	0.68 ± 0.03

with olivine and pyroxene, and they are moderately bright (Bus and Binzel 2002; DeMeo et al. 2009; Paton et al. 2011; Pieters and McFadden 1994). Also related with stony meteorites (and with S-class asteroids) are the Q-class, which show subtle variations of depth and width in the 1 and 2 μm bands (Bus and Binzel 2002; DeMeo et al. 2009). However, Q-class asteroids have a much lower slope of the whole spectrum, even more pronounced at redder wavelengths. Such spectral distinction is almost absent in the inner of the MB, and in fact the Bus-DeMeo classification (2009) reports less than ten samples, with only a few of them being well-catalogued in modern surveys.

According to our results and the spectra shown in Fig. 5, the distinction between the light- and dark-colored lithologies in Chelyabinsk (and Kilabo) can be tentatively connected to the distinction between S- and Q-type asteroids (overall similar shape but a noticeable decrease in slope and reflectance), as it has been done before between S- and C-type asteroids (Britt and Pieters 1994; Grier et al. 2004). The similarity between the Kilabo LL6 (light) and the Chelyabinsk LL5 #2 (light) spectra seems to indicate that those variations are not particularly dependent on the petrologic type 5–6, but more would be required to certify that. One possible explanation is that these differences are a consequence of the progressive darkening associated with impacts, shock metamorphism and brecciation, which affect the reflectance properties (Britt and Pieters 1994). That would be consistent with the optical darkening observed in the dark-colored lithology of Chelyabinsk due to higher degree of shock, and also with its definition as a breccia (Bischoff et al. 2013b). Other secondary processes have been associated with this overall variation. Space weathering can also have a deep influence in the reflectance properties (Pieters et al. 2000), and it has been suggested that the differences between OC meteorites and their S- or Q-type parent asteroids are a consequence of space weathering after surface exposition to the space environment for a long time (Gaffey et al. 1993). However, space weathering implies a stronger change in spectral slope than shock darkening (Kohout et al. 2014). Most NEOs could experience resurfacing due to the regolith displacement, originated by the differential gravitational pull at each surface point after close approaches to the terrestrial planets (Binzel et al. 2010). Also, small asteroids could have been produced in relatively recent impacts, and due to their low gravitational fields these would not have significant regolith on their surfaces. In fact, Q-type asteroids are relatively small, with its largest member being 1866 Sisyphus, which has a diameter of about 10 km, probably indicating that sample size is crucial to allow an unbiased distinction.

The classification of Chelyabinsk as an LL5-6 genomic breccia with an S4 degree of shock makes it an ideal candidate to test whether impact processing and induced metamorphism is behind the spectral variation reported between S- and Q-type asteroids. The shock-darkened lithologies, together with the shock veins filled mainly with melt troilite and metal and silicate fragments, are common all over our meteorite sample (Bischoff et al. 2013b; Kohout et al. 2014), where they occur as dark to opaque areas in the transmitted light mosaic (Figs. 1, 2 and 3). Shocked minerals are also to be expected in this meteorite, but the phases formed depend on the degree of shock and elemental availability (Xie et al. 2001). We mentioned before the detection of oldhamite, a mineral previously unknown in ordinary chondrites (Avril et al. 2013), in the sample studied with Raman spectroscopy. The interest of this mineral comes from the fact that it was considered as one of the components identified on asteroid (2867) Šteins in the interpretation of reflectance spectra taken by ESA's Rosetta spacecraft (Weissman et al. 2008). Therefore, our Raman identification of oldhamite indicates that the reflectance spectra of shocked S asteroids might have a minor, but still distinctive, depth in this region (Burbine et al. 2002), although the abundance due to shock would hardly be high enough to be apparent. The presence of merrillite and other high-pressure mineral indicates that Chelyabinsk experienced peak shock pressures higher than ~25 GPa, being consistent with the S4 shock stage already proposed (Bischoff et al. 2013b; Bischoff and Stöffler 1992; Kohout et al. 2014; Stöffler et al. 1991). The small shift in olivine spectra was previously reported for L6 chondrite Yamato 7304, and was attributed to residual stress after an incomplete recovery from the shock deformation, or a local deviation from the static pressure induced by the shock wave during its quick transit through the meteorite (Miyamoto and Ohsumi 1995). Also, we already reported the presence of the high pressure polymorph of olivine called ringwoodite (Trigo-Rodríguez et al. 2014b). Although it was identified long ago as quite common in shocked L-chondrites (Binns et al. 1969), it has been rarely found in LL chondrites (Bischoff 2002).

From the nanoindentations we obtained information about the mechanical properties of the main components of OCs (Moyano-Cambero et al. 2017). Both Young's modulus and hardness are much higher in Chelyabinsk than the values found on most CCs (Britt et al. 2015). The main reason for that is the difference in porosity between these two types of meteorites, as porosity is known to cause a reduction on both H and E_r (Pellicer et al. 2012). Indeed, some CCs can show porosities exceeding 30 % (Consolmagno et al. 2008; Macke et al. 2011), while the typical values for OCs are between 5 and 10 % (Consolmagno et al. 2008). Consistently, the microporosity of Chelyabinsk has been measured to range between 2 and 11 %, with a mean value of ~6 % (Kohout et al. 2014). Actually, the microporosity measured in Chelyabinsk does not vary much between the three lithologies, and therefore the differences in mechanical properties cannot be attributed to distinct porosities. As the composition and mineralogy of the lithologies are also very similar (Galimov et al. 2013; Kohout et al. 2014), there must be a different reason for the variations in mechanical properties. The results of shock processes, such as refinement of particle size, increase in the amount of structural defects, and the formation of shock-melt metal and troilite veins, introduce noticeable variations in the mechanical behavior. In spite of the higher E_r in the impact melt lithology, indicative of a higher resistance to elastic deformation, the formation of fractures and the presence of larger melt veins in the contours of this lithology (Popova et al. 2013) suggest that this lithology contributes to reduce the hardness of objects similar to the parent body of Chelyabinsk.

Mechanical properties play a key role in modeling the reaction of an asteroid to the impact of a solid projectile. In relatively hard and low porosity bodies such as Chelyabinsk, less impact energy is dissipated in the process of compacting pores, and therefore there is a better transmission of momentum between the two objects (Benz and Jutzi 2006; Bruck Syal et al. 2016; Hoerth et al. 2015; Holsapple and Housen 2012). Furthermore, lower hardness promotes the formation of impact craters and therefore the release of material from the body receiving the impact, which implies an even more efficient transmission and multiplication of momentum (Benz and Jutzi 2006; Holsapple and Housen 2012).

In order to deflect PHAs using kinetic impacts, a proper connection and understanding of the mechanical properties of meteorites and asteroids is crucial. The AIDA mission, with the goal of testing such a deflection mechanism and analyzing its consequences, will travel to the binary 800 m NEA Didymos and use DART to impact Didymos' 150 m companion (Michel et al. 2015a, 2016). Didymos has been characterized as an asteroid belonging to the S-complex (Pravec et al. 2006), and has been connected to L/LL-type metorites (Dunn et al. 2013). As a binary system, Didymos could have formed like a rubble pile, which would imply a considerable degree of shock and brecciation (Walsh and Richardson 2006). Chelyabinsk is therefore a possible good proxy for an asteroid like Didymos, and thankfully it is widely available due to the huge amount of meteorites recovered so far. Studying its mechanical properties will provide key information to properly model the effects of a kinetic impact in this type of asteroid.

The results provided here show that an impact on the light-colored lithology of Chelyabinsk would be more efficient than in the dark-colored, and therefore that a proper selection of the impact site is crucial even if the composition of the asteroid seems homogeneous. In Chelyabinsk the two lithologies are strongly mixed to be selected as targets, but it is not necessary the case for all NEAs.

However, it has to be taken into account that our results were obtained from small samples at microscopic scale, but we should expect significant macroporosity in a much larger rubble pile, which will probably have an important effect on the reaction to a kinetic projectile (Bischoff et al. 2006). Indeed, in those scenarios the cohesive strength that keeps the different parts together will play a significant role. The difference in scale between a meteorite sample and an asteroid are also important for both nanoindentations and UV-Vis-NIR spectra, so they need to be considered in future studies in order to extrapolate these results. Besides, the values obtained after quasi-static experiments like nanoindentations, are not identical to the results from dynamic experiments. With respect to hardness, for example, it has been seen that in dynamic experiments the results are generally larger than in quasistatic experiments (Anton and Subhash 2000). Nevertheless, trends relating hardness and compressive properties are similar between quasistatic and dynamic experiments (Subhash et al. 1999). Therefore, although additional experimental studies are required to understand the direct application to a large rubble pile asteroid, in the case of small compact bodies the connection is much more direct. Remarkably, as most OCs arriving to our planet possibly originate by the disruption and fragmentation of relatively large asteroids into smaller pieces, (Bottke et al. 2015; Michel et al. 2001, 2015b), that will be the case for most S- or Q-type NEAs.

5 Conclusions

Here we analyzed the shock processes in the Chelyabinsk meteorite, in order to propose appropriate evolutionary histories. According with the results presented before the main conclusions of this work are:

1. Chelyabinsk UV-Vis-NIR reflectance spectra are dominated by olivine and pyroxene bands, but due to the presence of shock darkened regions on this meteorite the overall reflectance of large samples will be lower than the one from other OCs, varying as a function of the amount of darkened lithology. Consequently, the collisional history has a deep influence in the reflectance of asteroids.
2. The different UV-Vis-NIR spectral behavior of the lithologies in Chelyabinsk resembles those of S and Q-type asteroids. Being relatively small and rare, Q-type asteroids could be the surviving fragments of strong collisional histories, which points towards a connection between the dark-colored lithology of Chelyabinsk, which is also the product of higher shock processes.
3. Shocked minerals exhibit distinctive drifts associated with residual stress due to the high pressure experienced by the body. Some examples are: jadeite, high P-merrillite, and oldhamite-like sulfides.
4. In such a processed compact object, momentum transfer and multiplication of a kinetic impact would be much more efficient than in the kind of porous body we expect to be the parent body of CCs.
5. The mechanical properties of Chelyabinsk are consistent with a low porosity material, and therefore an asteroid of OC composition would be deflected more effectively.
6. The different lithologies of Chelyabinsk show different mechanical behavior, being the light-colored the one which would more efficiently react to a kinetic impact, due to its lower hardness. Since their UV-Vis-NIR spectra can be easily distinguished, we would be able to select the best region to be impacted.
7. The impact melt veins in Chelyabinsk are more easily fractured than the other lithologies, and thus contribute to reduce the strength and induce fracture of a Chelyabinsk type asteroid.

To conclude, we could extract an important keynote: Shocked minerals participate in increasing the consistency of the PHA, which would become meteoroids tough enough for survival and contributing significantly to impact hazard. Despite of the history of collisions behind the Chelyabinsk meteorite, its impressive fall might be considered a serious warning about the threat that some surviving asteroid fragments imply for humankind. Therefore, it is important to be able to recognize the different regions and weakness of these asteroids, in order to select, before launching the projectile, the point which will more efficiently react to the impact of a kinetic projectile.

Acknowledgments Ministerio de Educaci6n y Ciencia (MEC) is acknowledged for providing AYA2011-26522 and 2015-67175-P grants (P.I. J.M.T-R.). J.L. is Serra Húnter Fellow and is grateful to ICREA Academia program. We thank Prof. Addolf Bischoff (Münster) for providing Chalyabinsk sections, and Marcos Rosado (ICN2, Barcelona) is acknowledged for analytical and technical assistance. This research utilizes spectra acquired by several authors with the NASA RELAB facility at Brown University that were compared with the obtained by our group under project AYA 2011-26522. N. Mestres/ICMAB, acknowledges financial support from the Spanish Ministry of Economy and Competitiveness, through the "Severo Ochoa" Program for Centers of Excellence in R&D (SEV—2015-0496). This study was done in the frame of a Ph.D. on Physics at the Autonomous University of Barcelona (UAB).

References

Anton, R.J., Subhash, G.: Dynamic Vickers indentation of brittle materials. Wear **239**, 27–35 (2000)

Avril, C., Malavergne, V., Caracas, R., Zanda, B., Reynard, B., Charon, E., Bobocioiu, E., Brunet, F., Borensztajn, S., Pont, S., Tarrida, M., Guyot, F.: Raman spectroscopic properties and Raman identification of CaS-MgS-MnS-FeS-Cr_2FeS_4 sulfides in meteorites and reduced sulfur-rich systems. Meteorit. Planet. Sci. **48**, 1415–1426 (2013)

Benz, W., Jutzi, M.: Collision and impact simulations including porosity. Proc. Int. Astron. Union **2**, 223 (2006)

Binns, R.A., Davis, R.J., Reed, S.J.B.: Ringwoodite, natural $(Mg, Fe)_2SiO_4$ spinel in the Tenham meteorite. Nature **221**, 943–944 (1969)

Binzel, R.P., Lupishko, D.F., Martino, M.D., Whiteley, R.J., Hahn, G.J.: Physical properties of near-earth objects. In: Bottke Jr., W.F., Cellino, A., Paolicchi, P., Binzel, R.P. (eds.) Asteroids III, pp. 255–271. University of Arizona Press, Tucson (2002)

Binzel, R.P., Morbidelli, A., Merouane, S., DeMeo, F.E., Birlan, M., Vernazza, P., Thomas, C.A., Rivkin, A.S., Bus, S.J., Tokunaga, A.T.: Earth encounters as the origin of fresh surfaces on near-Earth asteroids. Nature **463**, 331–334 (2010)

Binzel, R.P., Thomas, C.A., DeMeo, F.E., Tokunaga, A.T., Rivkin, A.S., Bus, S.J.: Lunar and planetary science XXXVII. In: 37th Lunar and Planetary Science Conference. Abstract 1491 (2006)

Bischoff, A.: Discovery of purple-blue ringwoodite within shock veins of an LL6 ordinary chondrite from Northwest Africa. In: 33rd Lunar and Planetary Science Conference. Abstract 1264 (2002)

Bischoff, A., Dyl, K.A., Horstmann, M., Ziegler, K., Wimmer, K., Young, E.D.: Reclassification of Villalbeto de la Peña-Occurrence of a winonaite-related fragment in a hydrothermally metamorphosed polymict L-chondritic breccia. Meteorit. Planet. Sci. **48**, 628–640 (2013a)

Bischoff, A., Horstmann, M., Pack, A., Laubenstein, M., Haberer, S.: Asteroid 2008 TC3-Almahata Sitta: a spectacular breccia containing many different ureilitic and chondritic lithologies. Meteorit. Planet. Sci. **45**, 1638–1656 (2010)

Bischoff, A., Horstmann, M., Vollmer, C., Heitmann, U., Decker, S.: Chelyabinsk—not only another ordinary LL5 chondrite, but a spectacular chondrite breccia. Meteorit. Planet. Sci. **48**, A61 (2013b)

Bischoff, A., Palme, H., Schultz, L., Weber, D., Weber, H.W., Spettel, B.: Acfer 182 and paired samples, an iron-rich carbonaceous chondrite: similarities with ALH85085 and relationship to CR chondrites. Geochim. Cosmochim. Acta **57**, 2631–2648 (1993)

Bischoff, A., Rubin, A.E., Keil, K., Stöffler, D.: Lithification of gas-rich chondrite regolith breccias by grain boundary and localized shock melting. Earth Planet. Sci. Lett. **66**, 1–10 (1983)

Bischoff, A., Scott, E.R.D., Metzler, K., Goodrich, C.A.: Nature and origins of meteoritic breccias. In: Lauretta, D.S., McSween, H.Y.J. (eds.) Meteorites and the Early Solar System II, pp. 679–712. University of Arizona Press, Tucson (2006)

Bischoff, A., Stöffler, D.: Shock metamorphism as a fundamental process in the evolution of planetary bodies: information from meteorites. Eur. J. Mineral. **4**, 707–755 (1992)

Bland, P.A., Artemieva, N.A.: Efficient disruption of small asteroids by Earth's atmosphere. Nature **424**, 288–291 (2003)

Borovička, J., Spurný, P., Brown, P., Wiegert, P., Kalenda, P., Clark, D., Shrbený, L.: The trajectory, structure and origin of the Chelyabinsk asteroidal impactor. Nature **503**, 235–237 (2013)

Bottke Jr., W.F., Broz, M., O'Brien, D.P., Campo Bagatín, A., Morbidelli, A., Marchi, S.: The collisional evolution of the main asteroid belt. In: Michel, P., DeMeo, F.E., Bottke Jr., W.F. (eds.) Asteroids IV, pp. 701–724. University of Arizona Press, Tucson (2015)

Britt, D., Yeomans, D., Housen, K., Consolmagno, G.: Asteroid density, porosity, and structure. In: Bottke, W.F. (ed.) Asteroids III, pp. 485–500. University of Arizona Press, Tucson (2002)

Britt, D.T., Abreu, N., Asphaug, E.I., Campins, H., Gertsch, L., Sercel, J., Zacny, K.: Boulder Physical Properties. Asteroid Redirect Mission (ARM): Final Report Draft for Public Comment, 35–43 (2015)

Britt, D.T., Pieters, C.M.: Black ordinary chondrites: an analysis of abundance and fall frequency. Meteoritics **26**, 279–285 (1991)

Britt, D.T., Pieters, C.M.: Darkening in black and gas-rich ordinary chondrites: The spectral effects of opaque morphology and distribution. Geochim. Cosmochim. Acta **58**, 3905–3919 (1994)

Brown, P.G., Assink, J.D., Astiz, L., Blaauw, R., Boslough, M.B., Borovička, J., Brachet, N., Brown, D., Campbell-Brown, M., Ceranna, L., Cooke, W., de Groot-Hedlin, C., Drob, D.P., Edwards, W., Evers, L.G., Garces, M., Gill, J., Hedlin, M., Kingery, A., Laske, G., Le Pichon, A., Mialle, P., Moser, D.E., Saffer, A., Silber, E., Smets, P., Spalding, R.E., Spurný, P., Tagliaferri, E., Uren, D., Weryk, R.J., Whitaker, R., Krzeminski, Z.: A 500-kiloton airburst over Chelyabinsk and an enhanced hazard from small impactors. Nature **503**, 238–241 (2013)

Brownlee, D.E.: The origin and properties of dust impacting the earth. In: Peucker-Ehrenbrink, B., Schmitz, B. (eds.) Accretion of Extraterrestrial Matter Throughout Earth's History, pp. 1–12. Springer US, New York (2001)

Bruck Syal, M., Owen, J.M., Miller, P.M.: Deflection by kinetic impact: sensitivity to asteroid properties. Icarus **269**, 50–61 (2016)

Burbine, T.H.: Asteroids: their composition and impact threat. Vestn. Ces. Geol. Ust. **77**, 243–252 (2002)

Burbine, T.H., Mccoy, T.J., Nittler, L.R., Benedix, G.K., Cloutis, E.A., Dickinson, T.L.: Spectra of extremely reduced assemblages: implications for mercury. Meteorit. Planet. Sci. **37**, 1233–1244 (2002)

Bus, S.J., Binzel, R.P.: Phase II of the small main-belt asteroid spectroscopic survey a feature-based taxonomy. Icarus **158**, 146–177 (2002)

Chapman, C.R., Morrison, D.: Impacts on the Earth by asteroids and comets: assessing the hazard. Nature **367**, 33–40 (1994)

Chen, M., Xie, X.: Raman spectra of orthopyroxene in two natural shocked H chondrites. Meteoritics **27**, 209 (1992)

Consolmagno, G.J., Britt, D.T., Macke, R.J.: The significance of meteorite density and porosity. Chem. Erde-Geochem. **68**, 1–29 (2008)

DeMeo, F.E., Binzel, R.P., Slivan, S.M., Bus, S.J.: An extension of the Bus asteroid taxonomy into the near-infrared. Icarus **202**, 160–180 (2009)

Dmitriev, V., Lupovka, V., Gritsevich, M.: Orbit determination based on meteor observations using numerical integration of equations of motion. Planet. Space Sci. **117**, 223–235 (2015)

Dunn, T.L., Burbine, T.H., Bottke, W.F., Clark, J.P.: Mineralogies and source regions of near Earth asteroids. In: Lunar and Planetary Science Conference 1997 (2013)
Fischer-Cripps, A.C.: Nanoindentation. Springer, New York (2004)
Gaffey, M.J.: Spectral reflectance characteristics of the meteorite classes. J. Geophys. Res. **81**, 905–920 (1976)
Gaffey, M.J., Bell, J.F., Brown, R.H., Burbine, T.H., Piatek, J.L., Reed, K.L., Chaky, D.A.: Mineralogical variations within the S-type asteroid class. Icarus **106**, 573–602 (1993)
Galimov, E.M.: Chelyabinsk meteorite—an LL5 chondrite. Sol. Syst. Res. **47**, 255–259 (2013)
Galimov, E.M., Kolotov, V.P., Nazarov, M.A., Kostitsyn, Y.A., Kubrakova, I.V., Kononkova, N.N., Roshchina, I.A., Alexeev, V.A., Kashkarov, L.L., Badyukov, D.D., Sevast'yanov, V.S.: Analytical results for the material of the Chelyabinsk meteorite. Geochem. Int. **51**, 522–539 (2013)
Goodrich, C., Bischoff, A., O'Brien, D.P.: Asteroid 2008 TC3 and the fall of Almahata sitta, a unique meteorite breccia. Elements **10**, 31–37 (2014)
Grier, J.A., Kring, D.A., Swindle, T.D., Rivkin, A.S., Cohen, B.A., Britt, D.T.: Analyses of the chondritic meteorite Orvinio (H6): insight into the origins and evolution of shocked H chondrite material. Meteorit. Planet. Sci. **39**, 1475–1493 (2004)
Hoerth, T., Schäfer, F.K., Hupfer, J., Millon, O., Wickert, M.: Momentum transfer in hypervelocity impact experiments on rock targets. Procedia Eng. **103**, 197–204 (2015)
Holsapple, K.A., Housen, K.R.: Momentum transfer in asteroid impacts. I. Theory and scaling. Icarus **221**, 875–887 (2012)
Horstmann, M., Bischoff, A.: The Almahata Sitta polymict breccia and the late accretion of asteroid 2008 TC3. Chem. Erde-Geochem. **74**, 149–183 (2014)
Keil, K., Bell, J.F., Britt, D.T.: Reflection spectra of shocked ordinary chondrites and their relationship to asteroids. Icarus **98**, 43–53 (1992)
Kieffer, S.W.: From regolith to rock by shock. Moon **13**, 301–320 (1975)
Kocherov, A.V., Korochantsev, A.V., Lorenz, C.A., Ivanova, M.A., Grokhovsky, V.I.: Recovery, laboratory preparation and current state of the main mass of the Chelyabinsk meteorite. In: 45th Lunar and Planetary Science Conference. Abstract 2227 (2014)
Kohout, T., Gritsevich, M., Grokhovsky, V.I., Yakovlev, G.A., Haloda, J., Halodova, P., Michallik, R.M., Penttilä, A., Muinonen, K.: Mineralogy, reflectance spectra, and physical properties of the Chelyabinsk LL5 chondrite—insight into shock-induced changes in asteroid regoliths. Icarus **228**, 78–85 (2014)
Lafuente, B., Downs, R.T., Yang, H., Stone, N.: The power of databases: the RRUFF project. In: Armbuster, T., Danisi, R.M. (eds.) Highlights in Mineralogical Crystallography, pp. 1–30. De Gruyter, Berlin (2015)
Llorca, J., Trigo-Rodríguez, J.M.: Raman spectroscopy of merrilite in Villalbeto de la Peña L6 ordinary chondrite. In: 37th Lunar and Planetary Science Conference. Abstract 1055 (2004)
Macke, R.J., Consolmagno, G.J., Britt, D.T.: Density, porosity, and magnetic susceptibility of carbonaceous chondrites. Meteorit. Planet. Sci. **46**, 1842–1862 (2011)
Metzler, K., Bischoff, A., Greenwood, R.C., Palme, H., Gellissen, M., Hopp, J., Franchi, I.A., Trieloff, M.: The L3-6 chondritic regolith breccia Northwest Africa (NWA) 869: (I) Petrology, chemistry, oxygen isotopes, and Ar-Ar age determinations. Meteorit. Planet. Sci. **46**, 652–680 (2011)
Michel, P., Benz, W., Tanga, P., Richardson, D.C.: Collisions and gravitational reaccumulation: forming asteroid families and satellites. Science **294**(5547), 1696–1700 (2001)
Michel, P., Cheng, A., Küppers, M., Pravec, P., Blum, J., Delbo, M., Green, S.F., Rosenblatt, P., Tsiganis, K., Vincent, J.-B., Biele, J., Ciarletti, V., Hérique, A., Ulamec, S., Carnelli, I., Galvez, A., Benner, L.A.M., Naidu, S.P., Barnouin, O.S., Richardson, D.C., Rivkin, A.S., Scheirich, P., Moskovitz, N., Thirouin, A., Schwartz, S.R., Campo Bagatín, A., Yu, Y.: Science case for the Asteroid Impact Mission (AIM): a component of the Asteroid Impact & Deflection Assessment (AIDA) mission. Adv. Space Res. **57**, 2529–2547 (2016)
Michel, P., Cheng, A., Ulamec, S.: The AIDA Team: Asteroid Impact and Deflection Assessment (AIDA) mission: science return and mitigation relevance. In: 4th IAA Planetary Defense Conference IAA-PDC-15, 1–3 (2015a)

Michel, P., Richardson, D.C., Durda, D.D., Jutzi, M., Asphaug, E.: Collisional formation and modeling of asteroid families. In: Michel, P., DeMeo, F.E., Bottke Jr., W.F. (eds.) Asteroids IV, p. 785. University of Arizona Press, Tucson (2015b)
Milani, A., Chesley, S., Chodas, P.W., Valsecchi, G.B.: Asteroid close approaches: analysis and potential impact detection. In: Asteroids III, pp. 55–69 (2002)
Miyamoto, M., Ohsumi, K.: Micro Raman spectroscopy of olivines in L6 chondrites: evaluation of the degree of shock. Geophys. Res. Lett. **22**, 437–440 (1995)
Morbidelli, A., Nesvorný, D.: Numerous weak resonances drive asteroids toward terrestrial planets orbits. Icarus **139**, 295–308 (1999)
Moreno-Ibáñez, M., Gritsevich, M., Trigo-Rodríguez, J.M.: New methodology to determine the terminal height of a fireball. Icarus **250**, 544–552 (2015)
Moyano-Cambero, C.E., Trigo-Rodríguez, J.M., Llorca, J.: UV-NIR spectra of the most reflective carbonaceous chondrite groups: CH, CR and R. In: 44th Lunar and Planetary Science Conference. Abstract 1533 (2013)
Moyano-Cambero, C.E., Pellicer, E., Trigo-Rodríguez, J.M., Williams, I.P., Blum, J., Michel, P., Küppers, M., Martínez-Jiménez, M., Lloro, I., Sort, J.: Nanoindenting the Chelyabinsk meteorite to learn about impact deflection effects in asteroids. Ap. J. (2017)
Nakamura, T., Noguchi, T., Tanaka, M., Zolensky, M.E., Kimura, M., Tsuchiyama, A., Nakato, A., Ogami, T., Ishida, H., Uesugi, M., Yada, T., Shirai, K., Fujimura, A., Okazaki, R., Sandford, S.A., Ishibashi, Y., Abe, M., Okada, T., Ueno, M., Mukai, T., Yoshikawa, M., Kawaguchi, J.: Itokawa dust particles: a direct link between S-type asteroids and ordinary chondrites. Science **333**, 1113–1116 (2011)
Oliver, W.C., Pharr, G.M.: An improved technique for determining hardness and elastic modulus using load and displacement sensing indentation experiments. J. Mater. Res. **7**, 1564–1583 (1992)
Ozawa, S., Miyahara, M., Ohtani, E., Koroleva, O.N., Ito, Y., Litasov, K.D., Pokhilenko, N.P.: Jadeite in Chelyabinsk meteorite and the nature of an impact event on its parent body. Sci. Rep. **4**, 5033 (2014)
Paton, M., Muinonen, K., Pesonen, L.J., Kuosmanen, V., Kohout, T., Laitinen, J., Lehtinen, M.: A PCA study to determine how features in meteorite reflectance spectra vary with the samples' physical properties. J. Quant. Spectrosc. Radiat. Transf. **112**, 1803–1814 (2011)
Pellicer, E., Pané, S., Panagiotopoulou, V., Fusco, S., Sivaraman, K.M., Suriñach, S., Baró, M.D., Nelson, B.J., Sort, J.: Localized electrochemical deposition of porous Cu-Ni microcolumns: insights into the growth mechanisms and the mechanical performance. Int. J. Electrochem. Sci. **7**, 4014–4029 (2012)
Pieters, C.M., McFadden, L.A.: Meteorite and asteroid reflectance spectroscopy: clues to early solar system processes. Annu. Rev. Earth Planet. Sci. **22**, 457–497 (1994)
Pieters, C.M., Taylor, L.A., Noble, S.K., Keller, L.P., Hapke, B.W., Morris, R.V., Allen, C.C., McKay, D.S., Wentworth, S.J.: Space weathering on airless bodies: resolving a mystery with lunar samples. Meteorit. Planet. Sci. **35**(5), 1101–1107 (2000). http://onlinelibrary.wiley.com/doi/10.1111/j.1945-5100.2000.tb01496.x
Popova, O.P., Jenniskens, P., Emel'yanenko, V., Kartashova, A., Biryukov, E., Khaibrakhmanov, S., Shuvalov, V., Rybnov, Y., Dudorov, A., Grokhovsky, V.I., Badyukov, D.D., Yin, Q.-Z., Gural, P.S., Albers, J., Granvik, M., Evers, L.G., Kuiper, J., Kharlamov, V., Solovyov, A., Rusakov, Y.S.: Chelyabinsk airburst, damage assessment, meteorite recovery, and characterization. Science **342**(6162), 1069–1073 (2013)
Pravec, P., Scheirich, P., Kušnirák, P., Šarounová, L., Mottola, S., Hahn, G., Brown, P., Esquerdo, G., Kaiser, N., Krzeminski, Z., Pray, D.P., Warner, B.D., Harris, A.W., Nolan, M.C., Howell, E.S., Benner, L.A.M., Margot, J.L., Galád, A., Holliday, W., Hicks, M.D., Krugly, Y.N., Tholen, D., Whiteley, R., Marchis, F., DeGraff, D.R., Grauer, A., Larson, S., Velichko, F.P., Cooney, W.R., Stephens, R., Zhu, J., Kirsch, K., Dyvig, R., Snyder, L., Reddy, V., Moore, S., Gajdoš, Š., Világi, J., Masi, G., Higgins, D., Funkhouser, G., Knight, B., Slivan, S., Behrend, R., Grenon, M., Burki, G., Roy, R., Demeautis, C., Matter, D., Waelchli, N., Revaz, Y., Klotz, A., Rieugné, M., Thierry, P., Cotrez, V., Brunetto, L., Kober, G.: Photometric survey of binary near-Earth asteroids. Icarus **181**, 63–93 (2006)

Righter, K., Abell, P.A., Agresti, D., Berger, E.L., Burton, A.S., Delaney, J.S., Fries, M.D., Gibson, E.K., Haba, M.K., Harrington, R., Herzog, G.F., Keller, L.P., Locke, D., Lindsay, F.N., McCoy, T.J., Morris, R.V., Nagao, K., Nakamura-Messenger, K., Niles, P.B., Nyquist, L.E., Park, J., Peng, Z.X., Shih, C.-Y., Simon, J.I., Swisher, C.C., Tappa, M.J., Turrin, B.D., Zeigler, R.A.: Mineralogy, petrology, chronology, and exposure history of the Chelyabinsk meteorite and parent body. Meteorit. Planet. Sci. **50**, 1790–1819 (2015)

Rivkin, A.S., Binzel, R.P., Sunshine, J.M., Bus, S.J., Burbine, T.H., Saxena, A.: Infrared spectroscopic observations of 69230 Hermes (1937 UB): possible unweathered endmember among ordinary chondrite analogs. Icarus **172**, 408–414 (2004)

Ruzicka, A., Grossman, J.N., Bouvier, A., Herd, C.D.K., Agee, C.B.: The meteoritical bulletin, no. 102. Meteorit. Planet. Sci. **50**(9), 1662 (2015)

Stöffler, D., Keil, K., Scott, E.R.D.: Shock metamorphism of ordinary chondrites. Geochim. Cosmochim. Acta **55**, 3845–3867 (1991)

Subhash, G., Koeppel, B.J., Chandra, A.: Dynamic indentation hardness and rate sensitivity in metals. J. Eng. Mater. Technol. **121**, 257–263 (1999)

Tapia, M., Trigo-Rodríguez, J.M.: Natural hazard associated to shock waves of meter-sized meteoroids. In: Trigo-Rodríguez, J.M., Gritsevich, M., Palme H. (eds.) Assessment and Mitigation of Asteroid Impact Hazards, pp. 199–217. Springer, New York (2017)

Trigo-Rodríguez, J.M., Blum, J.: The effect of aqueous alteration and metamorphism in the survival of presolar silicate grains in chondrites. Publ. Astron. Soc. Aust. **26**, 289–296 (2009)

Trigo-Rodríguez, J.M., Lyytinen, E., Gritsevich, M., Moreno-Ibáñez, M., Bottke, W.F., Williams, I., Lupovka, V., Dmitriev, V., Kohout, T., Grokhovsky, V.: Orbit and dynamic origin of the recently recovered Annama's H5 chondrite. Mon. Not. R. Astron. Soc. **449**, 2119–2127 (2015)

Trigo-Rodríguez, J.M., Lyytinen, E., Jones, D.C., Madiedo, J.M., Castro-Tirado, A.J., Williams, I., Llorca, J., Vítek, S., Jelínek, M., Troughton, B., Gálvez, F.: Asteroid 2002NY40 as a source of meteorite-dropping bolides. Mon. Not. Roy Astron. Soc. **382**, 1933–1939 (2007)

Trigo-Rodríguez, J.M., Moyano-Cambero, C.E., Llorca, J., Fornasier, S., Barucci, M.A., Belskaya, I., Martins, Z., Rivkin, A.S., Dotto, E., Madiedo, J.M., Alonso-Azcárate, J.: UV to far-IR reflectance spectra of carbonaceous chondrites—I. Implications for remote characterization of dark primitive asteroids targeted by sample-return missions. Mon. Not. Roy Astron. Soc. **437**, 227–240 (2014a)

Trigo-Rodríguez, J.M., Moyano-Cambero, C.E., Mestres, N., Bischoff, A.: A Raman Study of Chelyabinsk LL5-6 Chondrite Breccia: Investigating the Signatures of Shock-Induced Melting in Near-Earth Asteroids. In: 45th Lunar and Planetary Science Conference. Abstract 1729

Trigo-Rodríguez, J.M., Williams, I.P.: Dynamic sources of contemporary hazard from meteoroids and small asteroids. In: Trigo-Rodríguez J.M., Gritsevich, M., Palme H. (eds.) Assessment and Mitigation of Asteroid Impact Hazards, pp. 11–32. Springer, New York (2017)

Wachs, I.E., Routray, K.: Catalysis science of bulk mixed oxides. ACS Catal. **2**, 1235–1246 (2012)

Walsh, K.J., Richardson, D.C.: Binary near-Earth asteroid formation: rubble pile model of tidal disruptions. Icarus **180**, 201–216 (2006)

Weissman, P.R., Hicks, M.D., Abell, P.A., Choi, Y., Lowry, S.C.: Rosetta target asteroid 2867 Steins: an unusual E-type asteroid. Meteorit. Planet. Sci. **914**, 905–914 (2008)

Xie, X., Chen, M., Dai, C., El Goresy, A., Gillet, P.: A comparative study of naturally and experimentally shocked chondrites. Earth Planet. Sci. Lett. **187**, 345–356 (2001)

Xie, X., Minitti, M.E., Chen, M., Mao, H.K., Wang, D., Shu, J., Fei, Y.: Natural high-pressure polymorph of merrillite in the shock veins of the Suizhou meteorite. Geochim. Cosmochim. Acta **66**, 2439–2444 (2002)

Zolensky, M., Herrin, J., Mikouchi, T., Ohsumi, K., Friedrich, J., Steele, A., Rumble, D., Fries, M., Sandford, S., Milam, S., Hagiya, K., Takeda, H., Satake, W., Kurihara, T., Colbert, M., Hanna, R., Maisano, J., Ketcham, R., Goodrich, C., Le, L., Robinson, G., Martinez, J., Ross, K., Jenniskens, P., Shaddad, M.H.: Mineralogy and petrography of the Almahata Sitta ureilite. Meteorit. Planet. Sci. **45**, 1618–1637 (2010)

Zolensky, M.E., Ivanov, A.: The Kaidun microbreccia meteorite: a harvest from the inner and outer asteroid belt. Chem. Erde-Geochem. **63**, 185–246 (2003)

Asteroids, Comets and Meteorite-Dropping Bolides Studied from The Montsec Astronomical Observatory

Josep M. Trigo-Rodríguez, Toni Santana-Ros, Manuel Moreno-Ibáñez, Diego Rodríguez, Josep Sanz, Ivan Lloro, and Albert Sánchez

Abstract Multiple observational techniques are used to quantify the flux of meteoroids, asteroids and comets through near-Earth space. Telescopic CCD observations collect significant data of Earth-crossers being usually few hundreds of meters in diameter. These objects can be only observed during close approaches with Earth and are the most direct source of contemporary hazard to us. The 0.8 m Joan Oró Telescope located at Montsec Astronomical Observatory (Catalonia) allows a wide range of research in the minor bodies domain to be carried out. In this paper, two of its outstanding observational projects are described: (1) a follow-up of Near Earth Asteroids and Potentially Hazardous Asteroids during close approaches to our planet, and (2) a campaign collecting disk-integrated photometry of selected main belt asteroids which aims to enhance the Solar System science coming out from *Gaia* space mission. Additionally, a CCD-based all-sky camera together with video cameras from Catalonia stations record the appearance of large bolides that are linked with their parent asteroids using dissimilarity functions and backwards integrations of their respective orbits. With all these data it is expected to obtain not only a better understanding of the rate of meter- to tens of meters-sized asteroids impacting with Earth per unit of time, but also inferring new clues on the sources of these dangerous projectiles.

J.M. Trigo-Rodríguez (✉) • M. Moreno-Ibáñez • J. Sanz • I. Lloro
Institute of Space Sciences (IEEC-CSIC), Meteorites, Minor Bodies and Planetary Sciences Group, Campus UAB, c/Can Magrans s/n, 08193 Cerdanyola del Vallès (Barcelona), Catalonia, Spain
e-mail: trigo@ice.csic.es

T. Santana-Ros
Faculty of Physics, Astronomical Observatory Institute, Adam Mickiewicz University, Słoneczna 36, Poznań 60-286, Poland

D. Rodríguez
Guadarrama Observatory (MPC 458), Villalba, Madrid, Spain

A. Sánchez
Gualba Astronomical Observatory (MPC 442), Barcelona, Spain

J.M. Trigo-Rodríguez et al. (eds.), *Assessment and Mitigation of Asteroid Impact Hazards*, Astrophysics and Space Science Proceedings 46,
DOI 10.1007/978-3-319-46179-3_12

1 Montsec Astronomical Observatory Studies of Minor Bodies

The Montsec Astronomical Observatory was inaugurated in October 2008. This observatory houses a telescope of 80 cm in diameter, named after the famous biochemist Joan Oró (hereafter cited as TJO). The construction of the telescope and its facility came about through the efforts of Professor Joan Oró and other amateur astronomers involved in science popularization in Catalonia. Prof. Oró always fascinated about asteroids, comets and their meteorites and was pioneer in the study of these bodies and their implication on our existence (Oró 1961).

In the first years of operation the TJO was rarely used to the study Solar System minor bodies because non-sidereal follow-up was not implemented. In 2012 significant progress was made and reliable monitoring of asteroids was achieved, with continuous astrometric and photometric monitoring programs of near-Earth objects (NEOs) and periodic comets being carried out (see e.g. Moreno-Ibáñez et al. 2016). We have also cooperated in the follow-up of NEOs in the framework of the EURONEAR international cooperation (http://euronear.imcce.fr). The observing program is focusing on monitoring NEOs and potentially hazardous objects (PHOs) during close approaches to Earth. Among these hazardous objects, the asteroids delivered from the main belt are dominant in the population and known as PHAs. We will use both terms to make a distinction between them and the extremely dark dangerous bodies of cometary origin, but usually extinct or inactive. These bodies are contributors to Earth's impact hazard, and are also source of meteorite falls (Bottke et al. 2002; Morbidelli et al. 2002; Trigo-Rodríguez et al. 2015). The main aim is performing astrometric follow-up to improve orbits and obtaining additional physical information on these objects. Obviously, orbital characterization of these objects provides clues on their origin that is also useful in cosmochemical grounds (Alexander et al. 2012).

The TJO has proved to be an excellent facility for observing PHOs, as it includes a proper motion follow-up routine, to perform the observations of fast-moving asteroids. The methodology depends on the topic under study. Accurate astrometric measurements of 30 or 60 s target-guided exposures are reduced and reported to the Minor Planet Center (MPC). The observing program also can include ground-based photometry using standard John-son-Cousin filters, mostly in V, R, and I filters in order to get color indexes as a first clue on the nature of the object. Guided exposures are often stacked to achieve good signal/noise ratios in order to determine the presence or absence of cometary activity from the FWHM statistics and photometric growth curves.

Additionally, once the new *ARES* spectrometer will be operational, it is planned to obtain reflectance spectra of the brightest targets to be compared with our UV-NIR spectral database of primitive chondrites (see e.g. Trigo-Rodríguez et al. 2014). So far, the main goals are contributing to improve their orbits by performing accurate astrometric measurements, obtaining broadband photometry, rotational periods, and identifying cometary-like activity.

Repeated disk-integrated photometric observations of asteroids results in lightcurves that give clues on their main physical parameters: rotational period, variations in albedo, presence of satellites, etc. When the object has been observed under different viewing geometries (see chapter by Santana-Ros et al. for further details) it is also possible to derive a shape model including the body's spin state. Unfortunately, this is a highly time consuming technique for asteroids with main belt orbits or farther, as it requires gathering data during different apparitions. Where data mining is not possible for a given target, we are committed to gathering lightcurves during 3 or 4 years. However, data coming from high quality sky surveys is about to represent a great revolution in this field. In particular, the European Space Agency's *Gaia* mission will collect sparse-in-time photometric data of more than 300,000 asteroids. Using these measurements, it is expected that tens of thousands of asteroid will be modeled (see Cellino et al. 2006, 2009).

On the other hand, there might be some circumstances (e.g. not enough measurements, low lightcurve amplitude, etc.) in which *Gaia* data will not be enough to derive a unique model solution. As shown in Santana-Ros et al. 2015, in some cases, combining *Gaia* data with a single ground-based lightcurve can be enough to solve this issue. With that aim, the TJO is used to gather lightcurves of selected targets. These data will be included in a wider observation network called *Gaia*—Ground-based Observational Service for Asteroids (GOSA).

In this chapter two of the main observation projects in the TJO are described. Observations gathered within the NEA follow-up program allows the improvement of the orbits of NEOs and PHOs by providing accurate astrometric data to the Minor Planet Center (MPC). Lightcurves collected within the Gaia—GOSA collaboration network will contribute in enhancing the *Gaia* Solar System science.

2 2013–2016 NEA Follow-Up Program

The NEO and PHOs monitoring effort during these years involved a small network of observatories listed in Table 1. Several observational techniques are being used in order to cover different scientific cases. To infer color indexes we perform a photometric coverage using 30–120 s exposures, variable depending of the target magnitude to get good S/N ratios, in B, V, R and I filters. Such photometry is standardized to an aperture of 10 arcsec and usually to detect any possible cometary activity photometric growth curves are studied using increasing photometric apertures, as the growth curve of an extended object soon departs from that of a point

Table 1 Observatories involved in the present studies of minor bodies

Observatory	MPC code	Instrument
Gualba, Barcelona	442	SC 36.0 f/7
Guadarrama, Madrid	458	SC 25.0 f/10
Montsec, TJO, OAdM	C65	RCT 80.0 f/9.6

Fig. 1 PHA 2008 CM observed on 2.86 February 2014 at TJO (OAdM)

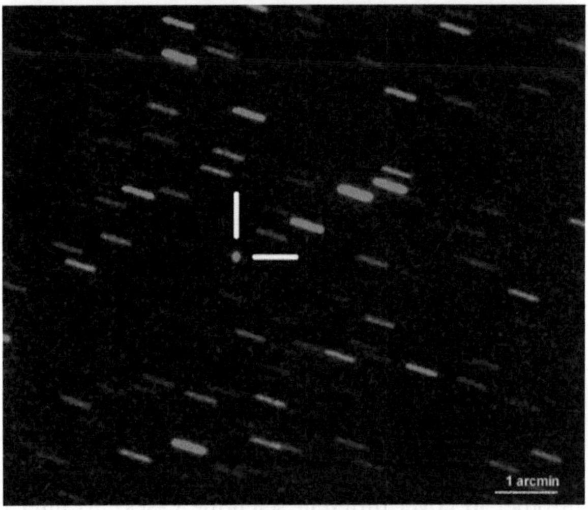

Fig. 2 PHA 3.5 km in diameter Apollo 85713 (1998 SS49) imaged by TJO on October 31, 2014 (OAdM)

like source. Using that approach we cover the photometric and full width at half maximum evolution of each NEO to be able to detect possible traces of cometary activity.

TJO tracking system works nicely and allows getting accurate images of asteroid and cometary targets whose astrometry is periodically reported to the Minor Planet Center (see e.g. Figs. 1, 2, and 3). Moderately-fast PHAs up to +18 are easily reachable using stacked images. For completeness of the NEOs and Main Belt (MB) asteroids monitored by TJO during the 2013–2015 period are listed in Table 2.

Fig. 3 C/2013 X1 PanSTARRS stacked image taken from OAdM on December 13.785, 2015 when the comet was in +13.1 R

Table 2 Main asteroidal targets followed by TJO during at least two nights between 2013 and 2015

Target	H	Classification
1862 Apollo (1932 HA)	16.25	Apollo (PHA)
3200 Phaeton (1983 TB)	14.6	Apollo (PHA)
85713 (1998 SS49)	15.6	Apollo (PHA)
97249 (1999 XT106)	14.7	MB asteroid
138852 (2000 WN10)	20.2	Apollo
204131 (2003 YL)	19.9	Apollo
214088 (2004 JN13)	15.3	Apollo
294739 (2008 CM)	17.2	Apollo (PHA)
345705 (2006 VB14)	18.6	Aten
412976 (1987 WC)	19.9	Amor
2014 MQ67	20.1	Amor

H is the absolute magnitude given by the JPL NEO homepage

In summary, we have been performing routine follow-up of NEOs during close approaches to Earth, together with observations of comets. By using photometric observations we are obtaining information about rotational parameters, color indexes, and albedo properties.

Another research program is focused in the follow-up of pristine comets approaching to the Sun. Our monitoring program of ground-based photometry is performed using standard Johnson-Cousin filters following the same methodology explained in (Trigo-Rodríguez et al. 2010). Due to our interest in learning about the development of cometary activity, we study the evolution of cometary commae. We observe comets far away from the Sun and during perihelion approaches, usually stacking guided exposures in order to achieve good signal/noise ratios (Fig. 3). TJO configuration currently uses a CCD camera that provides a square field of view of 12.3×12.3 arcmin^2 in which we are able to detect the appearance of the tail, the evolution of jet structures, and the comma brightness profile. Depending of the photometric profile we can conclude about the presence or absence of cometary activity at the different heliocentric distances. By building the respective photometric growth curves we are in conditions of detecting outbursts (Trigo-Rodríguez et al. 2008).

Since TJO telescope at OAdM was operational for minor bodies' tracking in 2012 we have been monitoring several comets (Trigo-Rodríguez et al. 2014, 2016). Astrometric and photometric observations are usually reported to the Minor Planet Center (MPC). Our photometric coverage focuses in B, V, R and I filters.

The measurements made using the R band have been proved to be the key to detect outbursts, because it increases in magnitude short after an outburst occurs due to the large particles released into the comma, while the magnitude in the other photometric bands remains stable while the particles are not fragmented into micron-sized dust capable to scatter solar light (Trigo-Rodríguez et al. 2010). Multiband photometry allows to get color indices of comets located at moderated solar distances and to get clues on their pristinity (Trigo-Rodríguez et al. 2016). Photometry is standardized to an aperture of 10 arcsec, and the photometric growth curves are studied using increasing photometric apertures to infer clues on the nuclear cometary activity. It is well known that the growth curve of an extended object soon departs from that of a point like source, allowing us to detect cometary activity at long heliocentric distances. Using that simple approach we plan to cover the comma development of comets from their first signs of sublimation until the intense activity suffered during perihelion approaches.

3 Gaia—GOSA: Enhancing *Gaia* Solar System Science

Gaia is probably the most ambitious mission ever undertaken by the European Space Agency. This spacecraft was launched on December 2013 and since mid 2014 it has been uninterruptedly surveying the sky. *Gaia* will keep gathering very precise astrometric and photometric data (as well as low resolution spectro-photometric measurements and radial velocities) at least until 2019. The final catalogue is expected to contain data of one billion stars and more than 300,000 minor bodies of the Solar System (Mignard et al. 2007).

Obviously, this huge amount of data received from Gaia is expected to induce a revolution in Solar System science. An excellent overview of the expected outcome and the influence in asteroid science was written by Tanga and Mignard (2012). In Table 3 we summarize some of the main results expected from analyzing the final *Gaia* Solar System catalogue, which release is planned for 2020.

The $\sim$50,000 asteroid models (more than 5 % of the known population) expected from the inversion of *Gaia* photometric data will provide us a more representative picture of the asteroid population. However, it has been shown in Santana-Ros et al. (2015) that the distribution of solutions will be burdened with a bias against different

Table 3 Expected impact in our knowledge of asteroids by using *Gaia* data

Property	Today	With *Gaia* catalogue
Rotation periods	~5000 asteroids	~50,000 asteroids
Shapes, poles	~400 asteroids	~50,000 asteroids
Astrometry	~1 milliarcsecond	~0.02 milliarcsecond
Mass	~10 asteroids	~50 asteroids

Table 4 A list with some of the Gaia—GOSA targets planned to be observed in 2016

Target	Classification
(7) Iris	Perturber
(42) Isis	Perturber
(99) Dike	Slow rotator
(207) Hedda	Slow rotator
(216) Kleopatra	Binary system
(241) Germania	Perturber
(243) Ida	Binary system
(384) Burdigala	Slow rotator
(882) Swetlana	Slow rotator
(1727) Mette	Binary system
(11429) Demodokus	Trojan with high-amplitude lightcurve

physical properties of asteroids, such as pole latitude or body elongation, as well as the number of *Gaia* measurements. On the other hand, it has also been proved that it is possible to reduce the bias by adding additional ground-based data. In order to collect as many additional ground-based data as possible during *Gaia* operations, a specific website called *Gaia*—GOSA (Santana-Ros et al. 2014) has been created. This service is available at www.gaiagosa.eu and has the purpose of coordinating a network of observers (including amateur astronomers) with the aim of gathering lightcurves of selected asteroids.

The Montsec Astronomical Observatory has granted us with 100 h of observational time in the TJO during 2016 which will be devoted to gathering data for the *Gaia*—GOSA project. A not exhaustive list of targets planned to be observed during 2016 is given in Table 4. Target selection has been done under scientific criteria. For instance, *Gaia* will obtain very precise astrometric measurements of asteroids being gravitationally perturbed by other larger bodies. By analyzing these data it will be possible to directly infer the mass of the so-called "perturber" (see Mouret et al. 2007 for details). Mass of an asteroid is a crucial physical property when studying the object, as it is required for obtaining a good estimation of its density. However, this effort will be done in vain if a detailed shape model of the asteroid is not available. Thus, collecting additional lightcurves of the "perturbers" will grant a greater detail of their shape models, allowing for a better estimation of the density.

Other particularly interesting group are the asteroids with long rotation periods (longer than 20 h), also called "slow rotators". The TJO is an optimal instrument for observing such objects. The Montsec Astronomical Observatory has developed

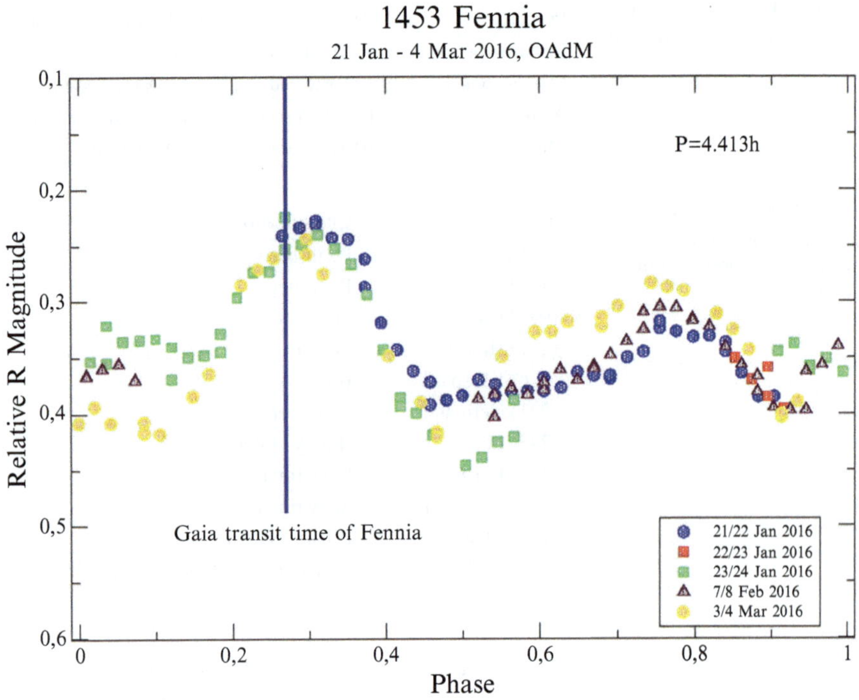

Fig. 4 Lightcurve of a Gaia—GOSA target obtained with the TJO

an excellent interactive application called MUR (Management for Users) where PIs can define their sequences for each target and night (for further details see Colomé et al. 2010 and Vilardell et al. 2013). Using this web tool it is possible to program the TJO robotic telescope to obtain a sparse photometric measurement of the slow rotator, and rapidly move to another object. The sequence can be programmed to repeat this process several times during the night, optimizing the observing time. After few good planned observing nights, it is possible to obtain a full lightcurve of these tough targets with a reasonable low use of telescope time.

Binary asteroids are also priority targets. By modeling these asteroid systems it is possible to directly derive their mass through Kepler's third law (see for instance Bartczak et al. 2014). Thus, some of the known binary systems are also included in our observational project. An example of a lightcurve of the asynchronous binary asteroid (1453) Fennia obtained with the TJO is shown in Fig. 4.

In order to obtain valuable data with the TJO and take advantage of its great potential, it is necessary to previously gather some know-how about MUR system. Next we provide a (very) short step-by-step guide for programming a TJO sequence for obtaining a lightcurve of an asteroid:

1. Define the target by providing the name, the type of object (minor planet) and the coordinate value. The expected coordinate format follows the Minor

Table 5 Illustrative exposure times for observations with R filter as a function of the asteroid brightness

Magnitude	Exposure time (in seconds)
12 or brighter	30
12–13	60–120
13–14	120–180
14–15	180–240
15 or fainter	300

Planet Center Orbit format. This database contains a file (MPCORB) which is regularly updated and can be downloaded from the IAU homepage http://www.minorplanetcenter.net/iau/MPCORB.html. In particular, characters 21–103 (or alternatively columns 4–11) are required.

2. Define the observing constraints. It is possible to obtain acceptable lightcurves even under poor sky conditions. For this reason we will allow the TJO to observe during bright nights, with poor seeing and spectroscopic sky conditions. The only requirements will be having the Moon at least 30° away from the target and collecting data during night time (i.e. Sun is more than 18° below horizon). It is also possible to specify observing windows if one requires data to be acquired during a particular span of time.
3. Configure the instrument (MEIA2). If no other special requirements exist, we will basically define the exposure time and the filter to be used. R filter is specially useful for bright nights. Using this filter, the illustrative value of the exposure time as a function of the asteroid brightness are given in Table 5.
4. Define the sequence. It is possible to combine different instrument configurations (and targets!) in a single sequence. This is an excellent feature of the TJO, making it a very flexible telescope which can be adapted to an endless number of observational requirements. As previously explained, this is particular interesting for optimizing your telescope time when observing slow rotating asteroids.

4 Search for PHAs Producing Meteorites

Additionally, we have been using the OAdM all-sky digital camera and other video stations set up by our institute and other research centers to record very bright bolides together with other video stations of the SPMN network: www.spmn.uji.es (see e.g. Trigo-Rodríguez et al. 2006). The standard way of linking meteoroids with a given comet or asteroid is through the similarity of their orbits. From the accurate astrometric analyses of the fireball images obtained from several stations we reconstruct the atmospheric trajectory, radiant and initial velocity (see e.g. Trigo-Rodríguez et al. 2006, 2008b). From these data the heliocentric orbit is inferred with the accuracy given by the measurement errors.

In order to search for the parent bodies of meteoroids, we use the so-called dissimilarity functions that allow comparing orbital affinity between bodies. These functions measure the "distance" between the orbit of the meteoroid and that of the potential parent. If this distance is below a predefined cut-off value, orbits are deemed to be similar. Porubčan et al. (2004) studied the similarity of meteoroid orbits with their parent bodies for a period of at least 5000 years before a generic association is claimed, a view later supported by Trigo-Rodríguez et al. (2007). The first dissimilarity criterion was proposed by Southworth and Hawkins (1963), who defined the D_{SH} dissimilarity function for two orbits A and B:

$$D_{SH}^2 = (e_B - e_A)^2 + (q_B - q_A)^2 + \left(2\sin\frac{I_{BA}}{2}\right)^2 + \left(\frac{e_A + e_B}{2}\right)^2 \left(2\sin\frac{\pi_{BA}}{2}\right)^2 \quad (1)$$

In this equation q and e are, respectively, the perihelion distance (expressed in astronomical units) and the orbital eccentricity. I_{BA} is the angle between the two orbital planes and π_{BA} the difference between the longitudes of perihelia as measured from the intersection of both orbits. A small value of D_{SH} implies a small difference between both orbits, which can then be regarded as similar. Usually, a cut-off value of 0.15 is adopted for D_{SH} (see e.g. Lindblad 1971a, b). Alternative dissimilarity criteria that are not necessarily equivalent and that are applied in current literature were defined later on by Drummond (1981), Jopek (1993), Valsecchi et al. (1999) and Jenniskens (2008).

In our continuous search for meteorite-dropping bolides in SPMN network imagery we have obtained the orbit of hundreds of bolides, not always associated with meteoroid streams or exhibiting sporadic origin. Some meteoroid orbits, when compared with the orbits of NEOs, show intriguing similarities, not by chance, if not extended over thousands of years (see Trigo-Rodríguez and Williams 2017). By integrating backwards their orbits and comparing the dissimilarity criteria we can establish significant associations that allow us to conclude that significant rocks can be detached from NEOs (Trigo-Rodríguez et al. 2007). The identified associations between large meteoroids and NEOs published in the literature by the SPMN network are listed in Table 6. A more exhaustive list is compiled in the chapter by Trigo-Rodríguez and Williams (2017).

Table 6 Near Earth Objects that probably deliver meteorites to Earth discovered by our team

NEO/PHA	N	Reference
2002NY40	3	Trigo-Rodríguez et al. (2007)
2007LQ19	2	Madiedo et al. (2014a)
2008ED69	4 (+MS)	Trigo-Rodríguez et al. (2009)
2008XM1	1 (+MS)	Madiedo et al. (2013)
2012XJ112	2	Madiedo et al. (2014b)

N is the number of bolides dynamically associated with the given NEO for orbital integration timescales of at least 10,000 years, while MS means that a meteor shower has been also associated with it

Fig. 5 A South Taurid, named SPMN141115D, exhibits several flares over TJO dome. This −9 absolute magnitude fireball appeared on November 14, 2015 at 23 h 40 m 28 ± 1 s UTC

During the last decades the number of networks recording fireballs has increased significantly, and so the number of meteoroid orbits (Jenniskens 2006). Then, the chance of identifying new mechanisms to deliver meter-sized projectiles to near-Earth space has increased significantly. At least we need to open the possibility that not only the classical delivery process from the Main Belt via planetary resonances is at work. An updated list of meteorite falls with the dynamic sources are described in Trigo-Rodríguez et al. (2015). For a detailed discussion about the physical processes delivering these rocks to Earth see Trigo-Rodríguez and Williams (2017). Figure 5 shows a South Taurid bolide from the 2015 outburst. A more complete example of South Taurid fireball recorded by OAdM all-sky, providing trajectory reconstruction and the derivation of its heliocentric orbit is discussed in chapter by Blanch et al. (2017).

5 Discussion and Conclusions

The main conclusions of this brief overview of our minor bodies studies using Montsec Astronomical Observatory (OAdM) instruments are:

1. The Telescope Joan Oró (TJO) is an excellent robotic instrument to perform routine follow-up of Solar System minor bodies. Astrometric and photometric data obtained with that instrument is particularly useful to gain insight on the dynamic evolution and physical properties of these objects.
2. Ground-based photometry of comets using standard Johnson-Cousin filters is a valuable tool to get information about their surface activity, and to predict their behavior.

3. Lightcurves collected with the TJO can be used to enhance the Gaia inversion results of certain asteroids. That also includes extreme cases like "fast rotators" or "slow rotators". This is possible because TJO is a very flexible telescope, as observers can define their own sequences in the MUR web tool, which allows for a huge number of possible configurations.
4. The OAdM also counts with other scientific instruments to study minor bodies. An all-sky CCD camera that, together with other fireball stations surrounding the observatory, allows obtaining accurate astrometric measurements of bright meteors and fireballs. These observations allow to provide a scientific explanation about the origin of very bright bolides, really helpful to popularize the interest of the science made at OAdM.
5. From the trajectory reconstruction, radiant, and velocity data of large bolides we can get orbital information of the large meteoroids penetrating on the Iberian Peninsula. The analyses of such results and the comparison with the orbits of NEOs provide new clues on the physical processes (disruption, tidal forces, etc) that these small bodies experience in near-Earth space.

Acknowledgments This work was financially supported by CSIC #201050I043, AYA2011-26522, and AYA2015-67175-P research grants. The Joan Oró Telescope (TJO) of the Montsec Astronomical Observatory (OAdM) is owned and funded by the Catalan Government and operated by the Institute for Space Studies of Catalonia (IEEC).

References

Alexander, C.M.O.'.D., Bowden, R., Fogel, M.L., Howard, K.T., Herd, C.D.K., Nittler, L.R.: The provenance of asteroids, and their contributions to the volatile inventories of the terrestrial planets. Sci. Express **337**(6095), 721–723 (2012). doi:10.1126/science.1223474

Bartczak, P., Michałowski, T., Santana-Ros, T., Dudziński, G.: A new non-convex model of the binary asteroid 90 Antiope obtained with the SAGE modelling technique. Mon. Not. R. Astron. Soc. **443**(2), 1802–1809 (2014)

Bottke Jr., W.F., Vokrouhlický, D., Rubincam, D.P., Bro, M.: The effect of Yarkovsky thermal forces on the dynamical evolution of asteroids and meteoroids. In: Bottke Jr., W.F., Cellino, A., Paolicchi, P., Binzel, R.P. (eds.) Asteroids II, pp. 395–408. University of Arizona Press, Tucson (2002)

Cellino, A., Delbò, M., Zappalà, V., Dell'Oro, A., Tanga, P.: Rotational properties of asteroids from Gaia disk-integrated photometry: a genetic algorithm. Adv. Space Res. **38**(9), 2000–2005 (2006)

Cellino, A., Hestroffer, D., Tanga, P., Mottola, S., Dell'Oro, A.: Genetic inversion of sparse disk-integrated photometric data of asteroids: application to Hipparcos data. Astron. Astrophys. **506**(2), 935–954 (2009)

Colomé, J., Casteels, K., Ribas, I., Francisco, X.: The TJO-OAdM robotic observatory: the scheduler. SPIE-The International Society For Optical Engineering, id. 77403 (2010)

Drummond, J.D.: A test of comet and meteor shower association. Icarus **45**, 545–553 (1981)

Jenniskens, P. Meteor Showers and Their Parent Comets, 790 pp. Cambridge University Press, Cambridge, UK (2006)

Jenniskens, P.: Meteoroid streams that trace to candidate dormant comets. Icarus **194**, 13–22 (2008)

Jopek, T.J.: Remarks on the meteor orbital similarity D-criterion. Icarus **106**, 603–603 (1993)

Lindblad, B.A.: A stream search among 865 precise photographic meteor orbits. Smiths. Contr. Astrophys. **12**, 1–13 (1971a)

Lindblad, B.A.: A computerized stream search among 2401 photographic meteor orbits. Smiths. Contr. Astrophys. **12**, 14–24 (1971b)

Madiedo, J.M., Trigo-Rodríguez, J.M., Williams, I.P., Ortiz, J.L., Cabrera, J.: The Northern X-Orionid meteoroid stream and possible association with the potentially hazardous asteroid 2008XM1. Mon. Not. R. Astron. Soc. **431**, 2464–2470 (2013)

Madiedo, J.M., Trigo-Rodríguez, J.M., Ortiz, J.L., Castro-Tirado, A.J., Cabrera-Caño, J.: Bright fireballs associated with the potentially hazardous asteroid 2007LQ19. Mon. Not. R. Astron. Soc. **443**, 1643–1650 (2014a)

Madiedo, J.M., Trigo-Rodríguez, J.M., Williams, I.P., Konovalova, N., Ortiz, J.L., Castro-Tirado, A.J., Pastor, S., de los Reyes, J.A.,., Cabrera-Caño, J.: Near-Earth object 2012XJ112 as a source of bright bolides of achondritic nature. Mon. Not. R. Astron. Soc. **439**, 3704–3711 (2014b)

Mignard, F., Cellino, A., Muinonen, K., Tanga, P., Delbò, M., Dell'Oro, A., Granvik, M., Hestroffer, D., Mouret, S., Thuillot, W., Virtanen, J.: The Gaia mission: expected applications to asteroid science. Earth Moon Planet. **101**(3–4), 97–125 (2007)

Morbidelli, A., Bottke Jr., W.F., Froeschlé, C., Michel, P.: Origin and evolution of near-Earth objects. In: Bottke Jr., W.F., Cellino, A., Paolicchi, P., Binzel, R.P. (eds.) Asteroids II, pp. 409–422. University of Arizona Press, Tucson (2002)

Moreno-Ibáñez, M., Trigo-Rodríguez, J.M., Rodríguez, D., Sánchez, A.: Astrometric and photometric follow-up of potentially hazardous asteroids using Joan Oró robotic telescope at Montsec Astronomical Observatory. 46th LPSC, LPI contribution, abstract #1138 (2016)

Mouret, S., Hestroffer, D., Mignard, F.: Asteroid masses and improvement with Gaia. Astron. Astrophys. **472**, 1017 (2007)

Oró, J.: Comets and the formation of biochemical compounds on the primitive Earth. Nature **190**, 389–390 (1961)

Porubčan, V., Williams, I.P., Kornoš, L.: Associations between asteroids and meteoroid streams. Earth Moon Planet. **95**, 697–712 (2004)

Santana-Ros, T., Bartczak, P., Michałowski, T., Tanga, P.: Gaia—GOSA, an interactive service for asteroid follow-up observations. EAS Publ. Ser. **67–68**, 109–112 (2014)

Santana-Ros, T., Bartczak, P., Michałowski, T., Tanga, P., Cellino, A.: Testing the inversion of asteroids' Gaia photometry combined with ground-based observations. Mon. Not. R. Astron. Soc. **450**, 333 (2015)

Southworth, R.B., Hawkins, G.S.: Statistics of meteor streams. Smiths. Contr. Astrophys. **7**, 261 (1963)

Tanga, P., Mignard, F.: The Solar System as seen by Gaia: the asteroids and their accuracy budget. Planet. Space Sci. **73**, 5–9 (2012)

Trigo-Rodríguez, J.M., Llorca, J., Castro-Tirado, A., Ortiz, J.L., Docobo, J.A., Fabregat, J.: The Spanish fireball network. Astron. Geophys. **47**(6), 26–28 (2006)

Trigo-Rodríguez, J.M., Lyytinen, E., Jones, D.C., Madiedo, J.M., Castro-Tirado, A., Williams, I., Llorca, J., Vítek, S., Jelínek, M., Troughton, B., Gálvez, F.: Asteroid 2002NY40 as source of meteorite-dropping bolides. Mon. Not. R. Astron. Soc. **382**, 1933–1939 (2007)

Trigo-Rodríguez, J.M., Madiedo, J.M., Gural, P.S., Castro-Tirado, A.J., Llorca, J., Fabregat, J., Vítek, S., Pujols, P.: Determination of meteoroid orbits and spatial fluxes by using high-resolution all-sky CCD cameras. Earth Moon Planet. **102**, 231–240 (2008a)

Trigo-Rodríguez, J.M., García-Melendo, E., Davidsson, B.J.R., Sánchez, A., Rodríguez, D., Lacruz, J., De los Reyes, J.A., Pastor, S.: Outburst activity in comets I. Continuous monitoring of comet 29P/Schwassmann-Wachmann 1. Astron. Astrophys. **485**, 599–606 (2008b)

Trigo-Rodríguez, J.M., Madiedo, J.M., Williams, I., Castro-Tirado, A.J.: The outburst of k Cygnids in 2007: clues about the catastrophic break up of a comet to produce an Earth-crossing meteoroid stream. Mon. Not. R. Astron. Soc. **392**, 367–375 (2009)

Trigo-Rodríguez, J.M., García-Hernández, D.A., Sánchez, A., Lacruz, J., Davidsson, B.J.R., Rodríguez, D., Pastor, S., De los Reyes, J.A.: Outburst activity in comets. II. A multiband photometric monitoring of comet 29P/Schwassmann-Wachmann 1. Mon. Not. R. Astron. Soc. **409**, 1682–1690 (2010)

Trigo-Rodríguez, J.M., Moyano-Cambero, C.E., Llorca, J., Fornasier, S., Barucci, M.A., Belskaya, I., Martins, Z., Rivkin, A.S., Dotto, E., Madiedo, J.M., Alonso-Azcárate, J.: UV to far-IR reflectance spectra of carbonaceous chondrites - I. Implications for remote characterization of dark primitive asteroids targeted by simple-return missions. Mon. Not. R. Astron. Soc. **437**, 227–240 (2014)

Trigo-Rodríguez, J.M., Lyytinen, E., Gritsevich, M., Moreno-Ibáñez, M., Bottke, W.F., Williams, I., Lupovka, V., Dmitriev, V., Kohout, T., Grokhovsky, V.: Orbit and dynamic origin of the recently recovered Annama's H5 chondrite. Mon. Not. R. Astron. Soc. **449**, 2119–2127 (2015)

Trigo-Rodríguez, J.M., Rodríguez, D., Sánchez, A., Moreno-Ibáñez, M.: Multi-band photometry to trace outbursts experienced by hyperbolic comet C/2013 X1 PANSTARRS. 47th LPSC, LPI contribution, abstract #1431 (2016)

Valsecchi, G.B., Jopek, T.J., Froeschle, C.: Meteoroid stream identification: a new approach—I. Theory. Mon. Not. R. Astron. Soc. **304**, 743–750 (1999)

Vilardell, F., Colomé, J., Sanz, J., Gil, P., Ribas, I.: Observing with the Telescopi Joan Oró. In: Highlights of Spanish Astrophysics VII, Proceedings of the X Scientific Meeting of the Spanish Astronomical Society (SEA), p. 958 (2013)